W9-AED-823

THE ORIGINS OF INVENTION

A STUDY OF INDUSTRY AMONG PRIMITIVE PEOPLES

BY

OTIS T. MASON

THE M.I.T. PRESS
MASSACHUSETTS INSTITUTE OF TECHNOLOGY
CAMBRIDGE, MASSACHUSETTS, AND LONDON, ENGLAND

Originally published by
Walter Scott, Ltd.
London, 1895

First M.I.T. Press Paperback Edition, August 1966

Library of Congress Catalog Card Number: 66-25002
Printed in the United States of America

ZUNI JEWELLER DRILLING TURQUOISE.

PREFACE.

—o—

At the celebration of the centenary of the American patent system in Washington (1891), I read a paper on the "Birth of Invention." The present volume is an expansion and illustration of the principles laid down in that paper. The history of the development of the inventive faculty is the history of humanity. In other respects we may resemble our friends the brutes, but here we part company intellectually, subdue and enslave them, and have dominion over the earth.

The term invention applies to four different yet related groups of phenomena:—

1. The things and institutions invented.
2. The mental acts involved.
3. The rewards and benefits of these acts.
4. The powers and materials of nature invoked.

I hold that all industries, arts, languages, institutions, and philosophies are inventions. The history of the mental acts is the account of an evolutionary series, beginning with taking notice and following examples, and ending with the highest co-operation in a great industrial establishment, with a symphony, with the writing of a dictionary, or with the framing of a government. The benefit or reward has also followed by analogy the processes of creation in Nature, from a single advantage accruing only to the inventor, up to a world-blessing conception.

As to the resources and powers of nature invoked, these have come into the service of man according to the law of ever-increasing complexity of structure for the performance of a greater variety of functions. The order of commanding kinetic energy has been the employment of—

1. Man-power in every pursuit.
2. Fire as an agent, in cooking, pottery, metallurgy, &c.
3. The power of a spring as in a bow or trap.
4. Beast-power, for burden and traction.
5. Wind-power, on sails, and mills, and in draught.
6. Water-power, as a conveyance, and a motor, and gravity or weight generally.
7. Steam-power, utilisation of an expanding gas.
8. Chemical power, in the arts of the civilised.
9. Electric power, motors, message-bearers, in mechanics and illumination.
10. Light as a mechanical servant, only beginning to be domesticated.

With Professor Payne I hold that the course of civilisation has been from naturalism to artificialism. And upon the lines of Mr. Spencer's division of activity into regulative and operative categories, it is the regulative side that exhibits the greatest differentiation and improvement. For instance, in simple tools, consisting of a working part and a manual part, it is the latter that has undergone enormous differentiations in applying the variety of kinetic energies.

I wish to express my gratitude to Professor Tylor, General Pitt Rivers, and Sir John Lubbock, without whose aid no one could write upon primitive technological subjects;

to the Smithsonian Institution and the Bureau of Ethnology at Washington for a thousand favours; to Mr. E. F. im Thurn, Mr. Man, and other modern travellers and ethnologists who, under the inspiration of the British "Notes and Queries," have vastly improved the material upon which studies are founded. I have been greatly aided by Professor Payne's *History of America*, and Mr. Henry Balfour's studies in the evolution of art, by M. Adrien de Mortillet's writings, and by the investigations of Mr. J. D. McGuire, Dr. Walter Hough, Professor Holmes, and Mr. F. H. Cushing.

O. T. M.

CONTENTS.

CHAPTER IX.

CHAPTER X.

CHAPTER XI.

CHAPTER XII.

CHAPTER I.

INTRODUCTION.

"Etenim omnes artes, quae ad humanitatem pertinent, habent quoddam commune vinculum et quasi cognatione quadam inter se continentur."—CICERO, pro A. Licinio Archia Poeta, section I.

IN this volume I desire to trace some of our modern industries to their origins, and to show how the genius of man, working upon and influenced by the resources and the forces of nature, learned its first lessons in the art of inventing. If the reader were to visit one of our great laboratories or mechanical establishments, he would see an army of skilled and intelligent men working together purposely to bring about larger results with less expenditure of effort. No one of these men would dream of doing all this work alone. A few years ago it would not have been possible for any number of men even to undertake it. There has been orderly procession, therefore, in the task to be done, and there has been growing complexity of organisation in the agency to be employed. In short, there is a close analogy between the natural history of the kingdoms of nature and the unfolding of the arts of life. The methods of studying the one may be successfully employed in a search for the true history of the other.

The term "invention" is here used in its plain, logical sense of finding out originally how to perform any specific action by some new implement, or improvement, or substance, or method. Fundamentally, it is a change in some one or all of these.

From the point of view here assumed, every change in human activity, made designedly and systematically, appears to be an invention. Not only mechanical devices, whose working models might have been stored in the vaults of prehistoric patent offices and the relics of which fill our museums, were inventions; but the processes of life, language, fine art, social structures and functions, philosophies, formulated creeds and cults,—all these involve over and over again the same activities of mind.

"Institutions," says Emerson, "are not aboriginal, every one of them was once the act of a single man, every law and usage was a man's expedient to meet a particular case." [1]

The foreshadowing of this faculty of finding out how is seen in the animal world. In certain exigencies even the invertebrates seem to have concentrated their intellectual activity upon methods of safety or escape. The conduct of one of these creatures in such emergencies is most instructive. First, it discovers a necessity, then follows a short period of confusion, finally the creature buckles down to hard thinking and experiment. The persistence of these humble inventors is often remarkable. Having conceived that a way of escape lies in a certain direction, it yields only to exhaustion or death. Sir Samuel Baker gives a picture of an immense elephant shaking a hēglik-tree three feet in diameter to secure the fruit. "The elephant butted his forehead against the trunk, the large tree quivered in every branch." [2] The creature had made a discovery and the shaking of the tree by the momentum of his body opened a new world of food supply. Had he gone further and made up a cushion to place between the tree and his head, that would have been an invention of the higher order rarely seen in the animal world.

This act of inventing involves the four causes of Aristotle,

[1] Emerson, *Essay on Politics*.

[2] *Ismailia*, New York, 1875, p. 230. See also *Rep. Brit. Ass.*, 1893, p. 861.

namely, the material, the formal, the efficient, and the final cause. Or, to be more explicit, every invention is made out of something, into some predetermined form, by means of certain apparatus and agencies, and to achieve some definite result. The change may be in any one of these. The inventor excogitates an alteration of causes or of movements. In primitive life and in the most cultured this is equally true ; only, in the latter, machinery performs the motions which in savagery must be effected by the human body.

Again, the term invention involves three sets of phenomena—the mental acts of inventing, of thinking out how ; the things invented, usually called inventions ; and the rewards bestowed on the inventor, nowadays called patents, but granted in some form during all the ages. In this work frequent allusion will be made to the growing intricacy of thought developed and demanded by this process through all history, terminating in the laboratories of invention with their coöperating experts, learned in every branch of science and mechanics ; but commencing with the relief of discomfort through a happy thought, by means of some slight modification or new use of a natural object.

Very similar to this coöperative invention in the laboratory or great mechanical establishment, is the united effort of a tribe, a community, a nation, a race, an age, the whole human species. The results of united effort, along the lines of activity, constitute the genius of each one of these. We speak of the genius of a man, meaning simply what he has invented, or of an age, having in mind all that it has done originally or found how to do in a new and striking manner. The unfolding of the genius of the ages has been the evolution of invention from the beginning.

The elaboration of rewards bestowed upon inventors from age to age should not be neglected. In this as in the other series of phenomena there has been increasing complexity and a sort of evolution. The public recognition and reward of invention may itself be said to have been invented. At first the public accorded really nothing. The man seized his own patent. His better bow or fish-hook got him more food, made him stronger, more acute, taught him that ingenuity is better than force, secured him admiration, respect, fear, obedience, homage, a larger number of wives, a more numerous following. The comforts, emoluments, distinctions, and power of able men were their first patents. Infringement upon such monopolies became dangerous and further originality was discouraged. Later in history, the tribe absorbed the benefit, then the state. Empty emoluments and public honours took the place of personal comfort. The inventor was crowned, or knighted, or medalled, or mentioned in the public *fêtes*. The history of the modern patent system would involve all of these.

"The education of the Indian," writes Morelet, "commenced early. When ten or twelve years of age, a *machete* is put in his hands and a load, proportioned to his years, upon his shoulders, and he is made to accompany his father in his excursions or his labours. He is taught to find his way in the most obscure forests through means of the faintest indications. His ear is practised in quickly detecting the approach of wild animals, and his eye in discovering the venemous reptiles that may lie in his path. He is taught to distinguish the vines, the juices of which have the power of stupefying fishes so that they may be caught by hand, as also those that are useful for their flexibility or for furnishing water to the wayfarer. He soon comes to recognise the *Leche Maria*, the precious balm with which he can heal his wounds, and the *guaco*, which neutralises the venom of serpents. He finds out the shady dells where the cacao flourishes, and the sunny eminences where the bees deposit their honey. He learns or is taught all these things early, and then his education is complete." [1]

But the chief object in view in this work is to follow out the lines of evolution or elaboration in the things invented. The shelter is the ancestor of the palace, the skin robe of

[1] Morelet, *Travels in Yucatan*, New York, 1871, p. 129.

the elaborate costume, the aboriginal roast of the costly dinner, the digging-stick of the steam plough, the carrying-strap of the burden trains and ships, and so on to the end of all the products of human activity in every direction. To trace out what constitutes an invention in this last sense is the history of original thought. What is really proposed is a study of those simple tools and processes out of which modern industrial life has grown. The finding out that a stone is hard and often sharp was a discovery, and the act lies within the capacity of very lowly creatures. But the slightest modification of that stone for industrial purposes was an invention, it was a step in that line of artificiality[1] which constitutes the progress of man. From this point of view, an invention, at first, was a slightly modified natural object or process. Every elaborative series exhibited in a museum should commence with such a specimen.

This barely modified natural object is susceptible of further changes, of added parts, of more complicated structure, of more diversified functions, passing in time from the simple to the complex, from the monorganic to the polyorganic. Exactly as one sees in the natural world in geologic time plant and animal forms becoming more and more highly organised, so in the constructions proceeding from men's minds and hands there has been a corresponding development and increased complexity and multiplied functions. In fact, the history of industry is the story of the greater diversity of materials used, of the more complicated thought in the mind of the inventors, of the perfection of tools and processes, which take the place of hands and feet and brain, and, lastly, of the final causes of the products of men's brains and hands. The play of these diversified motives and materials upon one another is one of the most interesting objects of human thought.

[1] "A change which has completely transformed human society, and to which the principal features which distinguish civilisation from savagery are traceable—the substitution of an artificial for a natural basis of subsistence" (Payne, *History of America*, New York, 1892, vol. i. p. 303).

Under the leadership of Gustav Klemm, the Germans have given to this theme the name of *Culturgeschichte*, and to its development his successors have devoted most serious consideration.

By analogy with the natural world those invented things or processes, those mental efforts involved in the act of inventing, those rewards of invention which obtained in the childhood of the race, also survive into the present, and may be seen in operation in every person's life; they are like those protophytes and protozoans, unicellular creatures which were once the only living things, which constitute now the majority of the life of the globe and which enter into the constitution of every higher organism in great numbers.

Necessity is commonly said to be the mother of invention. That is, all changes in human action are stimulated by man's needs. Now there are two classes of these, namely, those that act from within the individual, and those that affect him from without.

Of the former, hunger is the loudest. By hunger is meant the desire for food, or drink, or whatever enters the alimentary canal. The sense of fatigue and the desire for rest; the pain of monotony and the desire for change; the reproductive sense and many more, belong to these subjective stimuli.

The desire for warmth or cool shelter, or refuge from the storm, the sense of danger in the presence of savage beasts or men, in short, the discomforts which are produced by the want of harmony between a man and his environment constitute the second class of stimuli.

In the more lowly organised creatures, dwelling in the water, this disharmony is feeblest. The light, the temperature, the movements, the specific gravity even, are almost unchanging. In plants, which draw their nourishment from one spot, it is more varied. By an ascending scale terrestrial animals rise as they are wrought upon by the greater variety of stimuli.

Exactly as the inventive faculty, the things invented, and the rewards have passed through interesting evolutions, in which also the old ever survives into the new, so in the matter of stimuli there has been a parallel history. The pains of hunger are not the same in savagery and civilisation. The desire for house and clothing, and conveniences and art-products, and society and literature, and the explanation of things are childish in the one case, most exalted in the other. The evolution of human wants, therefore, is a part of the history of invention.

Just as there is an intimate relation between animals and plants on the one hand, and terrestrial phenomena and resources on the other, giving rise to phytogeography and zoögeography, so in the natural history of inventions there is the same relation never to be neglected. This correspondence or harmony between arts and industries and all that goes to make up environment enables the ethnologist to comprehend the proprieties of each region, and often to decide whether an art is indigenous or exotic. The study of this relationship between man's activities and the effect of his surroundings we may call technogeography, and the necessity of its careful study will appear at every point.

Nothing is more common than the assertion that men do not purposely invent in the lower civilisations, that they simply follow the leading strings and the mandates of Nature. The savage, it is said, does not invent, he simply borrows his clothing from the animals, his house from the trees and caverns, his food from many sources. He is an out-and-out imitator.

Now, nothing will be said here against Nature and her resources, her training school, and her wonderful teachers. Nor will it be denied that most men and women borrow everything and that all men and women follow suit in much that they do. This imitative process always supposes the existence of the thing to be imitated; the latter does not account for the origin of the process.

Furthermore, the whole amount of human progress

is undoubtedly to be accredited to human intelligence and volition. All Nature is clay in the hands of the potter. The suggestions of her three kingdoms are only the patterns hanging in the shop of the cabinet-maker or the modeller. The artificer and creator is man himself. Whatever may be one's theory regarding the manner, the place, the time of man's advent upon this planet, all agree that he was at first a houseless, unclothed being, without skill or experience, and that by the exercise of his faculties he has become the clothed philosopher. There is a sense in which the race may be said to have invented itself, proceeding against Nature often, with Nature often ; but always from naturalism to artificiality. Men were placed on earth to dress and keep it, to possess and subdue it. Through this wonderful faculty of invention the race has fulfilled its mission.

An objection to the opinion that physical nature is the most important element in civilisation is the well-known fact that the latter has progressed most rapidly, not in those favoured spots where food and raiments and shelter could be procured with the least effort, but in those regions where the quest for these things afforded the best school for the training of every faculty of the mind, where there were the greatest stimuli to exertion. In those places where Nature is too lavish the tribes of men have led a languid existence.

By laying too much stress upon the notion that the human race has borrowed all its plans and methods from Nature, one is apt to forget that the best of instruction has no effect on dull pupils, as every pedagogue will testify. The forms and movements of all things terrestrial were lying before the senses of animated nature for millenniums before our race arrived. How very few of them aroused the apperception of the brute, and stimulated him to those never-ceasing changes which constitute the life of progress. The profound teaching of Nature fell upon those who having ears, heard not.

One is perpetually hearing sociologists saying that men do not invent customs, but fall into them. Grant that the ninety and nine do follow suit, and in addition grant that each one of us follows his leader all but the thousandth time. It is the one act in a hundred or a thousand that each one originates, which constitutes the progress of the world. Again, we read that peoples do not invent civilisation, but borrow it; that one man left to himself would die, and that no people ever arose by itself. Borrow from whom? Where did the first lender get his stock?

It is inconceivable. It would be ungrateful to the ingenious minds that have brought a whole species of ignorant and inexperienced creatures to know and to conquer the world in an incredibly brief time. If we ignore or deny the existence of this adventurous spirit of climbing for the sake of being higher, of learning for the sake of knowing, of inventing for the sake of inventing, then, indeed, would æons on æons have been necessary for the evolution of our species, and man would have had to start farther back than Tertiary times to have drifted by mere imitation to achieve so much.

It is difficult to imagine the motives which actuate ethnologists in rendering their applause so grudgingly to this genius of invention. Mr. Wake says: "The ingenuity displayed by the Australians in overcoming the many difficulties they have to contend against in dealing with the hard conditions of nature is often, no doubt, very great. Great ingenuity is undoubtedly shown in the native weapons, one of which, the boomerang, would appear to be unknown, in principle at least, by any other race. It must be noted, however, that we do not know the progressive stages through which the boomerang has arrived at its present perfection, and that it may have been an accidental recognition of an operation of Nature."[1] Now, is not that too bad? The boomerang arriving at its present perfection! Accidental recognition of an operation of

[1] C. S. Wake, *J. Anthrop. Inst.*, London, 1872, vol. i. p. 75.

Nature! Was it not just such a history, of a humbler sort, as that of the rifle, the locomotive, the alphabet, the electric light? Recognition of the operations of Nature constitutes the genius of invention. The Australian or humble people just like him commenced this wonderful process. Those cunning little creatures, as Emerson called them, invented the boomerang. And there is not a patent office in the world that would refuse to grant them letters for the exclusive use thereof for seventeen years.

The civilised man passes his whole life in the midst of wheels and cranks and engines of iron. His eyes are on them every day. Now and then a new thought occurs to him in their motion. An improvement which would facilitate their action and lessen his pains or expense. That is called invention, for which he seeks a patent. The savage man passes his life away from wheels. He never saw a wheel until the new-comer showed it to him. But there are around him all sorts of suggestive things that take the place of wheels. He sees how he could improve them so as to facilitate their action, and so as to lessen his labour and multiply his gains. He makes the change. Is not that an invention also?

It surely ought not to fill this magnificent age with envy to admit that intimations at least of our times were heard long ago. The monorganic form of a tool from which a machine is a polyorganic evolution or elaboration ought, according to the nature of things, as they are now understood, to have come first. Furthermore, these old forms ought to have survived and do survive here and there, just as preglacial genera of butterflies may be seen even now flitting about Mount Washington, in New Hampshire. In the crude art of the French caves is found the prophecy of a people who in this last day should set the fashions of the Western world.

According to Le Bon, "The higher races have never been influenced by a foreign civilisation more rapidly than the lower races; and if they have sometimes adopted creeds, institutions, languages, and arts, different from those of their ancestors, it was not till they had slowly and profoundly transformed them and brought them into relation with their mental constitution."[1] Compare this with oft-repeated assertions that no race or people has ever raised itself to any higher culture. The race or people that did not lay at least one dressed stone on this stately edifice can not possibly have survived.

Schweinfurth says : " A people, as long as they are on the lowest steps of their development, are far better characterised by their industrial products than they are either by their habits or by their own representations, which are often incorrectly interpreted by others."[2] This is entirely in accord with what the present writer has said about the double history of the race, that written in words and that written in things and actions. The former is circumscribed in time and place and intelligibility ; the latter is universal, like the objects upon which it is based.

Again : " It is among the most secluded inhabitants, indeed among the rudest tribes, who are partly still addicted to cannibalism, aye, in the very heart of Africa, whither not even the use of cotton stuffs and hardly that of glass beads has penetrated, where we find the indigenous mechanical instinct, the delight in the production of works of art for the embellishment and convenience of life, the delight in self-acquired property best preserved."[3]

Captain Spicer, a whaler, who mingled with the Eskimo, told the writer that they often make invention a part of their sport. They go out to certain difficult places, and, having imagined themselves in certain straits, they compare notes as to what each one would do. They actually make experiments, setting one another problems in invention.

[1] Le Bon, *Pop. Sc. Month.*, New York, 1893, vol. xlii. p. 342.

[2] Schweinfurth, *Heart of Africa*, vol. i. p. 257. Quoted in *Artes Africanae*, London, 1875, p. x.

[3] Schweinfurth, *Artes Africanae*, London, 1875, p. ix. See also Man, *Andaman Islanders*, London, p. 26.

There is another error, equally illogical, into which a great many writers have fallen, of supposing that the ancients and prehistoric peoples were possessed of arts and mechanical appliances far in advance of aught we have nowadays. These are called "lost arts," and it is averred that they are now beyond the sagacity of man.

The answer to this argument is in the words of the wise man, "To everything there is a season, and a time for every purpose under heaven." The thing that hath been is the thing that may be, if it is desirable. The reason why arts are lost is that they have become antiquated by others higher in the scale, or because they were practised by a limited number who moved in a side current, whose secret died with them.

Sir John Lubbock quotes Mr. Wallace as saying that man is no longer influenced by "natural selection, and that his body has become stationary."[1] This is a mooted question; the oldest human skulls yet found are capacious enough, and there are no data at hand to show whether or not the brain is growing. But the work which this brain has to perform in making inventions has been growing, and therefore no one will doubt that it has increased in agility and knack. The brain may acquire knack as well as the hand. At first it had none of this quality, and had to peg away laboriously for hours and even days to comprehend and perform indifferently what it now does with celerity and ease.

In the higher walks of invention there is a perpetual rivalry between the mechanic and his work, between the scientist and his apparatus. In the lower levels of progress this emulation is often between the savage man and the material in which he works, or the tool with which he achieves his result. If one were to mark the history of sculpture he would notice at once a constant increase in the intractability of the material. This increase would also be coupled with a parallel improvement in the means of

[1] Lubbock, *Prehistoric Times*, New York, 1872, p. 591.

overcoming the resistance. Each success would embolden the sculptor to venture upon still finer rocks. As long as there were one mineral in sight that he could not work, this would be a standing menace to his ambition. The handier he became in mastering the stone that he had already attacked, the more eager would he be to find more difficult material to master.

Now these minerals are not scattered evenly over the earth. There is no flinty rock in the West Indies, but there is abundance of volcanic material and of granular rocks. What reckless waste of energy to spend one moment in trying to achieve therein such results as the chipped stone objects in use where quartz or flint abounded! The ancient Arawak wasted no time along that line, but from his rocks, that lent themselves peculiarly to the pecking and polishing process, elaborated the most beautiful stone implements in the world. There is abundance of evidence upon this matter. The ivory art of the Eskimo, the black slate art of the Haida, the red catlinite art of the Sioux, the jade art of China, the beautiful flint art of Western Europe in prehistoric times are in point, and, without doubt, it was in a certain sense the beautiful marbles of Greece and Italy that set up that duel between the man and his material which resulted in the grandest sculpture of the world. In fact, when national glory itself declined, the world still turned to these spots for artists, and sent thither its sons to breathe the divine afflatus.

A great deal that has been written about primitive industries and inventions is wide of the mark, because the writer has failed to take into account what may be called the knack of the age, or the tribe, or the particular method. He has described it as clumsy, and said that he could not for the life of him imagine how people could get along with such appliances. But they did. You will see a professional ethnologist sweating for hours to get a spark of fire with two sticks. The savage will do it for him in as many seconds. By and by the former acquires the knack, and then his trouble vanishes. Lafitau says the polishing of a stone axe requires generations to complete. Mr. Joseph D. McGuire fabricates a grooved jade axe from an entirely rough spall in less than a hundred hours. Every one who reads this will recall examples of this deftness; not only among jugglers and turners, but in the shop, on the farm, about the household there is always some one who has the knack of doing the thing.

> "The dean was famous in his time,
> And had a kind of *knack* of rhyme,"

says Swift, and there is no doubt that this is the quality which in the higher pursuits of life we call genius. Now this quick perception and dexterity of execution are not traits of higher civilisation alone; the savage and the barbarian possess them as well. Indeed, it is sometimes said that the substitution of unerring machinery has taken away the cunning from the human hand. The case is not nearly so bad as that, however. No change of apparatus can deprive the human race of geniuses, for the man of knack will be found excelling in the handling of the new machines. Now I have dwelt on this word in order to account for the earliest differentiation of trades. Doubtless in pristine civilisation every man had a multitude of functions, and every woman was mistress of all trades. But travellers tell us that, among the Eskimo, the Plains Indians, the fishing tribes, the Polynesian navigators, the Australian bushrangers, the man of knack takes the lead, is sent ahead freed from all burdens to spy out and slay the musk ox or other game, while the rest of the gang, men and women, come lumbering along with the conservative luggage. The bow-makers, the arrow-makers, the skin-dressers, the basket-weavers, the potters of the tribe, exalting their occupation and exalted by it in turn, find in this social differentiation the greatest opportunity and encouragement. The great procession of humanity drags along, too much encumbered with many cares to acquire excellence in any one occupation.

One of the greatest hindrances to more rapid progress among savages through the multiplication of inventions is their communistic system, their tribal intelligence and volition. House-building, canoe-building, hunting, fishing in common, borrowing indefinitely, parasitism are great impediments to personal ambition. When Turner told a Samoan about the poor in London, he replied, "How is it? No food! No friends! No house to live in! Where *did* he grow? Are there no houses belonging to his friends?"[1] But this very Samoan and many of his ingenious ancestors had been kept behind in the march of civilisation to support those who would not work.

In tracing the progress of invention or culture through modern aborigines and lowly tribes in the past, it is not necessary to make the description of monstrosities and of the deeds of human monsters the chief aim. These are atavistic, and exhibit the elements of destruction and decay. The people practising such things are in the suburbs of the world, passing away in the very nature of things. It is not out of such bloody conduct our present civilisation issued, but its progress was away from such things. Our culture is the offspring of parents whom it resembles. A people that practices infanticide and brutality to women has signed its death warrant. No cultured race ever arose out of such savagery as that. Among the most seemingly brutally savages there is a higher, purer society, the party of progress.

Mr. Wittich, who lived fifteen years among the Apaches, and had their confidence, says that no traveller has any chance in the world of seeing the best life of that people. The same is true there as would be true in any city of Europe. Let the Emperor of China announce that he will ride through London, and from the top of the omnibus let him take photographs and dictate to his stenographer his conceptions of the motley rabble around him. That would not be a history of the greatest social unit the world has

[1] *Cf.* Turner, *Samoa*, London, 1884, p. 160.

ever seen. No more is the statement of a cursory visitor an authentic account of a savage tribe; he passes only through the rabble, and misinterprets what he sees. Furthermore, we are not concerned here with the unoriginal moments of any man's life, nor with the stupid procession that never had a thought of their own, nor even with whole tribes or races of man after they have lost the divine genius of devising. The people that ceases to invent ceases to grow. Our concern is with the happiest moments of each, when he is in true sense a creator, with the cleverest thoughts of the best, and with the most beneficent contributions made especially by the lowest tribes to the general resources of the race. This will surely be a delightful quest, to ascertain how the world has improved under the guidance of the best and freshest minds.

The inventor in our own day is one who is from some motive seized with the notion of improvement. In the development of life on the globe, only those species and individuals survive which are best adapted to ever-changing environment. The progress of culture on the earth has followed the same law. The melancholy record of tribes and nations that have disappeared in historic times suffices to establish that. So that, in tracing forward or backward the inventions of mankind, we are always somewhere near the party of true success.

The progress of the world has been always toward grand results. It does seem, therefore, that an unseen hand has been holding a candle in the darkness to guide the successful races upward and onward. Also it seems that when a people got beyond the enlightening ray of this world-inviting beacon, they sooner or later declined and disappeared. A nation or an individual, in this regard, stands to this spirit of progress very much as the old-world people did to the oceanic currents. Columbus discovered a new world only when he was in the stream.

In the prosecution of this inquiry there are several kinds of witnesses to be interrogated: (1) the relics of bygone ages and peoples; (2) the operations of modern savages; (3) the publications of historians and travellers who were acquainted with savage tribes long ago; (4) the languages of cultured and uncultured races; (5) the makeshifts and contrivances of children and of the folk who never receive letters patent upon their devices.

Fortunately, in the life of our species the testimony of all these witnesses together affords a kind of "House that Jack built." The last verse, that is, the present industrial condition of human knowledge and industry, contains practically all that is in the poem. In any great modern city, and about its suburbs, it would be possible to get a tolerably accurate view of mankind in all ages.

The perusal of such a work as Mitchell's *Past in the Present* is enough to convince one of this assertion.[1] Indeed, a good patent attorney would go further, and point out to you in some intricate machine survivals of all the ancestral traits that have entered into it.

In the more remote parts of Iceland, many articles of bone and stone are still in use, which in more accessible districts, have been replaced by metal or earthenware. Mr. Anderson saw a wheelbarrow with a stone wheel, a steelyard with a stone weight, a hammer with a stone head, and a net with bone sinkers. At the same farm a quern was in use, also horn stirrups, harness fastenings of bone, to say nothing of bone pins and bone dice. The County Council of the district meets in a spacious cave in the lava.[2]

If the pristine artist or invention be represented by the letter a, then in the next epoch or improvement, going upward, this same implement or process or artisan will still survive, slightly modified, and may be represented by a'.

There will be in the second epoch also new men and methods and appliances, for which the letter b may stand, and the symbol of the whole artisan class, or the total of processes or the aggregate of devices will be $a' + b$. In the

[1] Arthur Mitchell, *The Past in the Present*, New York, 1881, Harpers.
[2] Tempest Anderson, *J. Soc. Arts*, London, 1892, vol. xl. p. 400.

third epoch it will be $a'' + b' + c$; and in the fourth $a''' + b'' + c' + d$, and so on. It has indeed been said that the latest of great modern inventors are an epitome of the genius of the world.

It is a fact, then, that our modern activities, with their results and methods and appliances, are the descendants of a long line of ancestors, that become more and more obscure and humble as we trace them backward. This being true, it is with the greater difficulty that the evidences of their existence and standing are brought together. As in other genealogies, names are dropped out or transposed, and the continuity of history is interrupted. In the case of inventions there is another method of checking off paternity which must not be neglected, namely, the study of peoples now living. For example, the wheeled vehicle exists as a native product in China, but not in Corea. The Coreans are for that reason behind the Chinese in this particular, and certainly represent a pre-Chinese civilisation. These have art-products unknown to the Manchu, and that relegates the latter to an elder day, other things being equal. Of course, in any case, before making such an assertion search should be made for evidences of degradation from higher standards.

In making up our minds concerning the status of an ancient people from certain things in their graves compared with similar things in possession of a modern tribe, it is also quite necessary to examine the tenure of the modern tribes by which they hold the things in evidence. There are at least three forms of ownership involved. The specimen of which we are speaking may be the product of a native art of a people in a rising scale. Or they may practice this art now while declining in culture. Or they may have borrowed it out and out, as the Navajos have borrowed weaving, still living in wretched hogans, and held back from the wild savagery of their cousins, the Apaches, by the possession of sheep which they borrowed from the Spaniards.

The pioneers in culture-history, such as Klemm and Waitz, Tylor and Lubbock, and more that could be mentioned, gathered with their best judgment into the works that have been our text-books, and with the greatest possible diligence, what travellers, explorers, missionaries, and settlers have said about native races. But the more careful observation of those who have looked deeper into these matters, have set aside much said by witnesses upon whom these great ethnologists have relied. If any one will take the pains to study the publications of the United States Bureau of Ethnology or the work of Mr. Man in response to the British *Notes and Queries*, he will see that half the labour is expended in correcting errors and disparaging misconceptions.

The Mediterranean race is the most mechanical of all, the blue-eyed and the brown-eyed variety must each settle for itself which shall bear the palm. The Semite is much less so. The Mongolian is, perhaps, more ingenious with his hands. The Africans and Papuans are more mechanical than the brown Polynesians; the Eskimo than the red Indians; and the Australians are the least clever of all. In each several division of humanity there are smaller centres of invention, owing both to natural ingenuity and to natural resources. In the higher walks of language, art, social structures, literature, science and philosophy, the peoples of Europe and Asia will need a new distribution for each classific concept. The Hebrew has never been excelled for sublime conceptions on religious topics, the Egyptian invented chronicles, the Greek perfected harmony and portraiture in art, the Romans laid the foundations for jurisprudence.

The regretful element in a study of this sort is that one must despair of seeing these older inventors at work in their descendants. The majority of human races had nearly quitted original research when they were discovered. Many, very many of them showed signs of undoubted decay. All of them were living on the ruins of civilisations superior to their own, or were in the possession of institutions and arts that they could not have devised. The wiser, younger, progressive stocks absorbed all the happy suggestions they had to offer, and left them to muse and to die among the ruins of ancestral genius. In a great modern factory old machines are at once sent to the scrap pile as soon as a new patent is issued, and whole chapters in the history of ingenuity have been torn up on the uprearing of a new and more advanced culture.

CHAPTER II.

TOOLS AND MECHANICAL DEVICES.

"What a plastic little creature man is! so shifty, so adaptive! his body a chest of tools, and he making himself comfortable in every climate, in every condition."—EMERSON.

AMONG inventions, the class of objects that are not an end in themselves, but which are used as means to ends, occupy a very prominent place. They are covered by such terms as "tools," "implements," "machines."

Many of these are the apparatus of special crafts, and should be considered among the inventions belonging to those crafts. But a great many of them have come down from remote antiquity, and belong to workmen of every trade.

The tool chest of the Andamanese, according to Man, would contain a stone anvil, stone hammers, chips, and cooking stones; one or more *Cyrena* shells for preparing arrow-shafts, for sharpening knives of cane and bamboo; and boar's tusks, for carving spoons, for knives in cutting thatch or meat, for scrapers in separating bast and bark in cord-making, for carving, and even for planes.

You would also find *Arca* shells for pot-making, *Pinna* shells for receptacles, and food plates and *Nautilus* shells for drinking-cups. The bamboo spear shafts, water holders, food receptacles, knives, netting-needles, tongs, &c., would call attention to the usefulness of that plant. Paint brushes from the drupe of the *Pandanus Andamanensium*[1] should not be overlooked.

[1] E. H. Man, *Andaman Islanders*, London, 1883, p. 156.

Under the head of general appliances for industrial processes may be included *tools*, *mechanical powers*, *metric apparatus*, *natural forces*, and *machinery*. M. Adrien de Mortillet has made a classification of simple tools which is adopted here, with additions and modifications.[1]

I. FOR CUTTING. EDGE TOOLS.

Working—

1. By Pressure. { Knives. Double-edge tools, shears. Planes.
2. By Shock. { Axes. Adzes. Chisels, gouges.
3. By Friction. Saws.

II. FOR ABRASION AND SMOOTHING.

Working—

1. By Pressure and Friction. { Scrapers, gravers, rasps, files, sandpapers, polishers, smoothers, burnishers, whetstones, grindstones.
2. By Shock. Bush-hammers.

In wood-working fire is an efficient element in abrasion.

III. FOR FRACTURING, CRUSHING, POUNDING.

Working—

1. By Pressure. Chipping and flaking implements.
2. By Shock. Hammers, pestles.
3. By Friction. Grinding apparatus, mills.

IV. FOR PERFORATING.

Working—

1. By Pressure and Friction. { Needles, prickers, awls, drills of all kinds.
2. By Shock. Punches, picks.

V. FOR GRASPING AND JOINING.

1. Tongs, pincers, vices, clamps, wedges.
2. Nails, lashings, glues.

---

[1] *Rev. Mensuelle de l'Ecole d'Anthrop.*, Paris.

Before entering more minutely upon the study of tools, a few words should be said concerning the composition of tools, their working parts and haftings. It is true that millions of ancient objects, in stone especially, lying in museums and cabinets have now no handles. But it is fair to assume that the great majority of them were once so furnished. Indeed, in their manufacture the artificer spent as much time and pains in getting them ready to be hafted as he did in finishing the working portions. The best guide in furnishing anew these objects with hand-attachments is the study of modern savagery.

These are to be studied both in their adaptation to the hand and in the method of their being fixed to the working part. The former, for convenience, may be called the grip, or handle; the latter, the attachment. The grip of an implement may be made to fit one hand or two, and to be held close to the object wrought upon, or at some distance. It is really this part that at last becomes a machine.

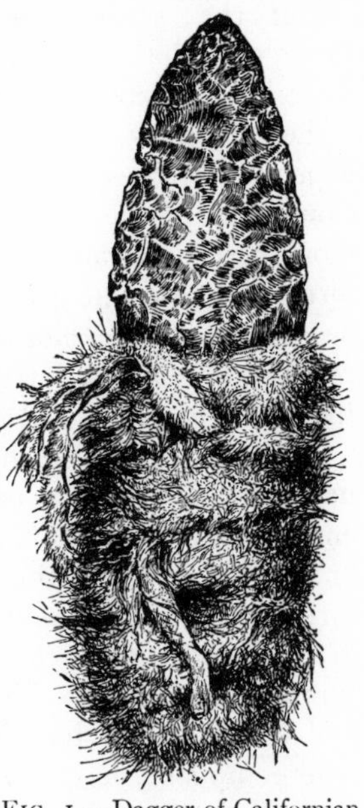

FIG. 1.—Dagger of Californian Indian. Grip, a long strip of Otter Skin bound around.

Many savages still use only the rudest kind of grip, merely smoothing the rough surface of the material or wrapping something about it, so as not to hurt the hand, but this is not true of all tribes.

The Eskimo men and women carve, from walrus ivory, musk-ox horn, and wood, the daintiest handles for their scrapers and other implements. They fit so exactly that the white man, with his much larger hands, is unable to use them. No modern sword grip is more convenient or more tastefully carved.

The Indians of the West Coast are not so particular, and yet on many of their tools there are grooves for the fingers. But a singular departure from this idea of convenience is to be seen on South American and Polynesian weapons, where for the sake of decoration the maker has carved a ridge that would be in the way of the hand.

But the great majority of haftings, shafts, handles, hilts, or grips of aboriginal implements were of some material separate from that of the working part, and attached thereto artificially. The form of this separate handle depended precisely upon the work to be done. The sagacious mind of the savage mechanic has nowhere worked to more perfect advantage. The economy of material and of form to acquire the greatest result with the least effort has been thoroughly explored. After the bare necessities of the case have been met, tribal genius, imagination, and good judgment have had full play.

To make a list of forms of aboriginal haftings it would be necessary to write a catalogue of the varieties of tools enumerated in the table at the beginning of this chapter. If one would examine the stock in a modern hardware or furnishing store, he would have to look over a great many kinds of tools before he could find a style of simple handle unknown to savages. He might begin with a cylindrical rod, and end with the handiest device from the Patent Office. There probably never was a more effective grip on a tool than the form used by the Eskimo women for their scrapers nor those on the Malay daggers or kris. A classification of haftings as to shape would commence with a mere stick or withe or fork of a sapling, and pass through a series of improvements ending with one in which the hand would be covered so that every finger and every muscle would have full play in every direction for pushing or pulling or rotary motion. This subject has never been worked out by a trained anthropologist.[1]

[1] Rau, "Chapter on Hafting in Aborig. Stone Implements," &c. *Smithson Cont. to Knowledge*. Murdoch, *Ninth An. Rep. Bur. Ethnol.*, many figures; Mason, *Rep. U. S. Nat. Museum*, 1889, many figures.

The methods of attaching the handle to the working part are more ingenious than the grip itself. The following are the principal types :—

1. Doubling a pliant hoop or sapling of wood about the working part.
2. Fastening the working part to a shoulder on the handle or to a forked stick.
3. Inserting the working part into a hole or groove or mortise in the handle.
4. Inserting the handle into or through the working part.
5. Binding the working part into a sling, which either encircles or covers it.
6. Seizing.
7. Gluing.
8. Rivetting.

In almost every section of North America occurs the "grooved axe," and there grow a great many varieties of wood, like ash or hickory, whose saplings will bend double without breaking and will easily split. The Indians were accustomed to take a piece of one of these saplings about six feet long and split it, so that in bending about the groove of the axe or adze or hammer, it would neatly fit. The hafting was completed by securely seizing the sides together near the working piece and at the grip. The method of this seizing will be presently explained. This style might have been seen in the United States anywhere between the two oceans.

In Matthews's "Mountain Chant" two young Navajos are sent out to chop poles for their tent. They had grooved stone axes, and for handles they bent flexible twigs of oak and tied them with fibres of yucca—that is, they doubled the twigs, inserted the grooved axe-head in the bend, and made all fast with yucca fibre.[1]

It is interesting to note in this account the transformation of a myth. While the story holds on to the oak withe it adopts the yucca binding. The Navajo moved southward

[1] Matthews, *Fifth An. Rep. Bur. Ethnol.*, p. 388.

into Arizona from Canada, and carried the memory of the oak while forgetting the old-time lashing of raw-hide.

Fitting a forked stick to the working part was thus accomplished. A young tree was selected from which a limb jutted out at the proper angle, having also the right size for the hand. The limb was split off with a goodly piece of the trunk attached, and this was trimmed to a shape so as to fit on the working part, which might be slightly let in, or laid flat with a shoulder on the haft. This process of onlaying and partly inlaying adapts itself to every type of handle used in savagery. The Eskimo even take old plane bits and iron axe-heads procured from whalers and so haft them. The boat-builders of the West Coast and the inhabitants of Australasia of every race make most varied and ingenious uses of the method. It has very great advantage to a savage whose grindstones are frequently of difficult access. The lashing or seizing can be readily done up and undone and the stone or metal working part quickly removed, sharpened, and replaced. The many ways of holding the parts together will quickly be explained.

Inserting the working part into the handle may be a much older and more primitive process. In the Swiss Lake dwellings are found good-sized blocks of antler, into the spongy end of which the poll of a small celt was driven. This block of antler was afterwards itself used as a handle, or again was inserted into another piece to serve therefor. The very same process is in vogue in America in our day wherever the antler or suitable material may be found. The tough exterior of antler and bone, and their spongy interior would almost suggest themselves to the most ignorant savage. While for small tools such as perforators, the rustic and the savage alike know that pith is soft and that the wood of some plants is very tough. This process may be seen in all stages of development among the working tools of Eskimo and Indians. Arrow heads, awl points, bone prickers and perforators, even scrapers and adze-chisels, may be found in abundance with their working part let in or driven into the handle. The parts are further secured by wrapping and by cement.

The Bongo method of hafting an axe—and this seems to have been the universal practice in true Africa—is to select a piece of wood that has a knot or gnarled place at one end and to drive the tang of the hoe or axe into a perforation through the knob. Fastened in this manner the wedge-shaped tang sticks more firmly in the handle at every stroke. On the other hand, spears and even many garden tools are furnished with a conical socket, into which the shaft is driven more firmly at every thrust.[1]

Says Kalm, the hatchets of the Delaware Indians were made of stone in shape of a wedge, with a groove around the blunt end. To haft it they split a stick at one end and put the stone between it ; they then tied the two split ends together. Some of these hatchets were not grooved, and these they held only in the hand.[2]

This is, in fact, a rude variation of the withe style of hafting. The blade is really inserted, however. There is a poor specimen of this kind of work in the United States National Museum from the Pueblos of New Mexico.

Lafitau describes a process which does not exist in modern savagery. I have found this writer's imagination or credulity playing tricks with his statements more than once and am

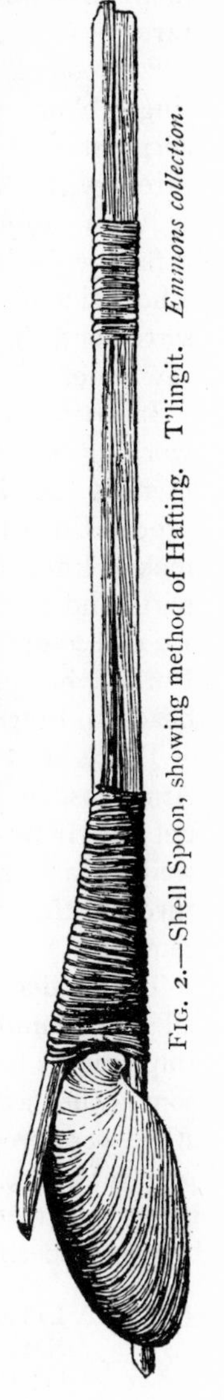

FIG. 2.—Shell Spoon, showing method of Hafting. T'lingit. *Emmons collection.*

[1] Schweinfurth, *Artes Africanae*, London, 1871, pl. vi. figs. 4, 5.

[2] Compare Kalm, *Travels*, &c., London, 1771, vol. ii. p. 37.

inclined to think the following method of insertion extremely rare.

"Choose a young tree," says Lafitau, "to split it with a single blow and insert the stone; the tree grows and incorporates it in such a manner that it is with difficulty and rarely withdrawn."[1]

A few examples occur in which the end of a stick is split, a ferrule or seizing stopping the rift at the point desired. The inside of the jaws were then trimmed out, the pole inserted and the outer ends tightly bound with green withe or raw-hide.

Inserting the handle into a perforation or a socket in the working part was not a common practice before the age of metals. Africa now affords the best illustrations of this process in rude metallurgy. But the Eskimo harpoon-maker knew how to mortise holes in his ivory working parts and to make the handle fit therein. Similar devices are not common among other races. The stone workers of Europe, however, were ingenious enough to drill stone axe heads and furnish them with handles.[2]

There is a "doughnut"-shaped stone found in both Americas, in Australasia, and in Europe whose function is not clearly made out. Sometimes it is called a digging-stick weight, and again a club head. But the handle passes through the stone and is held in place by an abundance of cement.[3]

The modern hammer, hatchet, adze, axe, and so forth have all good handles of hickory, but the ancient maker of stone implements fixed his edged and striking tools to handles in some other way. Though most beautiful perforated axes of stone were produced in the European stone age, they are too pretty for use. The working part with an eye for hafting came with metals.

The modern flail, the mediæval "morning star," are of a

[1] Lafitau, *Mœurs des Sauv. Amer.*, vol. ii. p. 111.

[2] Evans, *Anct. Stone Implements*, chaps. viii. and ix.

[3] Figured in Whymper's *Andes of the Equator*.

class whose method of hafting is well known in aboriginal workshops. I speak of the sling hafting. The Indians of the Plains sew up a round stone in green raw-hide, and attach the projecting portions to a stiff handle. The same tribes strengthen the attachment of their great stone mauls in a similar way. Indeed, the withe seems to furnish the rigidity and grip, while the raw-hide does the work of attachment. The long lines of the bolas and the sling are extensions of this method of having a flexible portion between the grip and the working part.

But the savage man's unfailing friend in holding together the parts of his tools is a seizing of some sort. It is so easy, so effective, so readily repaired, and it makes the handle stronger instead of weaker. Hence the Polynesian gentleman, when he goes out to visit or sits in the shade of his own vine and fig-tree, takes along a good quantity of cocoa fibre and braids it into sennit. If the reader never saw a roll of sennit, it will pay him to visit the nearest ethnological museum for this sole purpose. The uniformity of the strands, the evenness of the braid, the incomparable winding on the roll or spool, as one might call it, constitute one of the fine arts of Oceanica. But prettier still are the regular, geometrical wrappings of this sennit when it is designed to hold an adze blade and handle in close union. While speaking of this combining substance, it may as well be said that in the building of houses the framework is held together entirely by the braided sennit. The strakes of a boat are united by its means. In short, whatsoever is wrapped for amusement or seriously, and whatsoever is nailed or screwed or pegged or glued in other lands, is in this region united by means of this textile.

The peoples of the world who live north of the tree line, and many who dwell in more temperate zones, have discovered the virtue of raw-hide. The Eskimo spends many hours in cutting out miles of raw-hide string, or babiche, of all degrees and sizes. This he uses in holding together not only the parts of his implements, but in manufactures of every kind. It is a marvellous substance. Frost that will snap steel nails like glass has no effect upon it. When it is put on green and allowed to dry, it shrinks nearly one half, binding the parts immovably.

Further south, as well as in the Arctic region, the tough sinew is taken from the leg of the deer. It is shredded as fine as silk, spun into yarn, and then twisted or braided into cord. This has no end of uses, not only in tool making, but in all arts where the greatest possible toughness and pliability are demanded. It serves to make a secure ferrule on the awl handle, to strengthen the bow, to hold feather and head on the arrow. It has an economic use for every day in the year.

FIG. 3.—Hupa Dagger, Northern California, showing how the "leaf-shaped" blades were mounted in pitch.

All aborigines found out the art oi uniting the parts of tools by means of strings, made of the best textile the country afforded. Whatever deficiency they suffered in their materials or rude tools was met by string of some kind. The Fuegians are very clever in the manufacture of harpoons with long shafts. The barbed heads of bone are securely attached by string, and the Eskimo unites thus the many parts of his harpoon so ingeniously that if one be broken the pieces cannot be lost.

The poorest savages can make glue of some sort, and—which cannot be too often repeated in view of the frequent scandals heaped upon them—they will in Australia, or in Guiana, or in North America, tell you the best formula for glue that can be made on that spot. The coast tribes and the Shoshonean tribes of Western America produce excellent animal glue for holding together the fibre of the sinew backing of the bow.

The Eskimo makes cement of blood. The Utes and the Apaches, the Mohaves and the Pimas, always carry a stick, on the end of which is a mass of pitch or mezquit gum ready to heat and cement their arrow heads.

"The Hurons," writes Sagard, "with small, sharp stones extracted blood from their arms to be used to mend and glue together their broken clay pipes or pipe-bowls (*pippes ou petunoirs*), which is a very good device, all the more admirable, since the pieces so mended are stronger than they were before." [1]

For cements the Panamint Indians, of South-western California, used a glue made by boiling the horns of the mountain sheep, pitch gathered from the Nevada nut pine (*Pinus monophylla*), and a gum found upon the creosote bush (*Larrea Mexicana*). In its crude form the larrea gum occurs in the shape of small, reddish, amber-coloured masses on the twigs of the shrub, and is deposited there by a minute scale insect (*Carteria larrea*). The crude gum is mixed with pulverised rock, and thoroughly pounded. The product, heated before applying, was used to fasten stone arrow-heads in their shafts.[2]

The karamanni wax or pitch is prepared as follows: the basis is a resin drawn by tapping from a tree (*Siphonia bacculifera*), and is mixed with beeswax to make it more pliable, and with finely powdered charcoal to make it black. While in a semi-liquid state it is run into a hollow bamboo, or allowed to harden in the bottom of a buckpot. It is used as pitch to fill up crevices in woodwork, as, for instance, in boat-building, to fix the heads of arrows in their shafts, and in similar work.[3]

Quite similar in tenacity is the "black boy gum" of the Australians, used in great profusion in the manufacture of their implements.

Rivetting together the parts of a tool is by no means

[1] Sagard, *Le Grand Voyage du Pays des Hurons*, 1636, vol. i. p. 189.

[2] Coville, *Am. Anthropologist*, 1892, vol. v. p. 361.

[3] Im Thurn, *Indians of British Guiana*, London, 1883, p. 315.

unknown to savages. The same process is also applied to other sorts of joining. Metallic rivets were not employed, but little trenails or trunnels of bone or wood or antler. In some of the woman's knives brought home from Greenland the parts are united by means of little pegs or trenails of antler. The parts of harpoons are also thus joined. After the use of metal became common among these people, they came to be very expert at rivetting their knife-blades of various kinds upon the handles.

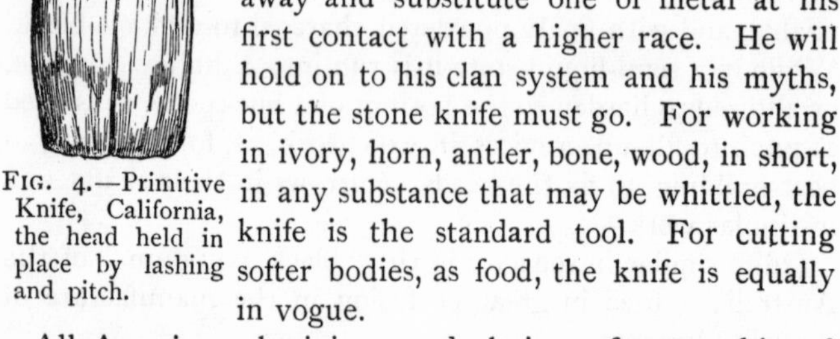

FIG. 4.—Primitive Knife, California, the head held in place by lashing and pitch.

And now it will be possible to follow the common tools of savagery in the order laid down in the classification above.

The jack-knife, the drawing-knife, and implements of that class are indispensable to the lowest grade of mechanic. When only stone is available, he fabricates his knife of stone; under other conditions, of the teeth of sharks and beaver, or of shells. But nothing demonstrates his absolute dependence upon the knife so convincingly as his willingness to throw the stone blade away and substitute one of metal at his first contact with a higher race. He will hold on to his clan system and his myths, but the stone knife must go. For working in ivory, horn, antler, bone, wood, in short, in any substance that may be whittled, the knife is the standard tool. For cutting softer bodies, as food, the knife is equally in vogue.

All American aborigines made knives of stone, chipped or ground, as the occasion or the natural resource demanded.

The African used his assegai for many purposes of the same sort, while throughout the Eastern Archipelago bamboo knives are in vogue, made while the stalk is green, and thus dried and charred to give them edge.

The Eskimo and Indians in whittling cut toward the body, and frequently make the handle of the knife long and curved so that the end will fit on the muscle of the forearm, to give a stronger grip and leverage. The modern curved knife only takes the place of one with stone blade, and it may now be seen throughout the whole intercontinental area from Lapland to Labrador.

The Polynesians had no other knife than a piece of bamboo cane. The serrated edge of the tool was formed in the extreme outer rind of the bamboo, and when the material has been recently split this edge is very sharp. And Ellis expresses his astonishment at the facility with which a large hog could be cut up with no other instrument.[1]

The readiness with which the peculiar structure of the cane and the bamboo has been seized upon everywhere for domestic knives, assists in the interpretation of the oft-repeated maxim that similar inventions spring from like environments and stress.

The shears of savages do not work like those of the civilised. There is not a pair of cutting edges, one working along the other. There is only one cutting edge, and the other piece is held at right angles. Indeed, there is no cloth or ribbon to be cut, only skins and human hair. The savage mother holds a bit of wood or leather against the child's head and haggles off the ends of the hair with a sharp stone, or a shell. The finishing touches are given with a fire brand. This practice was common among all American tribes.

For cutting the skins of animals the modern shears were preceded by the woman's knife,[2] called *ulu*, among the Eskimo. This consisted formerly of a blade of chert inserted

[1] *Polynes. Researches*, vol. iv. p. 346.

[2] Mason, "The Ulu or Woman's Knife," *Rep. U. S. Nat. Mus.*, 1890, pp. 411–416.

into a handle of ivory or wood, and glued fast. But even conservative Eskimo women obeyed the law of utility, and substituted iron blades on the advent of the whalers. All other women in the primitive world used similar shears, cutting skin as the modern saddler does, who has not a pair of shears in his shop.

The Algonkian Indians of North America secured splints of elm, birch, ash, and other hard woods of uniform thickness, by beating a log until the annual layers were loosened.

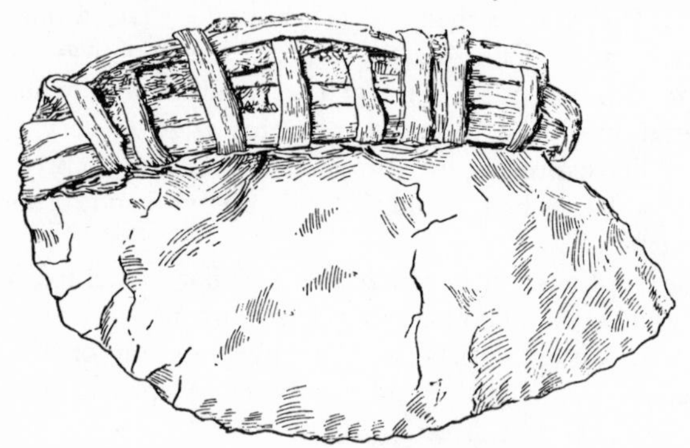

FIG. 5.—The "Woman's Knife," showing primitive form of grip.

They were then peeled off, scraped, and dressed into ribbons of the same width and woven into basketry.

For the jack-plane and the smoothing-plane, savagery has no mechanical substitute. There the set gauge to determine the thickness of the shaving is the thumb, which, in lieu of a better one, does tolerably well. The drawing-knife, the spokeshave, and such refined modern cutting tools, are all the lineal descendants of the primitive jack-knife, or curved knife, indeed, of the flake of flint or other hard stone struck off and used at the cutting edge. Lucien Turner, however, collected genuine little spokeshaves, with blades of chert, for dressing whalebone.[1]

The mechanic's edge tools in civilisation are axes, adzes,

[1] Collections of the United States National Museum.

and chisels of some sort. In general terms these work across the grain, with the grain, and into the grain. The lines are very feebly drawn in savagery. The very same stone blade is inserted into an antler and mounted on a helve for an axe, attached to a forked handle for an adze, and bound to the shouldered end of a straight handle for a chisel. The axe of savagery is a laborious tool, requiring great force and doing little execution. The adze is better, and in the culture areas where great trees abound near water, no aboriginal work is more attractive than the canoes tooled down with stone adzes. The chisel of savagery was seldom struck with a mallet. It was shoved from the workman after the manner of the modern trimming-chisel, and employed chiefly in connection with fire, as in hollowing out canoes. The invention of the tenon and mortise, the peculiar creation of the chisel, belongs to a culture-status in which domestic animals and extended commerce enter. Both in the East and the West Indies excellent adze and chisel blades were made of the great clam shells.

The Munbuttoo have an adze of iron which strongly calls to mind the socketted bronze celts of Scandinavia. A fork of a sapling serves for handle, one limb remaining long for the hands, the other cut short and inserted into the conical socket of the blade. "With this tool," says Schweinfurth, "Monbuttoo rough-hew their wooden vessels, subsequently smoothing and carving them more finely with a one-edged knife." [1]

The inhabitants of the Nubian part of the Nile valley use this mattock-like tool almost exclusively for all kinds of woodwork, while a real hatchet is never employed.

Saws are used by workmen in civilisation for cross-cutting and for ripping. The savage does not use the saw for the latter purpose. He gets out puncheons and planks by means of innumerable wedges distributed along a great log. Bone and harder substances he rips by boring a series of holes

[1] Schweinfurth, *Artes Africanae*, London, 1875, p. xxiii, fig. 11. In Finland iron axes preserve the peculiar shape of the ancient bronze blades.

through the substance in a straight line, and then breaking the pieces asunder with a blow. The rip-saw is in full force in China, Japan, and Corea. In ancient Egypt bronze saws were used, but the ripping was done single-handed.[1]

The cross-cut saw, on the contrary, is one of the oldest tools. There is no tribe of men who do not know how to haggle off a piece of wood by sawing with a jagged stone. This same method is used in separating antler, horn, ivory, and other industrial substances. The archæologists find among their collections blades of hard material serrated, and appearing to have been designed for saws. They will do the work excellently, and they seem to suit no other purpose. This tool must not be confounded with the stone-cutter's method of sawing stone and other hard substances by using sand and water.

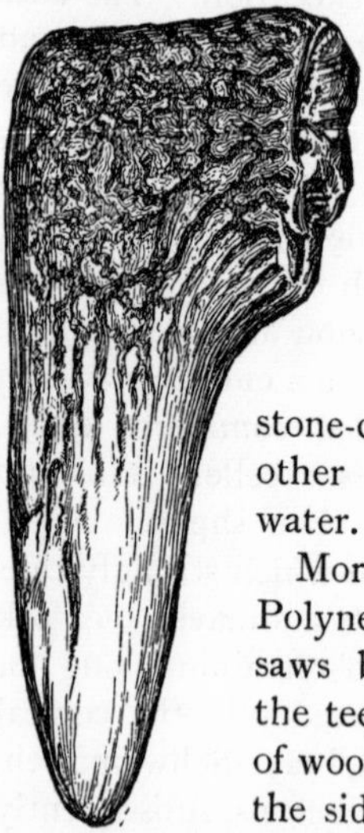

FIG. 6.—Wedge made from antler of elk (*Cervus Canadensis*).

Moreover, the ancient Mexicans and some Polynesian islanders knew well how to make saws by inserting bits of jagged stone and the teeth of sharks in a groove in a handle of wood or by sewing them with sennit upon the side of a thinner piece. The Australian saw-teeth are fastened to the handle with the "black boy gum."[2]

But the most efficient saw in savagery was a thin piece of stone, wood, or other soft substance used in connection with sand, to be described in the chapter on lithotechny.

The second class of common tools that have their ancestry in savagery are those that are used for abrading and smoothing surfaces. When the potter has finished shaping a vessel, the

[1] Wilkinson, *Anct. Egyptians*, New York, 1854, Harper, vol. ii. p. 118, fig. 1.

[2] Wood, *Unciv. Races*, Hartford, 1870, vol. ii. p. 35, with fig. on p. 36.

surface is corrugated and covered with finger prints. By the use of bits of leather, or gourd, or stone, she scrapes away these inequalities, and leaves the surface without a mark upon it.

The box-maker, the boat-builder, the fabricator of war implements, the worker in bone and horn and ivory, take away the inequalities from the surface of their industrial products in two ways—by scraping and by grinding, as is done to-day. The cabinet-maker with his wood rasps and his steel scrapers has his counterpart in the savage worker with scraping tools and grinding tools of stone. The Fijian war-club maker, the American boat-builder, the African metal-worker, grind and scrape away a deal of their material in bringing the article into shape. The North American Indians use sandstone, or fish skin, or grass ; the South Americans, the palate bones of certain fish, and the rough leaves of trumpet wood, *Cecropia peltata*, or of the *Curalitta Americana ;* the Polynesians employ pumice and coral ; and each location has its peculiar method of procedure.

When Europeans first opened trade with the South Sea Islanders, steel fish-hooks were among the things pressed upon the attention of the natives. But these last, or the fish, we had better say, like the mother-of-pearl hooks better. But the metal points were sharper, so nails and wire were in great demand. Perceiving in the nails a close resemblance to the scions from the root of the breadfruit tree, the fishermen actually planted some, expecting them to grow. There were no files to be had, so the nails were formed into shape and ground and bent by the use of stone. The introduction of the file wrought as much change in native art here as it did in the New World.

All of these processes of breaking, boring, sawing, cutting, grinding, and polishing are shown by Professor Putnam in his paper on the manner in which bone fish-hooks were made in the Little Miami Valley. A series of partly finished examples were taken from a grave in the Madisonville Cemetery, near Cincinnati.[1]

[1] *An. Rep. Peabody Mus.*, Cambridge, 1887, pp. 581–586, 11 figures.

Engraving, or ornamentation answering to the graver's art, was produced on softer substances by means of a blunt pointed, hard tool, and the design traced out by a series of creases on the surface. This is done on wood, bone, and pottery. But most of the decoration of this class was accomplished by scratching away the material with chips of flint or other hard substances. The Eskimo used to rely upon the hard tooth of the beaver, the Polynesian wrought with sharks' teeth, and in other places hard shells and gravers of flint were employed.

The Indians of Central America are expert in the engraving and painting of calabashes. With a pointed instrument they work out designs upon the surface of a dish and give relief to the ornamentation by roughening the intervals. In painting them the blue is made with indigo, the red with anotto, and the black with indigo mixed with lemon juice. The colour is fixed by means of a greasy substance formed by boiling an insect called *aje*.[1]

For giving a polish to surfaces, grass containing silex, very smooth stones, ochres laid on buckskin strips, or the hard hands were quite sufficient. Experiments lately made in the United States National Museum demonstrate that the objects mentioned are quite adequate to the result, with patience and knack. The archæologist is frequently puzzled in studying prehistoric methods of working, because all traces of chipping and sawing are obliterated by the polisher. But, in a great collection of polished objects like that of Commodore Douglas, in New York, or the jade objects in the British Museum, it is hard to believe that every one of them was first battered into its present shape.

Akin to the burnishing and polishing of the surface of different wares is the whole genus of greases, oils, varnishes, and other devices for filling the grain of the substance and giving a better shine. The idea of preserving

[1] Morelet, *Trav. in Cent. Am.*, New York, 1871, p. 314. Juarros mentions the *aje* among the drugs of Vera Paz, lib. i. c. 3.

wood by the use of paint and oils hardly entered the savage's mind. The study of paint as a purely decorative matter belongs to æsthetology. But the investigation of surfacing would be deficient if it did not include inquiry concerning paints and varnishes and burnishing powders.

The oil used by the Guiana Indians to anoint their bodies and their weapons is prepared from the crab-wood tree (*Carapa guianensis*). At the proper season the nuts are gathered, boiled and put away until half-rotten. They are then shelled and kneaded into a coarse paste. Troughs of bark, cut in form of a steel pen, are filled with the nut-paste and fixed in a sunny place, slanting, and with the point over a vessel. The oil oozes from the paste and drips into the vessel below. Sweet-scented substances are added to overcome the rancid odour. Palm oil is also obtained by crushing and boiling the seed. The oil rises to the surface and is skimmed off with pads of cotton.[1]

The calabashes of the Sandwich Islanders are dyed in the following manner: When the fruit has grown to its full size they empty it by placing it in the sun. The dried contents are removed through an aperture made at the stalk. In order to stain the shell, bruised herbs, ferruginous earth and water are mixed and poured in until it is full. They then draw with a piece of hard wood or stone on the outside of the calabash, rhombs, stars, circles, waves, &c. After the colouring matter has remained within three or four days, they are put in an oven and baked. When they are taken out, the figures appear in brown or black on the outside, while those places where the outer skin had not been broken retain their natural bright yellow colour. The dye is emptied out and the calabash dried in the sun; the whole of the outside appears perfectly smooth and shining, while the coloured figures remain indelible.[2]

It is difficult to find a better example of the specialisation going on throughout all history of men in all grades,

[1] Im Thurn, *Indians of British Guiana*, London, 1883, p. 314.

[2] Ellis, *Polynesian Researches*, London, 1859, vol. iv. p. 372.

operated upon by the resources at hand and yet developing the local or tribal technique.

"The split-cane of the Rotang (*Calamus secundiflorus*) is buried in the leaf-mould in the bottoms of brooks by the Niam Niam until it becomes thoroughly blackened. This dyed material, mixed with the splints of the natural colour, is wrought into all sorts of geometric patterns."[1] The Indians of Washington State and Oregon have discovered the very same fact, and use splints of root, or sprouts, or straws, blackened in the same fashion. The Indian women bury the split roots of the spruce in marshes to get the dark-brown splints for basketry.

The Andamanese paint in water and in oil colours. White clay mixed in water is daintily laid on the body as well as on bows, baskets, buckets, trays, &c. This work is done by women. Oil colours are made by mixing ochres with fat of pig, turtle, iguana, dugong, oil of almond, &c. It is applied to the person as ornament or otherwise.[2]

Finally, the whetstone and the grindstone must find a place in the tool-chest of the primitive man. And they are abundant. Constant reports are sent to the Smithsonian Institution of the finding of huge masses of sand-rock whose surfaces show marks of constant use as grindstones. When it is remembered that every edged tool of stone has been many times ground, the number of these implements reported will not appear astonishing. The whetstone is only a portable grindstone, and those gathered in museums show by their surfaces and grooves what a variety of uses they have served.

Whetstones are found in shell-heaps, graves, and mounds all over the earth, and they are of the best material the locality affords. They are an empirical result of the highest order. Among modern savages the whetstone is universal. In its ancient forms the great variety of grooves and worn places testify to the many kinds of implements to which

[1] *Cf.*, Schweinfurth, *Artes Africanae*, London, 1875, p. xii. figs. 12–14.
[2] *Cf.*, Man, *The Andamanese*, p. 184.

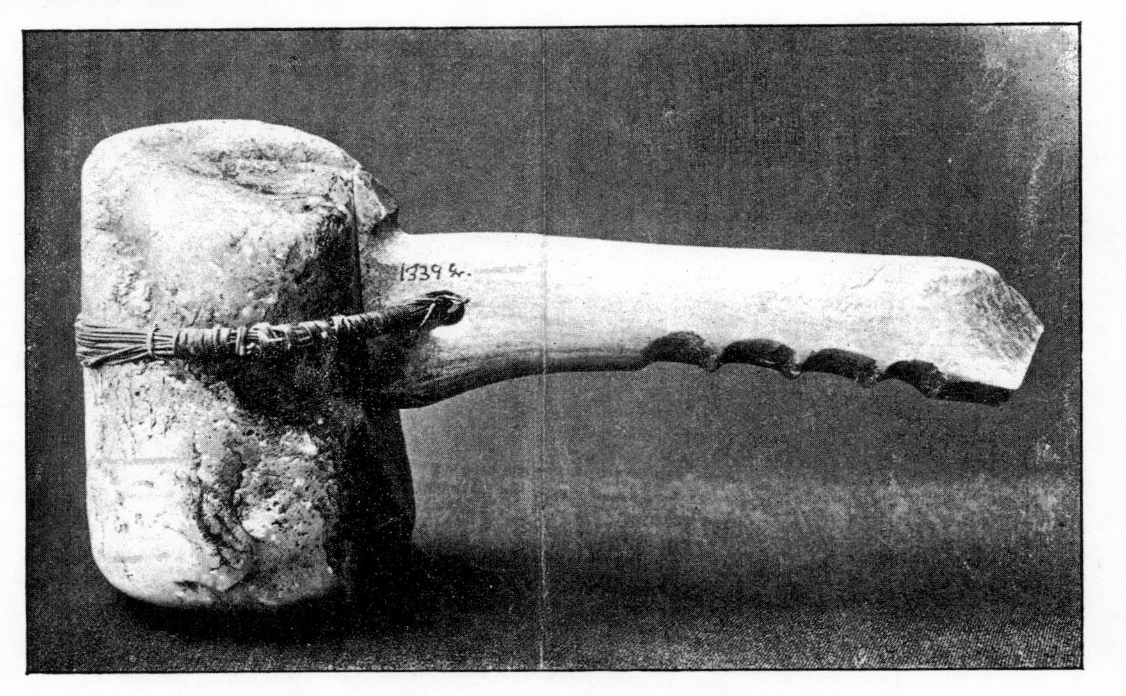

PECTOLITE HAMMER WITH HANDLE OF CARIBOU ANTLER. (*U.S. Nat. Museum.*)

they once gave point and edge. The Eskimo collections of our museums abound in good hones. The Andamanese wood-worker holds the blade of his adze over the inner side of his left foot and renews the edge with his hone. Many of the stone axes and hammers seen in collections show marks of having also been used as grindstones.

An implement of the greatest importance in the early history of mankind, universal in its use, found on ancient camp sites everywhere, is the hammer stone. It will be minutely studied in the chapter on stone-working. It seems strange that with all the ingenuity that our race can exercise it is yet necessary to abrade granite in the same way that the ancient Egyptians are represented as doing it, in the same way that primitive man did it, namely, by pecking and battering away the surface a few grains at a time.

But every man and woman in savagery needs a hammer, each in their several industries. The Indian women of North America with hammers of stone break dry wood for fires, crush bones to extract the marrow, pound dried meat into meal for pemmican, drive down pegs for setting the tent, beat the hides of animals to make them pliable. In this last operation they are imitated all over the tropical world by their sisters who hammer cloth out of the bark of trees.

The savage man uses his great hammers in driving wedges, in breaking off stone in the quarry, in mining, and as a pestle in pulverising various materials.

The North-west Coast Indians use a very graceful hammer, which is grasped in the middle like a dumb-bell. The pounding end is flattened out, while the other extremity is usually ornamented by carving. Hafted hammers are common in Eskimo land, in the canoe region of the Pacific Coast and in the buffalo country, each region adopting a characteristic method depending on the work to be done and upon the natural resources.

Prehistoric hammers and hammer heads are among the

commonest objects in collections. Those that are used as millstones or pestles are described in the proper place. The object in each case, whether with paint or with foodstuffs, is to crush and to pulverise without mixing any of the detritus of the apparatus with the product. The stone-chipping and flaking tools, developed in savagery and almost lost in modern times, save by the glazier and the gun-flint maker, will be described particularly in the chapter on stone working.

The making of holes by means of a punch struck by another body is the product of the metallic age. The African smith is not only acquainted with the art of engraving on the surface of his knives and assegais with punches, but he also makes holes by the same process. The other savages of the world do not perforate in this manner, but employ such tools as the needle or awl, thrust through soft substances ; the hand perforator, working like a reamer or a gimlet, and the drill operated by a string in a reciprocating motion.

There is no end to the sharp-pointed tools employed by both sexes among lowly peoples. They use them for sewing clothing, tents, utensils, for making basketry and other textiles. They have little stilettoes or prickers of bone no bigger than a needle, and others as strong as a marlinspike. Each one is a device exactly adapted and studied out for its work, so that the archæologist, finding a similar implement in some ancient *débris*, at once begins to set up in his mind the industrial life of a departed people.

With the two palms a drill is rotated after the fashion of the cook in mulling chocolate. It consists of two parts, a shaft of wood, with a point of hard substance lashed to the lower end. A beautiful specimen of this sort is in the United States National Museum, with a delicate point of the Alaskan jade. This would be capable of boring almost any stone object.

From this form, having a point fastened at the end of a shaft, have been invented the bow-drill, the two-handed strap-drill, the pump-drill, and the top-drill. The distribution of these three forms of drills is discussed under the chapter on fire. The same method of changing vertical or horizontal motion into rotary motion would be available alike in creating fire as in boring holes. Mr. Hough, who has studied the fire problem thoroughly, is decidedly of the opinion that the mechanical drill is older than the fire drill —in short, that the heat developed in boring holes led up to the creation of heat by this means.

The Samoan drill,[1] used in boring the pearl-shell shanks of fish-hooks, is precisely the same as the pump-drill used by the Pueblo Indians of the United States. In the Samoan example the crossbar or handle does not seem to have been perforated for the shaft.

The Hawaiians were acquainted with the rotary drill for boring.[2] In the island of Lombok Wallace saw the primitive gunsmith at his work.

"An open shed with a couple of small mud forges were the chief objects visible. The bellows consisted of two bamboo cylinders, with pistons worked by hand, having a loose stuffing of feathers thickly set round the piston, so as to act as a valve. An oblong piece of iron on the ground was the anvil, and a small vice was set on the projecting root of a tree outside. The apparatus for boring the barrels was a strong bamboo basket, spheroidal in shape, through the bottom of which was stuck upright a pole about three feet long, kept in its place by a few sticks tied across the top with rattans. The bottom of the pole had an iron ferrule and a hole in which four-cornered borers of hardened iron can be fitted. The barrel to be bored is buried upright in the ground, the borer is inserted into it, the top of the vertical shaft is held by a cross-piece of bamboo with a hole in it, and the basket is filled with stones to get the required weight. Two boys turn the bamboo around. The barrels are made in pieces about eighteen inches long, which are

[1] Minutely described in Turner's *Samoa*, London, 1884, p. 169.

[2] Brigham, *Cat. Bishop Mus.*, Honolulu, 1892, vol. ii. p. 31.

first bored small, then welded together upon a straight iron rod."[1]

The last type of common tools whose evolution commenced with early man to be mentioned here is the series of gripping implements. Tongs, pincers, vices, and all such things are represented in the aboriginal tool chest. All these devices are temporary expedients for holding two or more objects firmly together until they can be made fast by sewing or lashing, or they are designed for holding on to hot objects or small objects while they are being wrought. The words "vice," "tongs," "nippers" cover the three classes.

FIG. 8.—The Primitive Vice. Eskimo method of making a ladle. (*U. S. Nat. Museum.*)

In the collection brought home by E. W. Nelson from Alaska there is a very primitive vice just as effective for the work in hand as one made with a screw would be. The woodworker is about to make a dipper out of a thin spruce board. He rolls one end of the board into a cylinder after thoroughly boiling it, leaving six inches of the other end still free and unbent to be fashioned into a handle. To hold the bent end fast and tight to the part of the board against which it rests until it could be secured by sewing with whalebone or tough fibre, two sticks a little longer than the board is wide or the cup is deep are laid parallel to each other, one without and one within the cylinder, and their projecting ends tightly lashed together with fine, wet spruce

[1] Wallace, *Malay Archipel.*, New York, 1869, p. 179, with figure.

root. In drying the root contracts and holds the surfaces together water-tight. A block of wood is then fastened in one end of the cylinder with wooden pegs, and the dipper is completed. Several pieces that are in the United States National Museum have been made in the same fashion, and doubtless with a vice as crude and effective as Mr. Nelson's specimen. The capability of raw-hide and sinew for shrinking and holding things together so that they could not budge was well known and constantly utilised all over North America. These and other savages also knew that twisting a cable shortened the length and served as a press.

The Bongo smith uses a smooth gneiss boulder for his anvil, another smaller one for a hammer, with the cunning hand of the operator for a handle. For pincers he splits the end of a stick of green wood, seizes the hot mass between the jaws, and holds them firmly together by an iron ring slipped along the stick. The same tongs are mentioned by Speke among the Wanyamuesi.[1]

In the enumeration of the chest of tools belonging to savages we must not omit the teeth, which among seamstresses and other craft people could not be dispensed with. Every osteologist has noticed how the teeth in the crania of savages are worn to the socket, and we are frequently told that this arises from the large quantity of sand in the food. Basket-makers all the world over use their teeth in peeling and cutting their strands or filaments, and the Eskimo boot-maker uses her jaws for crimping irons. Whoever has seen an Eskimo boot neatly puckered all around the edge of the sole will not be surprised at the brevity of the good woman's teeth when he comes across her skull in the museum.

An original and very simple press is found among the Haida of Queen Charlotte Sound. Bancroft says, "After a sufficient supply of solid food for the winter is secured, oil, the great heat-producing element of all northern tribes, is extracted from the additional catch, by boiling the fish in wooden vessels, and skimming the grease from the water or

[1] Schweinfurth, *Artes Africanae*, London, 1875, pl. v. fig. 8.

squeezing it from the refuse. The arms and breasts of the women are the natural press in which the mass, wrapped in mats, is hugged. The hollow stalks of an abundant seaweed furnish the natural bottles in which the oil is preserved for use as sauce, and into which nearly everything is dipped before eating." [1]

The subject of the knots used by savages would require a book. The arrow-maker, to begin with, has great faith in tucking the ends under. So has every implement user who desires to separate the parts readily. The manipulator holds his left thumb on the end of a string, and in wrapping simply covers up this end. At the finish the last end is tucked under and concealed so as seldom to get loose. The different hitches and knots of the sailor are all well known to the uncivilised. On Polynesian spears and nets will be observed the whole series of ties that one would see on a ship.

The Arctic peoples have developed an entire series of tools and implements that have been made to take specialised forms by reason of the snow and ice. They put diminutive snow-shoes on the bottoms of the long staves which they use for canes or alpenstocks. From huge plates of bone taken from the scapula or the jawbone of the whale, or from slabs which they split from driftwood, they construct shovels, lining the cutting edge with thin plates of walrus ivory. To the back a handle is securely lashed by means of raw-hide. This is for removing the soft snow. But against the hard ice and frozen snow they have also a remedy in the form of a pick of walrus tusk. This may be lashed to a straight handle to form a crowbar, or at an angle to constitute a pickaxe. These are held to the handle by walrus hide as tight as a tire on a wheel by wrapping when the skin is green. The shrinking binds the parts so tightly together that the whole tusk of a huge walrus is worn quite out before the lashing comes loose.

They make tiny scoops and strainers for dipping the

[1] Bancroft, *Native Races*, New York, 1874–76, vol. i. p. 163, quoting many original authorities.

broken ice from a seal hole, and paper-knife clothes whisks to scrape the snow from clothing. The eyes are protected by snow goggles, which are cups of wood with narrow slits cut across the bottoms and inverted over the eyes. At once these devices keep the annoying snowdrift out of the eyes, and prevent the brilliant reflection of the snow from blinding the hunter. They put under their boots ice creepers also made of ivory, and precisely similar to those worn in Europe. The trowel for cutting out blocks of snow and building up the cunning, dome-shaped habitations must not be overlooked.

Having to do his work with gloved hands, the Eskimo has

FIG. 8.—Wrench for straightening wood or bone. (*Emmons collection U. S. Nat. Museum.*)

thought out an ingenious series of toggles, swivels, detachers, "frogs," buttons, any one of which will do its work, and some of them enable the hunter to make fast and cast loose frozen lines after a whole day's drive. He also has an ingenious wrench for winding up his sinew-backed bow.[1]

It is time to turn to the primitive knowledge of mechanics. By the mechanical powers is meant that series of devices which enables one man to do the work of several by the interchange of time and direction and momentum, namely, the *inclined plane*, *the wedge*, *the lever*, *the wheel and axle*,

[1] For tools of the Eskimo, systematically described, see Murdoch, *Ninth An. Rep. Bur. Ethnol.*, Washington, 1892, p. 150–190, with many illustrations.

the *pulley*, and the *screw*. One does not expect to find all of these full fledged in the lowest savagery, but the intimations of them all are to be looked for among very primitive folk. It is not true that any mechanical power has been lost. The great engineering feats of the megalithic epoch were performed with powers well known in our day, acting through co-operation.

The screw, the pulley, and the wheel and axle, are known to savages only in a very rudimentary way. Dr. Boas represents a plug used by the Baffin Land Eskimo to thrust into a spear wound on a seal to prevent the escape of blood. A sort of "thread" is cut on this wooden plug, and if the object be entirely a product of native thought, is the most primitive example of the screw.[1]

The Eskimo also approached a knowledge of the power of the screw in the tightening apparatus on the back of their bows and in their wolf traps. They know that tremendous power was accumulated by winding a cable of sinew by means of a lever. A very ingenious device, involving the lever of the third kind, and coming as near to the screw as we shall be able to find in savagery, is the cassava strainer of the Guiana Indians. After the roots are ground or grated, the pulp is placed in a long woven bag or cylinder, in which the warp and weft of tough splints run spirally and diagonally, so that when the two ends are forced together the cylinder becomes short and wide, and when they are pulled apart, it becomes long and slender. As soon as the squeezer is drawn into its shortest length and filled with pulp, one end is suspended from a tree overhead, and one end of a log of wood is thrust through the lower loop of the squeezer, the other extremity of the log resting on the ground. The woman then sits on the log, and by her weight gradually elongates the bag and squeezes the poisonous juice out of the mixture, the interstices in the woven fabric of the press acting at the same time as a sieve. These cassava squeezers are to be seen in

[1] Boas, *Sixth An. Rep. Bur. Ethnol.*, Washington, p. 480, fig. 402.

most museums, together with the graters, which are nothing more than flat blocks of wood into whose surfaces little bits of flinty rock have been firmly set. The whole apparatus

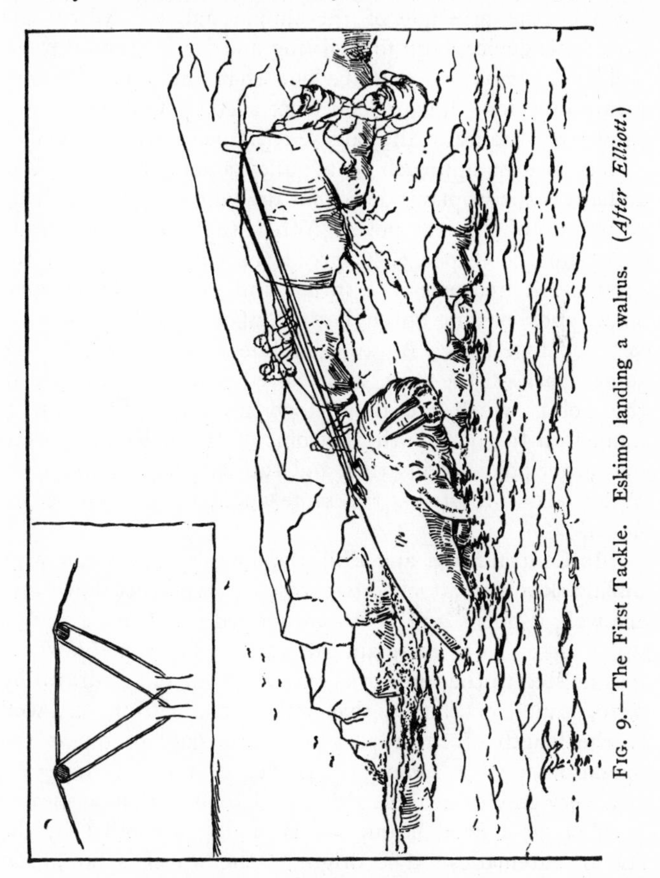

FIG. 9.—The First Tackle. Eskimo landing a walrus. (*After Elliott.*)

is entirely aboriginal, and the basket work of the press constructed with exceeding neatness and skill.[1]

The pulley may exist, and did primarily exist, without

[1] *Cf.* E. im Thurn, *Indians of British Guiana*, London, 1883, p. 260 *seq.*

the wheel, in the form of the "dead-eye." Any line drawn around a fixed object, as a tree, and pulled in one direction for the purpose of moving an object in another direction, involves the principle of the simple pulley. All savages know this device, both for hoisting and for horizontal work.

The Eskimo have gone beyond that, and know how, by means of a long line, to construct a compound pulley and draw from the water the carcase of immense sea mammals.

The nearest approach to a pulley among the American Indians is the woman's device for drawing the skin covering to the top of the tent poles. When the women are ready to set up the teepee, they spread the covering out on the ground. Three poles are thrust under the covering, their small ends passing through the orifice and being loosely fastened together. A raw-hide line is made fast to the upper part of the tent, and passed over the juncture of the poles, which are then stood upright. The tent is hauled up to the top, the bottoms of the poles are spread out, other poles are inserted, and the covering is stretched. When about to strike, the same apparatus lets the cover down.

"In Central Syria and Philistia, for raising water, a large buffalo-skin is so attached to cords that, when let down into the well, it opens and is instantly filled; and being drawn up, it closes so as to retain the water. The rope by which it is hoisted to the top works over a wheel, and is drawn by oxen, mules, or camels, that walk directly from the well to the length of the rope and return, only to repeat the operation until a sufficient quantity of water is raised."[1] It is very easy to imagine this wheel to be either a sheave, a roller, or a fixed beam, one becoming the other by the law of eurematics. The origin of the wheel is not made out. The precise mechanism of those we do see on Egyptian, Assyrian, and Grecian chariots and waggons is not clear to the minds of modern wheelwrights. The other wheel, used as a mechanical convenience in changing the

[1] Thomson, *The Land and the Book*, New York, 1880, vol. i. p. 20.

direction of a force or as a mechanical power, is still more difficult to follow up.

The roller is older than the wheel. One day, Mr. Henry Elliott came near catching a company of men inventing the roller. A crew of Eskimo rowed to a gravelly beach in one of their skin canoes. The craft was heavily laden, and they had either to get into cold water, to lift all the freight ashore and then carry the boat so that the gravels would not cut the very thin and delicate sealskin bottom, or they had to set their wits to work. As on many another occasion the inventive spirit predominated, and they placed a row of inflated seal-skin floats in front of the umiak, and rolled her high and dry up on the beach by this means. The very recent adoption of the pneumatic tire on bicycles and racing sulkies, after this explanation may leave the impression that Solomon was not altogether wrong when he said, "There is no new thing under the sun."

Long before the roller was invented, the pole road afforded an easy and slippery method of conveyance. Im Thurn describes the portage of a boat in the interior of Guiana. "We were obliged to carry our boat across the portage, which is about a quarter of a mile long, up and then down a very considerable hill. Our men laid rollers all along the path, then harnessed themselves by a rope attached to the bows of the boat, and drew her merrily over in a very short time." [1] The same method is in vogue in all mountainous countries for getting logs down to the level, and Robinson Crusoe would not have been compelled to dig canals if Daniel Defoe had been a South American Indian.

The windlass, the capstan, the winch, are modern appliances to convert time and momentum. The ancient engineers had rollers and chutes and greased ways. Even in savagery they could remove very heavy logs to the seaside, and stones weighing hundreds of tons were brought to the places where they were to be set up. Co-operation in great

[1] Im Thurn, *Among the Indians of British Guiana*, London, 1883, p. 61.

FIG. 10.—PRIMITIVE ENGINEERS. THLINKET INDIANS LANDING CEDAR LOG FOR COMMUNAL HOUSE. (*After Elliott.*)

labour took the place of invention ; but it must not be forgotten that this working together was an invention in social order of the highest value.

The inclined plane is found everywhere in ancient and modern engineering. The Pacific Coast Indians, in erecting their totem posts, and in laying up great crossbeams, use skids, guys, shore poles, and the parbuckle, besides their own main strength. In Africa, Corea, and in North-western United States, the porters draw their loads up on their backs by a strap which also act as a parbuckle.[1]

The lever and the wedge are well-known devices to savages. It has been previously mentioned that with wedges the California Indians felled trees, the British Columbia Indians split out immense planks, the metallurgists broke off masses of ore, and the engineers lifted great weights. The wedge was also understood in tightening the lashing of haftings, and in working clamps for holding objects together.

"I was interested," says Sir Samuel Baker, "in the mechanical contrivance of the Lobore for detaching the heavy metal anklets, which, when hammered firmly together, appeared to be hopelessly fixed in the absence of a file. The man from whose ankle the ring was to be detached sat on the ground. A stick of hard, unyielding wood was thrust through the ring, and both of its ends rested on the ground. A man stood on one end, and a stone was placed on the other end of this bottom stick. A lever of tough wood rested on the top of this stone as a fulcrum, one end passing through the ring. When the long arm was pressed down, it opened the jaws of the manacle, and released the man's foot.[2]

That system of counting and weighing and measuring, which lies at the basis of all tool-using, now demands our serious attention. To begin with, the sense of number is universal, and is found in a rudimentary state among the

[1] Mason, "Human Beast of Burden," *Rep. U. S. N. Museum*, fig. 42.
[2] *Ismaïlia*, New York, 1875, p. 268.

animals, but they have no notation nor any mechanical invention for recording numbers. Most of the tribes of men have adopted the quinary notation. But the only numerals in use among the Andamanese are those denoting "one" and "two," and they have no word to express specifically any higher figures, but they indulge in some such vague terms as "several," "many," "numerous."[1]

Among the North American savages the universal method of keeping account was by means of tally sticks

FIG. 11.—Land and water transportation in Mexico, showing tally. (*From Codices.*)

or shells or stones or notches, one for each unit being laid away or kept after some fashion. In the United States National Museum is an old census of a tribe of Comanches. It is simply a collection of bundles of straws, one for men, one for women, and one for children. Besides this example are many bundles of gambler's counters, which are simply short sticks tied together. One of the most charming things Mr. Wallace ever wrote is telling how the rajah of Lombok took the census.[2]

[1] Man, *Andamanese Islanders*, London, Trübner, 1883, p. 32. L. L. Conant, "Primitive Number Systems," *Proc. Am. Assoc. Adv. Sc.*, Salem, 1892, vol. xli. p. 270.

[2] Wallace, *Malay Archipel.*, New York, 1869, chap. xii.

Memory-helping devices for numbers, such as notched sticks or knotted strings, have a wide distribution. The message-sticks of Australia, the *rush* of the Pelew Islands, had their counter parts everywhere. The Maoris, says Tregear, used notched pieces of wood for this purpose, specially for recording genealogies. In China, the invention of memorising by knotted cords is attributed to the Emperor Luy-jin. Turner in his account of Nui (Ellice Group) says, "Tying a number of knots on a piece of cord was a common way of noting and remembering things among the South Sea Islanders." In Hawaii the tax-gatherers, although they can neither read nor write, keep very exact accounts of all the articles of all kinds collected from the inhabitants throughout the island. This is done by one man; the register is a line of cordage, distinct portions of which are allotted to various districts, which are known from one another by knots, loops, and tufts of different shapes, sizes, and colours. Each taxpayer has his part in this string, and the number of dogs, hogs, pieces of sandalwood, &c., he has to furnish is well defined."[1]

In every patent office there is an examiner of instruments of precision. The very mention of a standard yard or metre, of square feet or acres, of cubic inches or centimetres, of delicate balances and platform scales, of gallons or bushels, of degrees and their subdivisions, of clocks and chronometers and calendars, of pounds, shillings, and pence, awakens in the mind a consciousness of the nicety with which things are measured or weighed or paid for in our times. Only the astronomer, the chemist, the physicist, the microscopist, the great banking houses, know to what a degree of finesse all of these devices for getting the correct

[1] E. Tregear, *J. Polynes. Soc.*, 1892, vol. i. p. 127, referring to Keats; *Pelew Is.*, pp. 367, 392; Erman, *E. Travels*, vol. i. p. 492; Goquet, vol. i. p. 161, 212, and vol. iii. p. 322; Klemm, *Cultur-Geschichte*, vol. i. p. 3, and vol. iv. p. 396; Charlevoix, vol. vi. p. 151; Long, *Erep.*, vol. i. p. 235; Talbot, *Disc. of Lederer*, p. 4; Humboldt and Bonpland, vol. iii. p. 20; Marsden, p. 192; Tyerman and Bennett, *Journal*, vol. i. p. 455.

figures have attained. It will be interesting to note how, in the earliest industries the places of all these diversified measuring apparatus were filled. The correct metric or chronometric data within the exigencies of each tribal life will give a fair idea of the status of that tribe.[1] It is well known that the history of navigation is almost the history of clocks, that speed in trains is allied to red glass and signalling, that the accuracy of the cubit is the gauge of the quality of ancient architecture, and, in a general way, the history of metrology is the history of civilisation. A separate book on this subject would be worthy of preparation, only the data are so meagre.

Metric apparatus and instruments of precision include all devices covered by what in the school arithmetics is denominated "tables of weights and measures." The measuring appliances involved, and their numerical values in different ages constitute the science of metrology. This alone has had a very interesting elaboration. The lowest peoples have their standards of measuring and comparing quantity. Out of these have grown the modern processes.[2]

The scale or balance was known in America before the Discovery. The Peruvians made beams of bone, suspended little nets to each end, supported the beam at the middle by means of a cord, and used stones for weights. The transition from the balance to the "steelyard" is not easy to make out.

The standards of compound arithmetic were very low among the Andamanese. About forty pounds was a man's load, and anything above that would simply be more than

[1] A photographer, who lived fifteen years among a tribe of savages, told the author that the unit of capacity for small quantities was universally the double handful.

[2] For an elaborate study of the origins of metric standards, the reader is urged to consult William Ridgeway, *The Origin of Metallic Currency and Weight Standard*, Cambridge, 1892, Univ. Press, 418 pp., 8vo; *Mound Builders*, Cleveland, 1883; McGee, *American Antiquarian*, April, 1881, reviewing Petrie's *Inductive Metrology*, commented on by Dr. Brinton, *op. cit.*, p. 447.

a man's load. Size was rated by well-known natural objects, seeds, fruits, nuts, &c. Capacity was counted by handfuls, basketfuls, bucketfuls, canoefuls. There is no prescribed form or dimensions for any object. No tallies were kept nor counters, and this is very low down, because all American tribes knew the use of tallies. Distance was spoken of as a bowshot, or as from there to there, indicating the limits. Fifteen miles, about, was a day's journey, and over that was said to "exceed a day's journey."[1]

Those ancient manufacturers and builders had no government standards of measuring their work, but referred everything to their bodies. This system was far more accurate among rude peoples, where anthropometric differences between the sexes and between individuals were very slight. Many witnesses confirm the opinion that every weapon, or chunkey pole, had its proportion to the owner. Dr. Matthews says that the Navajo pole for the Great Hoop Game was twice the span long, and Mr. Dorsey found that the Omaha arrow had to measure from the inner angle of the elbow to the tip of the middle finger, and thence over the back of the hand to the wrist-bone. I have examined many hundreds of quivers, and have always found the arrows to be of the same length, while those of the tribe resemble in general appearance, but vary slightly in length for each man. Dr. Dorsey found the Naltunne, on Siletz Agency, in Oregon, using the double arms' length, the single arm's length, half the span, the cubit, the half cubit, the hand length, the hand width, the finger width (1, 2, 3, 4, 5), from the tip of the elbow across the body to the end of the middle finger of the other hand. In most of these cases the starting-point is the meeting of the tips of the thumb and index finger.[2]

Among the Aztec or Nahuatl and the Maya, the two most cultivated stocks of North American aborigines, Brinton finds no words for estimating quantity by gravity,

[1] *Cf.* Man, *Andaman Islanders*, London, 1883, Trübner, p. 116.

[2] *Science*, New York, 1892, vol. xx. p. 194.

no weighing terms. For extension the human body and, largely, the hand and the foot furnished standards of measuring. Among the Mayas the footstep or print or length of the foot was very familiar, and frequently in use by artisans, as well as the pace or stride.

Quite a series of measures were recognised from the ground to the upper portions of the body, to the ankle, to the upper portion of the calf, to the knee-cap, to the girdle, to the ribs or chest, to the mammæ, to the neck, to the mouth, to the vertex. Other measures were the hand, finger-breadths, the span, half around the hand, as in measuring for a glove, the cubit, the fathom. Journeys were counted by resting-places.

In Aztec metrology, the fingers appear to have been customary measures. The span was not like ours, from the extremity of the thumb to the extremity of the little finger, nor the Cakchiquel, from the extremity of the thumb to that of the middle finger; but like that now in use among the Mayas, from the extremity of the thumb to that of the index finger. There were four measures from the point of the elbow—to the wrist of the same arm, to the wrist of the opposite arm, to the ends of the fingers of the same arm, to the ends of the opposite arm, the arms extended always at right angles to the body.

The Aztec arm measures were from the tip of the shoulder to the end of the hand; from the tip of the fingers of one hand to those of the other, from the middle of the breasts to the end of the fingers. The *octocatl* or "ten foot pole," approximately, was the standard of length employed in laying out grounds and constructing buildings. The road measure of the Aztecs was by the stops of the carriers, as in Guatemala. The Aztecs were entirely ignorant of balances, scales, or weights. The plumb line must have been unknown to the Mexicans also.[1]

[1] D. G. Brinton, *Essays of an Americanist*, Philadelphia, 1890, pp. 433-451. This whole paper should be consulted. Charles Whittlesey, *Metrical Standards of the Mound Builders.*

Federal money and the metric system as applied to the mechanism of exchange are modern returns to very primitive modes of reckoning values. The basis of money is at times a shell, a bead, a robe, a skin. The purchasing power of the unit is fixed in each case. And among certain tribes there is a table of moneys, such as two elk teeth equal one pony, eight ponies equal one wife. The principal involved does not seem to be different from that of our own standards, namely, to have some rare and portable object for standards.[1]

The Bongo make iron spade-shaped disks, which represent their coined money.[2] The hoe-and-spade currency is widespread in Africa. Crosses of copper, and ingots of native iron hammered out from nuggets of iron ore pass for currency. Furthermore, to give to these objects the further semblance of coinage the manufacturers put a certain twist or mark on the object, which is in effect a tribal mark, and suggests the coins of the realm. These marks are not government stamps, however, and they do not raise the objects above the rank of tokens.

Although the native canoe-builders in the Louisiade Archipelago work with adzes made of hoop-iron, the payment for their work is made in stone axes, ten to fifty of these being the price of a canoe. The stone axe is still the accepted medium of exchange in large transactions—pigs, for instance, and wives are valued in that currency. It is only fair, by the way, to mention that the purchase of a wife is stated by the natives not to be such in the ordinary sense; the articles paid are, they say, a present to the girl's father. In Mowatta, sisters are specially valued, as they can be interchanged with other men's sisters as wives.[3]

Almanacks and clocks, how indispensable to all our activities! They were never absent from human traffic.

[1] Stearns, "Ethno-Conchology," *Rep. U. S. Nat. Museum*, 1887, with bibliography; also Cushing, *Am. Anthropologist*, Washington, 1892, vol. v,

[2] Schweinfurth, *Artes Africanae*, London, 1875, pl. iv. p. 14, 15.

[3] Trotter in *Proc. Roy. Geog. Soc.*, p. 795, Nov., 1892.

The Andamanese have natural calendars, partly in the sky, partly in nature around them. Having no numeration, they did not count the moons in a year, but noted the cool season, the hot season, the rainy season, in their proper order. The year was also divided into twenty minor seasons, named for the most part after trees which flowering at successive periods, afford the necessary supply to the honey bees. These flowers are used to name the children born while they are blooming, and these names, added to the prenatal name conferred by the parents, constitute the denomination of the person until maturity or marriage.

The phases of the moon and its connection with the tides were both designated by appropriate terms. Of the starry host they take little notice, confining their special observations to Orion and the Milky Way.

They knew the four points of the compass, and the prevailing winds by name, and distinguished certain meteorologic phenomena. So much for the calendar.

As to clocks, they had no mechanical device for marking time of day, but had thirteen separate expressions for known parts of the twenty-four hours. But these were extremely vague, and the divisions over-lapped one another. For that matter, clocks and watches are extremely modern devices.

The day's journey is often mentioned as a fixed distance. This is only true within wide limits, and it scarcely ever exceeds ten miles for marching. "The Indians, finding that their wives were so near as to be within one of their ordinary day's work, which seldom exceeded ten or twelve miles, determined not to rest till they had joined them." [1]

In these journeys the Canada Indian hunters are said to stand a stick in the snow and make a mark along the shadow as they pass some well-known spot. The women and old men coming later note the angle between the former and the present position of the shadow, and they are thereby enabled to regulate their future speed.

The Zuñi Indians know well that the light of the rising

[1] Hearne, *Journey*, &c., London, 1795, Strahan, p. 185.

sun falls on the same spot but two days in the year, and that at noon the shadow of a pillar lengthens and then shortens back to the same spot in the same period. They have a pillar dedicated to astronomical observations. On many houses in the Pueblo there are scores on the wall opposite windows, or loop-holes for the purpose of recording the movements of the sun. There are also pillars to be seen in other parts of the world which could possibly be dedicated to the same end, since such a feat is performed by at least one tribe.

"Each morning, just at dawn, the Sun priest, followed by the master priest of the Bow, went along the eastern trail to the ruined city of Ma-tsa-ki by the river side, where, awaited at a distance by his companion, he slowly approached a square, open tower, and seated himself just inside upon a rude ancient stone chair, and before a pillar sculptured with the face of the sun, the sacred hand, the morning star, and the new moon. There he awaited with prayer and sacred song the rising of the sun. Not many such pilgrimages are made ere 'the suns look at each other,' and the shadows of the solar monolith, the monument of Thunder Mountain, and the pillar of the gardens of Zuñi lie along the same trail; then the priest blesses, thanks and exhorts his father, while the warrior guardian responds as he cuts the last notch in his pine-wood calendar, and both hasten back to call from the housetops the glad tidings of the return of spring. Nor may the Sun priest err in his watch of time's flight; for many are the houses in Zuñi with scores on their walls or ancient plates embedded therein, while opposite a convenient window or small porthole lets in the light of the rising sun, which shines but two mornings of the 365 on the same place. Wonderfully reliable are these rude systems of orientation, by which the religion, the labours, and even the pastimes of the Zuñi are regulated." [1]

[1] F. H. Cushing, *Century Magazine*, quoted in *Nature*, London, 1892, March 17, p. 464.

In the Moki village of Wolpi, Arizona, there are means of telling noon and midnight. Fewkes says: "When the sunlight through the kibva [sacred chamber] entrance fell in a certain place on the floor and indicated noon time each of the four priestesses made a single *baho*, consisting of two willow twigs equal in length to the distance from the centre of the palm of the hand of the middle finger." Again, "At 12.15 the head priestess ascended the ladder and minutely examined from the roof the position of the stars. She looked anxiously for some star in the constellation of Orion or the Pleiades, but the stars she sought were hidden by a cloud, and she at last decided what she had in mind by observing a bright star in the western sky. Then she went down the ladder and announced that the time had come for the midnight ceremony." [1]

The ancient Polynesians had thirteen months in their year, regulated by the moon, and once in a while dropped out a moon. They had separate names for every night in the lunation, and twenty-seven separate names for time of day during each twenty-four hours. [2]

In the long voyages which they undertook about six hundred years ago, they made excellent use of the stars both for direction and time of day. In another chapter some mention will be made of fire as a time measure, but the near kindred of these Polynesians anticipated the hour-glass by boring a small hole in the bottom of a cocoanut cup, and placing it in a vessel of water, noting the time it took the cup to sink.

The reader well knows that the primitive folk were good meteorologists. That they knew something about natural thermometers and barometers and hygrometers may be gathered from the story of Gideon's fleece. Mr. Ling Roth contributes the following charming bit from the Malay:—

[1] Fewkes, "A Tusayan Dance," *Am. Anthropologist*, 1892, vol. v. pp. 109, 117. For time of day among the Navajo, see Matthews, "Mountain Chant," *Fifth An. Rep. Bur. Ethnol.*, p. 389.

[2] Ellis, *Polynes. Res.*, London, 1859, Bohn, vol. i. pp. 86–9.

"When the natives of Borneo are selecting the site for a new village a piece of bamboo is stuck in the ground, filled with water and the aperture covered with leaves. A spear and a shield are placed beside it, and the whole is surrounded by a rail. The latter is to protect the bamboo from being upset by wild animals, and the weapons are to warn strangers not to touch it. If there is much evaporation by the morning the place is considered hot and unhealthy, and is abandoned."[1]

The evolution of machinery cannot be ignored in this connection. A machine in this view is a contrivance for changing the direction and the velocity of motion or force. It cannot create force any more than a tool can. On the contrary, it consumes a vast amount of force in its own working. By means of a tool the entire force exerted is brought to bear upon the material. The machine, by the waste of a portion of the force enables the workman to apply his efforts more rapidly, more powerfully, or in ways unattainable by hand.[2]

All power at first was hand-power, the machinery of the world was moved only by human muscles. In the chapter on animals will be treated the gradual enlistment of domestic beasts in the service of man. Besides these, winds and water currents, vapours and electric currents and chemical force have been domesticated for human uses. The study of these is essential to a knowledge of industrial progress. Muscular power is the basis of all power, just as human backs will be shown later to be the basis of the carrying trade.

The Zuñi or Nicobar woman's simple potter's wheel, which is nothing more than the turning of her vessel about in a dish or basket as the work goes on, is only a little more rude than the fashion in the interior of China

[1] Ling Roth, *J. Anthrop. Inst.*, London, 1892, vol. xxii. p. 31.

[2] "A machine is a combination of materials arranged by man so as to enable determinate motions to be obtained" (Shaw, *J. Soc. of Arts*, London, 1885, vol. xxxiii. p. 395).

of putting a lump of clay on the top of a revolving shaft, which they turn with one hand while the pot is formed with the other.

"The potter's wheel was known in the world from high antiquity. The Egyptian artisan turned the wheel by hand. The Hindu potter goes down to the river-side when a flood has brought him a deposit of fine clay, when all he has to do is to knead a batch of it, stick up his pivot in the ground, balance the heavy wooden table on the top, give it a spin and set to work."[1]

The spindle with its whorl is a free wheel and axle, with the principle of the fly-wheel fully developed, and the twister, well known to savages, is a still simpler fly-wheel. The Zuñi Indians make a block of wood about 8 in. × 3 in. × $\frac{1}{2}$ in. Near one end a hole is made $\frac{3}{4}$ in. in diameter, and the stick is notched just outside this hole. This is the fly wheel. A stick with a head cut on it is thrust through the hole and serves for handle. One end of the material to be twisted is tied to the notch on the fly-wheel, and the other end to some fixed object. The twister holds to the handle and causes the fly to revolve by the motion of his hand.

The regular spindle serves for yarn-making, thread-making, and twine-making, and the product is wound on the shaft, which is twirled in a small vessel, rolled along the thigh, or sent spinning in the air, held up by the thread caught in a hook on the upper end. Here the operation stops, and the writer does not know of any primitive people to whom it occurred to fix the two ends of the shaft as journals in bearings. The nearest approach to such a device is the Eskimo drill; in which the piece held in the mouth furnishes the upper socket, the perforation being made the under socket and the bow or strap applying the power. The true wheel and axle reverses this process, and does its work where the Eskimo applies his force.

Crank motion applied to the potter's wheel is of very recent date. Dr. Smith, long resident in Siam, informed

[1] Tylor, *Anthropology*, New York, 1881, p. 275.

ZUNI WOMAN SPINNING WOOLLEN YARN. (*Photo in U.S. Nat. Museum.*)

the writer that the potter first gives an immense impetus to a fly-wheel, and then works the clay while the wheel is turning. The next progress forward is placing the heavy fly-wheel low down where the potter may keep it in motion with his toes. "So doth the potter, sitting at his work and turning the wheel about with his feet, he fashioneth the clay with his arm."[1]

In polishing the basket lacquer work, the Shans use a crude lathe. A bamboo basket is coated with lac or with lac mixed with ashes of straw. When the lac is dry, the basket is turned on a very simple lathe, the wheel of which revolves backwards and forwards, the principle of the crank being apparently unknown. The workman uses a treadle, which turns the wheel one way, and it is brought back in the opposite direction by a long bamboo which acts as a spring.[2] The reader should compare with this the exceedingly crude Moorish lathe in which the operator works a bow drill in one hand and uses his toes to assist the other hand to holding the cutting tools.

"There are strong grounds," says Shaw, "for considering the fire drill or twirling stick, first revolved between the hands of one or two operators, as one of the earliest examples of machinal motion, and that a long time must have elapsed before the introduction of continuous, instead of alternating rotary motion." But Mr. Shaw forgets the fly-wheel on the spindle, called usually the whorl. The spinning of fibre is as old as the fire sticks. Indeed, it would not appear that the fire sticks are among the oldest of human devices. Men had fire very long before they knew how to create it.

"It is extremely probable that the first continuous motion was employed in connection with the grinding of corn."[3]

[1] Ecclus. quoted in Tomson, *op. cit.* vol. i. p. 34; excellent plate, p. 35, showing two types of burden-bearing and the fly-wheel turned with the foot.

[2] Ernest Satow, *J. Soc. Arts*, London, 1892, vol. xl. p. 186.

[3] Shaw, *J. Soc. Arts*, London, 1885, vol. xxxi. p. 395.

Shaw arranges corn-grinders as:—(1) Simple stone pounder; (2) Mortar and pestle, worked (*a*) by slaves, (*b*) by bondsmen, *c*) by cattle; (3) flat cylindrical stone with vertical spindle. But in reality there have been two series, the mortar series and the grinding series, the order of which last would be (1) rubber and flat nature rock; (2) metate and muller; and (3) the rotary mill driven first by hand and after by animals, winds, and water.

The employment of the wind to separate chaff from grain is an appliance in primitive agriculture or harvestry. The utilisation of the wind in locomotion will be studied in the chapter on primitive transportation. The Indians of the Plains, who dwelt in skin lodges, understood the use of the fly and extra pole on the tent to utilise the wind in creating a draught and drawing the smoke out of the dwelling. The sail is also used in the Arctic regions to aid in driving the sledge over smooth ice. But no savage had any conception of a windmill, or invited the air to participate in doing mechanical work.

If I were permitted to coin a word, I should call all the combined arts that relate to the getting, preserving, and utilising water, hydrotechny; but that would furnish rather a long term for the study of these arts—hydrotechnology—though it is not lacking in euphony. The spring, the well, the city reservoir, and waterworks; the open stream, the canal, the locomotive; the tide wheel, the overshot, the turbine—all of these indicate progress in hydrotechny as related to aliment, to transportation, to irrigation, and to manufactures. The world's progress has followed the water, and water has never been absent from men's minds.

No aborigines, unaided by domestic animals, have displayed so much patience and ingenuity in the storage and conducting of water as the Indians of the arid region of the United States. Throughout the Pueblo region, says Mr. Hodge, works of irrigation abound in the valleys and on the mountain slopes, especially along the drainage of the Gila and the Salado, in Southern Arizona, where the inhabitants engaged in agriculture to a vast extent by this means. The arable tract of the Salado comprises about 450,000 acres, and the ancient inhabitants controlled the watering of at least 250,000 acres. The outlines of 150 miles of ancient main irrigating ditches may be readily traced, some of which meander southward a distance of fourteen miles. In one place the main canal was found to be a ditch within a ditch, the bed being 7 feet deep. The lower section was only 4 feet wide, but the sides broadened in their ascent to a "bench" 3 feet wide on each side of the canal. Remains of balsas were recovered, showing that the transportation of material was also carried on. Remains of flood gates were found by Mr. Cushing, and great reservoirs for storage of water, one example being 200 feet long and 15 feet in depth.[1]

In Mexico and Peru, especially in the latter, this art reached its highest perfection. "Higher up in the Andes irrigation was carried out on a far more extensive scale. Partly by tunnelling through the solid mountains, partly by carrying channels round their sides, the waters of the higher valleys, where the supply was abundant, were made available for the cultivation of others where it was deficient : and in the district between the Central and Western Cordilleras, to the northward and westward of Cuzco, such channels were extensively constructed to irrigate, not only the valleys, but the llama pastures on the mountain sides."[2]

In the evolution of hydrotechny the curious invention of the Bakalahari negroes has a place. The women dig tiny wells in the wet sand. They then fasten a bunch of grass to the end of a reed and bury it in the pit. By means of the reed they suck water into their mouths and discharge it into ostrich shells, using as a guide to the stream a stalk of grass. When twenty or thirty shells have been filled they are placed in a net, carried home and buried in the earth for future use.[3]

[1] F. W. Hodge, *Am. Anthropologist*, July, 1893, pp. 323-330.

[2] Payne, *History of America*, Oxford, 1892, p. 380.

[3] Livingstone, *Trav., &c., in S. Africa*, New York, 1858, p. 59, ill.

The wheel and bucket are in common use through the eastern continent. For lifting water out of shallow wells or sources of supply, a wheel may be used whose diameter is a little more than the vertical distance from the water to the point of discharge. On the rim of the wheel are buckets resembling those in an old-fashioned mill-wheel. The apparatus is worked by a draught animal. But, in more elaborate specimens of the same sort, the machine is set in a running stream, which, working against paddles on the rim, revolves the wheel and lifts the water. The Chinese make an enormous apparatus of this sort, and fasten bamboo buckets diagonally on the outside of the rim. These descending are plunged mouth first under the water, and ascending retain it until they pass the centre of motion, when they discharge into a trough. Thomson speaks of enormous wheels at Hums, on the Orontes, the diameter of some being 80 or 90 feet.[1]

The *nà'urah*, or Persian water-wheel, common throughout Western Asia, consists of a clumsy cog-wheel, fitted to an upright post, and made to revolve horizontally by a beast attached to a sweep. This turns a similar one perpendicular at the end of a heavy beam, which has a large wide drum built into it, directly over the mouth of the well. Over this drum revolve two rough hawsers, or thick ropes, made of twigs and branches twisted together, and upon them are fastened small jars or wooden buckets. One side descends as the other rises, carrying the small buckets with them, those descending empty, those ascending full. As they pass over the top they discharge into a trough. The buckets are fastened to the hawsers about 2 feet apart. The hawser is made of twigs, generally of myrtle, because it is cheap, easily plaited, and its extreme roughness prevents its slipping on the drum.[2]

[1] Thomson, *The Land and the Book*, New York, 1880, vol. i. p. 21.

[2] *Ibid.*, vol. i. p. 19. An excellent full-page plate represents a camel working the *nà'aurah*. The harness is extremely primitive. Compare figure vol. iii. p. 8.

In matters of engineering the starting-point backward is itself in a remote past. Watkins, in his "Beginnings of Engineering" says: "Of the races to be considered I will mention in what seems to me to be their order of importance, Chaldea, Babylon, Egypt, Assyria, Phœnicia, Etruria, Palestine, Moab, Persia, India, China, and the Incas. To this aggregate every form of engineering was known which did not require the application of the generated forces. They built canals for transport and irrigation, reservoirs and aqueducts, docks, harbours, and lighthouses. They erected bridges of wood and stone, as well as suspension bridges; laid out roads, cut tunnels, constructed viaducts, planned roofs for their massive buildings; tested the strength and discovered the weakness of their building materials; instituted elaborate systems of drainage; planned fortifications; designed engines of attack and floating bridges; devised methods for the transport of heavy objects—in fact, covered to a greater or less degree all departments of hydraulic, bridge and road, sanitary, military, and mechanical engineering."[1]

Assuredly even these enterprises were the mature results of still earlier efforts, which it would be delightful to trace. In the earliest engineering feats two facts must be sharply kept before the mind, to wit, that time was no object, and that there were no private buildings. Suppose that every labouring person in London should be immediately withdrawn from all private work, and that they all should be organised to labour for ten years upon some government building as a memorial of the reign of Her Gracious Majesty. One million hand labourers would erect a pyramid containing fifteen thousand milliards of tons of earth, and the mechanics would put on the top of it a structure larger than all the monuments in Egypt combined.

The only puzzle the modern student can have is to conceive how the ancient engineer made and moved his crib-

[1] Watkins, "Beginnings of Engineering," *Trans. Am. Soc. Civ. Engineers*, vol. xxiv., No. 5, 1–76, p. 800.

work. It is within the ability of a company of savage Indians to hammer down any great stone into any form. It is customary for them as a tribe to all engage in the same operation in hauling logs, or seines, or boats, or stones. The problem is somewhat like that of Archimedes, "Given a rope long enough, and a crib-work strong enough," and any modern savage people will undertake to set up the monuments of Brittany.

"The usual method of removing the iron open rings worn on the ankles by the Madi requires a number of men. A rope is fastened to each side of the ring, upon which a number of men haul in opposite directions until they have opened the joint sufficiently to detach the leg." [1] In pictures of Egyptian stoneworkers great companies of men are seen hauling together on some heavily-weighted sledge, and in Constantinople one may see any number of men from eight to twelve carrying a heavy tierce of wine in slings attached to four parallel bars.

The Khasi Hill tribes of India still erect megalithic monuments. The slabs of sandstone are quarried near by where they are to be set up by means of wedges. Some of these weigh twenty tons. They are moved on a cradle made of strong curved limbs of trees, roughly smoothed and rounded, so as to present little surface to friction. In dragging and setting up the slabs all the members of a community are under an obligation to assist on such an occasion, and are not paid for their labour, beyond receiving in the evening a little food or liquor at the dwelling of the family who sought the aid.[2] This is exactly like the "barn-raisings" familiar to all American farmers.

"The block" (of stone) "is detached by means of wedges introduced into natural fissures and artificially drilled holes. Two or three stout logs are placed under the slab at right angles to its axis and equi-distant. Under these are fastened four bamboo trunks, two on either side parallel to the axis

[1] Sir Samuel W. Baker, *Ismailia*, New York, 1875, p. 269.

[2] Austen, *J. Anthrop. Inst.*, London, 1872, vol. i. p. 127.

of the stone, and beneath these bamboos series of smaller bamboos like the rounds of a ladder. The whole forms a gigantic crib-work, or carrying frame. Three or four hundred men can unite their efforts thus in picking up the whole and carrying it to its destination. In two or three hours the stone may be transported a mile. It is set up by means of guy ropes and lifting, and planted in a hole previously prepared." [1]

A curious fact in engineering is recorded by that most careful of observers, Rev. J. O. Dorsey, regarding the Omaha tribal circles. He says, "The circle was not made by measurement, nor did any one give directions where each tent should be placed; that was left to the women" (§ 9). "Though they did not measure the distance each woman knew where to pitch her tent." She also knew the proper distances apart for safety, on the one hand, or for the convenience of dressing hides on the other (§ 11).[2]

[1] A. L. Lewis, *Materiaux pour l'Hist. de l'Homme*, Toulouse, 1876, 2 S., vol. vii. p. 185, 3 figs., which explain the process well.

[2] Dorsey, *Third An. Rep. Bur. Ethnol.*, Washington, 1883, pp. 219, 220. Parkman speaks of a whole village of fifty tents being set up in half an hour.

CHAPTER III.

INVENTION AND USES OF FIRE.

"Quo modo homines ignis usum primum intellexerint non nostrum hoc loco dicere, immo hercule nihil certum inveniri potest."—M. H. MORGAN, *Harvard Studies*.

IT is scarcely possible to conceive of man without fire. Very early in history he discovered the Promethean spark, and a train of blessings came with its advent. The light and warmth of the sun were let into his cheerless dwelling. Forests and jungles, with their poisonous malaria, noxious insects, venomous serpents, and ravening beasts, were subdued or quickly removed. Life was prolonged by the cooking of food and by the ability to preserve it for future use through drying, smoking, roasting, &c. In one Indian house in Guiana there were fires under each hammock as well as for cooking, and, in the open, the hunter sleeps secure from ravenous beasts so long as his fire is burning.

In old archæological sites in Europe, representing the remains of the Cave men of the Mousterian epoch in France and Belgium, are found flints that have been cracked by fire, fragments of charcoal, burnt bones that have been split for the marrow. In the low prairie lands of the United States the settlers, in digging wells, come upon piles of charred logs. The latter are, doubtless, the result of natural causes, but the former are the relics of human industries, and belong to the earliest history of Europe, whenever that may have been.

The study of fire in its relation to human invention, and

as a tool in doing work, must include many topics, among them the following :—

1. Its natural available sources, from which it could have been procured before men learned to create it artificially.

2. The methods of its artificial creation, together with such apparatus and appliances as were invented to make it burn.

3. The preservation and manipulation of fire after it had been kindled.

4. The use of fire in domestic operations, in cooking food, preserving food, and in giving warmth.

5. Illumination.

6. Fire in doing mechanical work and in the aid of man in peace and war.

7. Fire as a timekeeper.

8. Fire as an object of worship.

In the allegorical panel of tile-work designed by Bracquemond for the Centennial Exposition in Philadelphia, and presented by the Havilands to the Smithsonian Institution, the genius of fire is represented as already subdued by man, rising from the centre of the picture, bearing in one hand a vase, in the other a freshly-cast bronze image. About him are factories and foundries sending forth clouds of smoke and steam. Huge trains of freight and passenger cars cross the foreground. The scene is laid in daylight, so there is no need of artificial illumination. But every typical use of fire conceivable in 1876 has been wrought into this bold design.

Strangely in contrast with this bustling scene is the time spoken of by Sir John Lubbock, when fire had not been kindled on this earth. "In what precise manner," he says, "Nature communicated this secret to our species, is now difficult to determine. Even the few lowly tribes of our day that were devoid of fire-making apparatus, had found at least some way of keeping the smouldering spark alive." [1] However, there were fires and great conflagrations kindled

[1] Lubbock, *Preh. Times*, New York, 1872, p. 558.

without the aid of man by lightning stroke, by the smouldering furnaces within the earth, by chemical action, and by friction of falling rocks and the chafing of limbs and stems in the dense forests. Man in his pristine state was witness of these, and, following the current of activity in other matters, had doubtless learned to dread and rudely to worship this element before he subdued it to his use. To light his torch or his domestic fire at the hearth of beneficent nature was his first step in its subjugation, his first lesson in its manipulation.[1]

The T'lingit family of Indians in South-eastern Alaska say that the raven gave them fire. Flying through the inky darkness once upon a time he came upon a great medicine man, who had the sun and moon and stars and the divine spark in a box, which he hid away in his sacred chest. To secure this spark it was necessary for the raven to incarnate himself into the womb of the old man's daughter, and to be born of her. The baby waxed strong, much to the delight of the grandparents, and when he demanded the contents of the chest to play with that was granted. No sooner did the boy open the little box than in a twinkling the heavenly bodies sprung into the sky. The raven assumed his wonted form, seized the glowing coal in his beak, and sped away to the T'lingit home. The preservation of the fire in the box, and bearing it from tribe to tribe, has no allusion to fire-making, but it does preserve the reminiscence of the fact that the race had fire long, long before the days of the fire stick, that coals were carried from house to house, and that tribes lost fire and had to get a new coal in the best way they could.

The traditions of the Polynesian race clearly point to

[1] George Goodfellow, of the United States Geological Survey, says that many fires were started by falling boulders in the great earthquake in Arizona, May, 1887, *Pop. Sc. Monthly*, June, 1888. See also "L'Art du feu, est'il une caracteristique de l'homme?" By M. Duncan, *Bull. Soc. d' Anthrop. de Paris*, 2 S., 1870, vol. v. pp. 61–86; 90–114; 141–145, discussion by Broca, Ploix. Letourneau, &c.

three events in the history of fire. First was a time when the islanders ate everything raw ; in the second place came ownership of fire through the earthquake god ; and, finally, the creation of fire from rubbing two pieces of dry wood together, for, said the earthquake god Mafuie, you will find the fire in every wood you cut.[1]

On the authority of O. F. Cook, who lived among them, the Golas, of West Africa, put out all the fires in a village when lightning ignites any substance, and immediately kindle new fires therefrom.

Mr. A. G. Theobald, who lived many years in the jungles of Southern India, assured Mr. W. T. Hornaday that fires often occurred in the Animallai forests from the rubbing of the bamboo stems in a high wind.[2]

Likewise spontaneous combustion must always have been an active agent in the kindling of natural fires. It is not improbable that in the innumerable experiments going on always, the savage unwittingly brought together the substances that ignited.[3]

The reader will find references to the methods of fire-making in classical authorities given by M. H. Morgan, with mentions of the works of Weiske (1842), Lasaulx (1843), Holle (1879), Milchöfer (1886), Kuhn (1886), Hough (1890). This charming subject has not escaped the serious attention of all distinguished writers upon culture-history. When, to the affirmation that man is the only tool-using animal, it is suggested that many animals perform work by means of natural objects, as implements, the rejoinder is safe that man is the only fire-making animal.

The work of Dr. Walter Hough in the Department of

[1] Turner, *Samoa*, London, 1884, p. 209. Tregear, *Trans. N. Z. Inst.*, vol. xx. The Origin of Fire, pp. 369–399. An excellent *resumé*.

[2] Private letter.

[3] Coal veins in West Virginia and Pennsylvania, I am told by Dr. Walter Hough, are often ignited by spontaneous combustion of the pyrites decomposed by percolating water. Drifts in a coal vein frequently enter on a burned-out area, nothing but ashes remaining.

Ethnology of the National Museum is especially devoted to the methods of savagery in the creation of fire. The following classification is based on the presumed order of invention :—[1]

I. FIRE-MAKING BY RECIPROCATING MOTION.

1. *Simple, two-stick apparatus*—Indians of two Americas, Japanese, Veddahs, Australians, Somalis, Kaffirs, &c.

2. *Four-part apparatus*—Eskimo, some Indian tribes, Hindoos, Dyaks.

3. *Weighted drill, with spindle whorl*—Iroquois, Chukchis.

II. FIRE-MAKING BY SAWING.

Malays, Burmese.

III. FIRE-MAKING BY PLOUGHING.

Polynesians, Australians, Papuans, Americans, Africans.

IV. FIRE-MAKING BY PERCUSSION.

1. *With pyrites or stone containing iron, and flint*—Eskimo and Northern Indians, Fuegians, Prehistoric Europe.

2. *With flint and steel*—General.

V. BY COMPRESSION OF AIR.

Dyaks and Burmese.

The creation of fire by the friction of sticks is a process in the evolution of which may be observed some of the nicest co-operations between Nature and human effort. There are just three ways in which one stick of wood may be rubbed upon another—by moving with the grain, commonly called ploughing; by moving across the grain, called sawing; and by twirling. The first-named two methods seem to have had arrested development, that is, they reached their perfection in rudimentary form, and after that were improved no more. Besides, they have been kept within very narrow geographic limits. The method by ploughing is found among the Indo-Pacific races, including

[1] Walter Hough, *Proc. U. S. Nat. Mus.*, 1888, pp. 181–184; *Rep. U. S. Nat Mus.*, 1887–88, pp. 531–587; *id.*, 1890, pp. 395–409; *Am. Anthropologist*, 1890, vol. iii. pp. 359–371; *id.*, vol. vi. pp. 207–210.

the Australian, the Polynesian, and the Papuan, and sporadically in America. As these three races have done much borrowing from one another, it would be far from easy to say which of the three would be entitled to the patent on the fire-plough. There is no special adaptedness in any tree or wood peculiar to one of the regions mentioned above, so that we might declare that region to be the fire-plough's native home. The case resolves itself into one of racial knack and genius. The Polynesians have carried the apparatus through many degrees of latitude, employ no other method, and in many places are still using it. Let us provisionally give the invention to them.

From first to last the fire-plough consists of two parts, namely, a stout stick of thoroughly dried hibiscus, three feet or more long and two inches in diameter. This forms the hearth or stationary part. A smaller stick of the same kind of wood, about a foot long, cut at the lower extremity in shape of a wedge, with its edge forming a very obtuse angle, constitutes the working part or plough. The fire-making is accomplished by violently rubbing the end of the plough backward and forward on the stationary piece, cutting a groove running with the grain of the wood for a distance of four or more inches. Instead of rubbing off woodmeal, the plough disengages extremely minute ribbons or shavings, which the friction succeeds in heating above the point of ignition. The groove is not used a second time, and the plough receives a fresh trimming after each effort.

The natives of New Britain, in whom the Papuan element greatly predominates, rub a sharpened piece of hard stick against the inside of a dried piece of split bamboo. This has a natural dust that soon kindles. They use soft wood when no bamboo can be procured, but it takes longer to ignite. The flame is fed with grass.[1]

In Samoa the blunt-pointed stick is taken between the clasped hands and shoved along the groove at an angle of forty to forty-five degrees, slowly at first, with a range

[1] W. Powell, *Wanderings, &c.*, p. 206.

of six inches, till the wood begins to be ground off and collects in a heap at the end of the groove. The speed is then accelerated until the brown dust ignites.[1] The Australian process and apparatus are evidently under foreign influence, existing only in circumscribed areas. They rub a knife of wood along a groove made in another stick previously filled with tinder.[2]

The Hawaiians obtained fire by the plough. A small stick, the *aulimá*, is held in the hands and rubbed in a groove in a larger stick *aunakí*. The *aulimá* is of some hard wood, while the *aunakí* is of hau or some soft wood. In five seconds the rubbed wood is charred, and in about a minute the dust which collects at the bottom of the groove ignites, and the flame is caught in a bit of tinder, or a fuse composed of twisted kapa or cotton cloth.[3]

Fire-making by tranverse friction is a Malay device. It is true that Malays use all other methods, and the saw may be found elsewhere, as in both India and Further India ; but to the Malay the fire-saw belongs. Again, it is safe to assert that the saw belongs to a bamboo area. It is true there are bamboos where there are no fire-saws, but fire-saws do not work well where there are no bamboos. " For this method two pieces of bamboo are used, a sharp-edged piece like a knife is rubbed across a convex section in which a notch is cut, nearly severing the bamboo. After sawing across for awhile the bamboo is pierced, and the heated particles fall below and ignite." [4]

Eliminating the ploughing and the sawing method, we are brought to the method by twirling. Now, this may not have antedated the other processes in time, but it has had a more interesting history.

[1] W. Hough, "Fire-making Apparatus," *U. S. Nat. Mus. Rep.*, 1888, p. 520.

[2] R. Brough Smith, *Aborig. of Victoria*. Quoted by Hough, *ut supra*, p. 571.

[3] Brigham, *Cat. Bishop Mus.*, Honolulu, 1892, vol. ii. p. 31.

[4] Wallace, *Malay Archipelago*, p. 332. Quoted in Hough.

The simplest possible device for this operation is a rod of dry wood, and a partly decayed and very dry lower piece. The rod is the vertical, moving element, the soft piece is the horizontal, stationary element ; the former is twirled in some fashion, the latter remains on the ground and is held firm by means of the foot or the knee. Mr. Hough enumerates the tribes that use this method.[1]

The vertical spindle and the horizontal hearth or socketed piece, with its side notches for the escape of the wood meal in which the fire first appears, being the starting-point, the next thing was to make some additions to the manual part in order to shorten time, to decrease effort, and to render more certain the result. The bottom piece seems to have been improved by the Eskimo by making it broader and cutting a step or wide rabbet along one border. The wood meal and fire could then fall on this extended part. In snowy countries the horizontal step with the hearth attachment was a decided improvement. Another trick of the ingenious Eskimo was to make his sockets along the middle of a tolerably broad piece and then to cut his notches for draught and the escape of the woodmeal along the median line of the stick instead of outward. Living on the ice where there was little wood to be had, this also was just the thing to do.

FIG. 13.—The simplest form of Fire-drill. (*After Hough.*)

The spindle or vertical piece could be improved in several ways. The writer has often seen Mr. Hough making fire with the common fire-stick, and learned that it is necessary to keep the border of the lower edge of the vertical whittled

[1] Fire-making Devices, &c., *Rep. U. S. Nat. Mus.*, 1888, pp. 531–587.

away to prevent binding. This border simply polishes a ring about the fire-socket. To improve the spindle further, some sort of string or strap could be wrapped about it once or twice and pulled backward and forward as rapidly as possible. If a bowstring be used, similar to that of the jeweller, and a socketed rest be held on the top of the vertical shaft after it has been whittled to a point, then one man can still operate the combination and get a speedy result. If the reader will look at a collection of Eskimo men's tools in any museum he will see often a cavity about one quarter of an inch in diameter and one-tenth inch deep somewhere along the middle of the handle. This is to convert the knife grip, bow stretcher, or other tool, into a top rest for the drill, either the fire-drill or the perforator. A still further complication of this same pattern requires the co-operation of two persons. Instead of the bow to operate the string, one holds the rest at the top, and the hearth or horizontal, while the other pulls the string backward and forward with his two hands. In other words, one of them furnishes the bearings of the drill, while the other furnishes the power.[1]

Among the Indians of the South-west the pump-drill is very common for boring, and, at least among the Iroquois, this form of fire-drill is reported. The parts of this apparatus are the vertical shaft, the fly-wheel or spindle-whorl, the hand-piece by the up-and-down motion of which the drill is worked, and the string. The hand-piece, or grip, is a stick held in the hand and attached at its extremities to the cord or string, which also passes over a notch in the top of the spindle. In the best forms this hand-piece is perforated, and the shaft passes through it loosely. In the ruder forms the hand-piece simply rests against the spindle. The apparatus is put in motion by twisting the string once or twice about the shaft and kept in operation by moving the hand up and down. The sacred Hindu fire-drill is on the plan of the two-handed cord drill, and is, in fact, the climax of this sort

[1] Murdoch, *Ninth An. Rep. Bur. Ethnol.*, Washington, 1892, pp. 289–291, fig. 282.

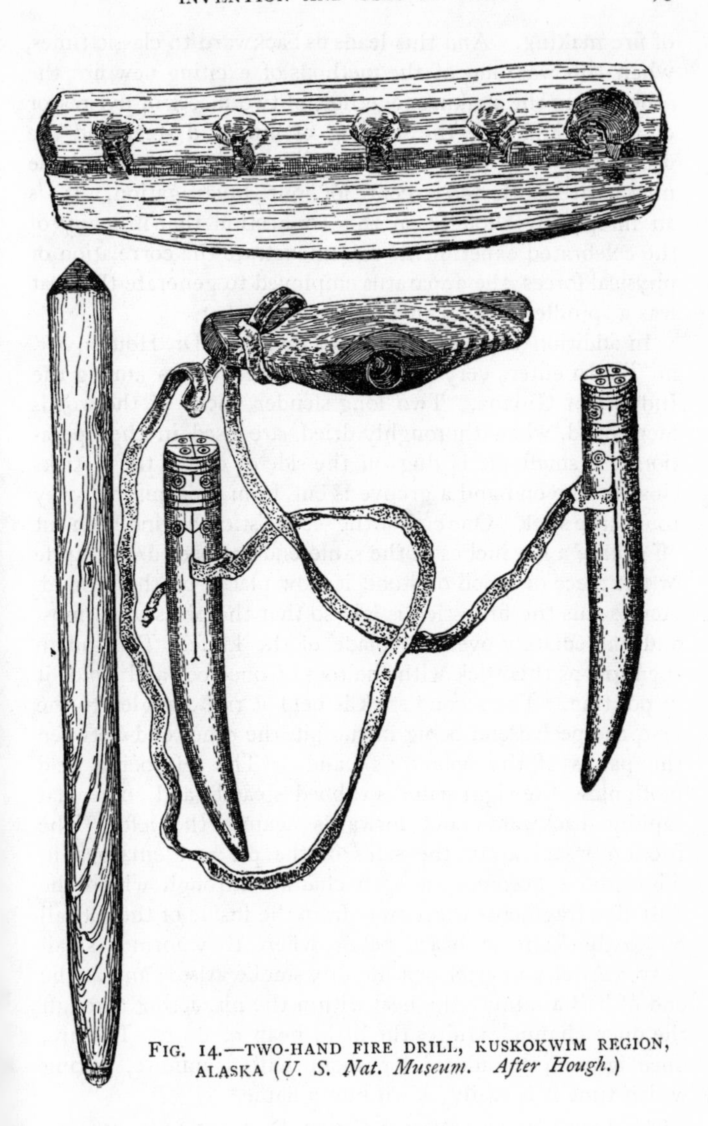

FIG. 14.—TWO-HAND FIRE DRILL, KUSKOKWIM REGION, ALASKA (*U. S. Nat. Museum. After Hough.*)

of fire-making. And this leads us backward to classic times, when this was one of the methods of exciting new fire, the other being the striking together, of two pieces of pyrites, or of pyrites and flint, the striking of flint and steel, and the use of a lens or concave mirror. The flint and steel and the mirror are both too recent for our investigation. It is an interesting thought in this connection that in many of the celebrated experiments to demonstrate the correlation of physical forces, the apparatus employed to generate the heat was a spindle working in a horizontal socket.

In addition to the descriptions given by Dr. Hough, Mr. im Thurn enters very minutely into the process among the Indians of Guiana. Two long slender sticks of the woods mentioned, when thoroughly dried, are used in the operation. A small pit is dug on the side of one of these sticks close to one end and a groove is cut from this pit half-way round the stick. One end of the second stick having been cut off square a few inches at the same end are peeled. A knife or flat piece of wood or stone is now placed on the ground. Across this the first stick is laid so that the pit is uppermost and immediately over the blade of the knife. The Indian then grasps this stick with the toes of one foot and holds it in position. The second stick is held at right angles to the first, the peeled end being in the pit, the other end between the palms of the operator's hands. The left being held motionless, the right palm is rubbed steadily and somewhat rapidly backwards and forwards against the left. The friction wears away the sides of the pit and enlarges it. The groove becomes an open channel through which the dust-like fragments worn away from the inside of the pit fall on to the knife or board below, where they form a small heap. After a quarter of a minute, smoke arises; and at the end of half a minute the heat within the pit, acting through the open channel, ignites the little heap of dust. The fire, once ignited, smoulders for about half a minute, during which time it is easily blown into a flame.[1]

[1] Im Thurn, *Indians of British Guiana*, London, 1883, p. 257; excellent figure, 17.

Two straight pieces of boughs thirty centimetres long, and of the thickness of a lead-pencil, serve most negro tribes for the generation of the spark by means of friction. The Bongo generally carry the sticks in the quiver. Into the end of one stick a score is cut, in which the other piece is vertically inserted. By means of friction the upper piece is made to twirl around between the palms of the hands, while the lower piece is held tight with the foot. A hard support (a smooth spear-head answers the purpose best) must be always provided. A small heap of glimmering wood-dust is the result after the brief manipulation; tinder and straw effect the rest.[1]

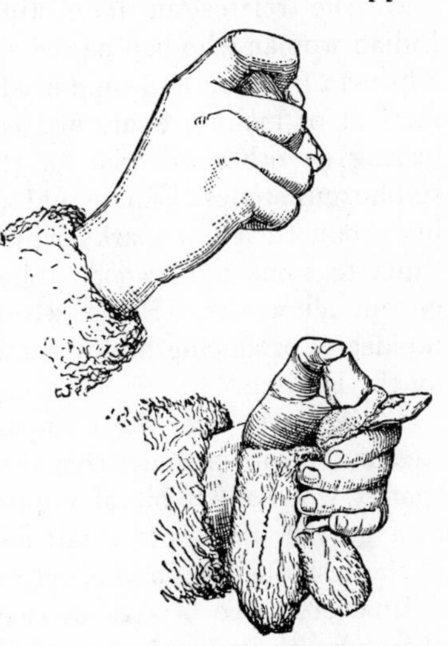

FIG. 15.—Striking fire with two pieces of pyrites. Central Eskimo. (*After Hough.*)

If flint and steel are a comparatively modern invention owing to the late appearance of the latter, this method of creating fire had an ancestor that may be older than any fire sticks. This progenitor was flint and pyrites, or two pieces of pyrites, the latter mineral deriving its name from its striking fire. A very complete strike-a-light sent from Cape Bathurst consists of flint, pyrites, and tinder done up in dainty little bags, with leather pad to guard the fingers from injury. The Eskimo, from Smith Sound to Behring Strait, used this method. Evans points also to Fuegia and the European archæological sites for the antiquity of this method.[2]

[1] Schweinfurth, *Artes Africanae,* London, 1875; S. Low, pl. iv. fig. 16.
[2] See the pyrites and steel in Hough, *Rep. U. S. Nat. Museum*, 1888.

"Skipertogan is a small bag that contains a flint and steel, also a pipe and tobacco, as well as touchwood, &c., for making a fire. Some of these bags may be called truly elegant; being richly ornamented with beads, porcupine-quills, moose-hair, &c., a work always performed by the women; and they are, with much propriety, greatly esteemed by most Europeans for the neatness of their workmanship." [1]

Hearne relates an interesting episode of a Dog Rib Indian woman who had passed seven months without seeing a human face and had supported herself meanwhile. "Her method of making a fire was equally singular and curious, having no other material for that purpose than two hard sulphurous stones. These by hard friction and long knocking produced a few sparks, which at length she communicated to some touchwood. She did not suffer her fire to go out all winter. Hence we may conclude that she had no idea of producing fire by friction in the manner practised by the Eskimo." [2]

To light the fire, the Campas of Peru use the flint and steel and a sort of tinder that they make themselves from a spongy wood and a bit of impure copal, which in the form of a grayish mass and small density, is found at the foot of the resinous trees and is very inflammable.[3]

In every tribe of savages that have fire, those who are charged with its management have learned to promote the flame by supplying fresh air. At the critical moment when the smoke bursts from the wood-meal at the bottom of the fire drill, the fire-maker knows that he does not dare to blow upon the spark with his breath, for fear of scattering the dust. He simply waves his hand back and forth in front

pp. 571–577; with excellent figures. Also Murdoch, *Ninth An. Rep. Bur. Ethnol.*, p. 291.

[1] Hearne, *Journey*, &c., London, 1795; Strahan, p. 48. See also Harris, *Voyages*, vol. i. p. 816, fol.; New England Canaan, p. 23.

[2] Hearne, *Journey*, &c., London, 1795; Strahan, p. 261.

[3] Ollivier Ordinaire, *Rev. d' Ethnog.*, 1887, vol. vi. p. 271.

of the tiny hearth with sufficient rapidity to clear the smoke away as it rises. When he applies his spunk or torch of shredded bark or kindling material to the Promethean spark he keeps up the motion of his hand until the fire has well set in the kindling. Then only can he trust his breath or dare to whirl his material around his head. Of the three processes involved in the act the hand is the ancestor of all fans and blowers ; the mouth easily suggested the bellows, great and small ; and the whirling deadwood or torch was the first attempt to compel the wind to blow our fires. In this example, however, Mahomet goes to the mountain.

The people who weave or make basketry have no trouble to manufacture fans. Indeed, the hunting tribes of North America used the large birds' wings, not to keep the hearth tidy as our grandmothers did, but to encourage the fire. In tropical countries palm-leaf fans are used quite as much in coaxing the flame as in cooling the face. Indeed, im Thurn tells us that the women who cook the cassava in Guiana smooth the upper surface of the cakes on the griddle with the same fan that urges the fire along.[1]

From the hand to the fan and from the fan to the blower is only a march of two steps, to say the least. The fire maker, the fire nurse, and the locomotive engineer or mining ventilator, are grandfather, father, and son.

The labial tube, supplemented by the nasal openings, enabled the savage fireman, or firewoman rather, to excite the live coal into a flame. Further along it occurred to the pyrotechnician to move the eyes a little from the blinding smudge by means of a tube of cane or something like it, and the Japanese cooks at least keep up the practice. This method also allows the air to be more effectually driven to the very spot where it is wanted. This tube or mouthpiece will also be the nozzle of the coming bellows. It is difficult to imagine just how the inventor of long ago first attached an artificial buccal cavity to this tube and thus

[1] *Indians of British Guiana*, London, 1883, p. 262.

perfected his apparatus. Doubtless one bag sufficed at first and he did not get quite as far as the continuous current all at once. The African blacksmith represents a later patent. His boy sits on the ground between two goat skins, the necks of which are securely attached to the ends of reeds. The other ends are fixed like the mouths of mail sacks or travelling bags, and the boy has his thumb and fingers attached to the flat pieces of wood along the borders. The tubes at the lower ends of the skins connect with a main reed or pipe under the fire. When all is ready the boy opens one hand wide and raises it as high as the skin will permit, then closes the hand tight and presses down, holding the other hand closed and down meanwhile. When the first hand is nearly down, the second is deftly caught up to repeat the process. With a little song and much style the boy manages to get considerable fun out of the exercise. The valve in the bellows is too far along for our study, and as for compound blowpipes and other appliances for forcing air in large or small quantities through a tube, they are only modern variations of quite ancient devices.

In the matter of draught, the tall chimney is absent from all ancient cities. The savage built his fire in the middle of his hut just as our English ancestors made theirs in their halls, only, if you will look carefully at the picture of a wigwam, you will see that two poles are longer than the rest and each one of them is attached to a fly of hide, which enables the inmates to place a barrier against the wind, and create such a draught as to suck the smoke out of the tent. The Roman baths, perhaps, and the Oriental kangs teach us how a closed tube first conducted the smoke away. But savages had their appliances as well.

Every step in the process of inventing the chimney may be seen in the Pueblo region of the United States. Imagine a room built up of stone spalls and adobe, with ceiling of stout poles and brush and mud. A low rectangular opening is the door to this kennel, and in one corner of the roof a hole allows the smoke from the fire on the floor beneath to ascend. This is the first lesson. Now these people are great potters, and a long time ago some one discovered that a water jar whose bottom had been worn out or broken through would do excellent work in urging the smoke out of the hole in the roof. Nay more, the good women also found out that two, three, four, are better than one, and piled them up and stopped the chinks with mud and smoothed up the inside. There were thus tile chimney-pots in America before there were any in London.

Quite as ingenious has been the evolution of the smoke-stack and the chimney-jamb inside the rooms, crawling down the wall over the fire. It only needs the brass andirons to complete the metamorphosis.[1]

Though William the Conqueror introduced the curfew bell into England, the practice of preserving fire by covering it up with ashes is much older than his time. Indeed, Dame Nature, long before there were any housewives, piled ashes over lava streams and kept the fire for indefinite periods. Dr. M. H. Morgan[2] has collected references from Homer, Ovid, Virgil, and other classic authors mentioning the practice of covering fire, and also the rights of hospitality in this regard. But they were moderns compared with the antiquity of the practice, as can be shown.[3]

The diligent search through the round of preservative substances for the best in this department of labour is in a line with the whole body of aboriginal activities. About a hundred years ago Hearne made some acute observations on this matter in North-western Canada. "Westward to procure birch-rind for making two canoes and some of the fungus that grows on the outside of the birch-tree, which is used by all the Indians in those parts for tinder. There are two

[1] Victor Mindeleff, *Tenth An. Rep. Bur. Ethnol.*, pp. 162–67, figs. 51–74, relating to chimneys. The whole work must be studied for the history of Pueblo architecture.

[2] *Harvard Studies in Class. Philol.*, Boston, 1890, Ginn, vol. i. p. 16.

[3] See also Planck, *Die Feuerzeuge der Griechen und Römer*, &c., Stuttgart, 1884.

sorts of these funguses which grow on the birch-trees; one is hard, the useful part of which much resembles rhubarb; the other is soft and smooth like velvet on the outside, and when laid on hot ashes for some time and well beaten between two stones is somewhat like spunk. The former is called by the northern Indians *jolt-thee*, and is known all over the country bordering Hudson Bay by the name of *Pesogan*, it being so called by the southern Indians. The latter is used only by the northern tribes, and is called by them *Clalte-ad-dee*. The Indians, both northern and southern, have found by experience that by boiling the pesogan in water the texture is so improved, that when thoroughly dried, some part of it will be nearly as soft as sponge. Some of these funguses are as large as a man's head; the outside, which is very hard and black and much indented with deep cracks, being of no use, is always chopped off with a hatchet. Besides the two sorts of touchwood already mentioned, there is another kind of it in those parts that I think is infinitely superior to either. This is found in old decayed poplars, and lies in flakes of various sizes and thicknesses, some is not thicker than chamois leather. It is rather surprising that the Indians, whose mode of life I have just been describing, have never acquired the method of making fire by friction, like the Esquimaux." [1]

"With the decayed or thoroughly dry trees the Indians always kept up their annual holy fire; and they reckon it unlawful and productive of many temporal evils to extinguish even the culinary fire with water. In the time of a storm, when I have done it, the kindly women were in pain for me, through fear of the ill consequence attending so criminal an act. I never saw them to damp the fire, only when they hung up a brand in the appointed place as a threatening symbol of torture with death or when their kinsman dies. In the last case, a father or brother of the deceased takes a fire-brand, and, brandishing it two or

[1] Hearne, *Journey*, &c., London, 1795, Strahan, p. 278.

three times round his head, with his right hand dips it into the water and lets it sink down."[1] The Eskimos keep a blubber lamp burning the year round.

The Andamanese, and perhaps some other very uncultured peoples, know not the art of kindling fire at all. They do not seem to have any traditions on the subject like the Promethean myth, but hold that their people have had fire from the Creation. Whether they procured their first supply from volcanoes, still active on one of their islands, or from forest fires, the Andamanese can teach us a lesson in its preservation. From E. H. Man we learn that when they all leave an encampment with the intention of returning in a few days, besides taking with them one or more smouldering logs, wrapped in leaves, if the weather be wet, they place a large burning log or faggot in some sheltered spot, where owing to the character of the wood invariably selected on these occasions, it smoulders for several days, and can be easily rekindled when required. Decayed pieces of the *Croton argyratus*, and two species of *Diospyros* (bastard ebony or marble-wood), and a fourth, called by them *chôr*, are chiefly used as fuel. All labour of splitting and chopping is saved, and it is only necessary to beat a log of this description on a stone or other hard substance a few times before it breaks. In each hut that is occupied there is invariably a fire, the object of which is to keep the owner warm, to drive away insects, and to cook food; while the smoke is useful in preserving the stores of provisions, which are placed above it. Fires are generally kindled by fanning the embers with a frond of the *Asplenium nidus*, and they are extinguished by pressing the burning logs against some such object as a tree, or stone. If more than a few hours' absence from home is anticipated, besides a supply of provisions, a smouldering log is entrusted to some one member of the party, whose duty it is to kindle it into a blaze whenever a fire is required.[2]

Among the Taveta, East Africa, the fires are not allowed

[1] Adair, *American Indians*, p. 405.

[2] Man, *Andaman Islanders*, London, 1883, Trübner, pp. 82, 137.

to go out in the village. When a family fire goes out, it is rekindled by getting a blazing faggot from a neighbour. In the history of the tribe they had always preserved the fire. They use fire-sticks and flint, and procure tinder from the Mahwale tribe.[1]

In the course of time these spunk knots and smouldering logs became joss-sticks in China, and all sorts of fuses and slow matches in western nations.

For the handling of fire-brands the naked hand sufficed, as it does with most of our backwoodsmen. But among eastern savages tongs are employed in lifting the hot stones used in cooking. These are generally a long strip of bamboo bent in the middle so as to bring the ends together as in a pair of tweezers. The art of stone-boiling was well known among the North American savages, and always they supplied themselves with tongs, either by bending a pliable stick or splitting a larger one half way its length. These rude appliances, and a stick for stirring the fire, supplemented with a skewer for moving the food, include the outfit of the ancient cook.

Just how it first occurred to the primitive folk that cooked meat would last longer and digest more quickly than raw meat is unknown. The ever-ready guesser will say that a lucky accident was the teacher. But lucky accidents give no lessons to those who are not already alert. The only truth that can be arrived at is in the study of the cookery of modern savages.

The most abject peoples on the earth cook their food. The only exceptions at all worthy of mention are the Eskimo, "who," says Collinson, "seldom cook their food, the frost acting as a substitute for fire."[2] Any one who has eaten freshly-killed beef and a slice of beef after hanging a month in a freezing atmosphere, will take sides with the Eskimo.

[1] French-Sheldon, *J. Anthrop. Inst.*, 1892, vol. xxi. p. 370.

[2] Collinson, "H.M.S. Enterprise, &c.," *J. Roy. Soc.*, London, 1855, vol. xxv. p. 201. According to Murdoch, the Eskimo prefer cooked food, and never eat raw meat unless compelled.

Surely, there can be no bacteria in that climate. Provisions keep for millenniums, for that matter, and raw meat there is tenderer than cooked meat here.

The rudest savages that have come to a knowledge of fire, before they apply this element to vessels of any kind, depend upon roasting or baking alone. If we are to credit the assertion of a cursory visitor, "the Australians never take trouble to cook their food, but merely tear off the exterior skin of the animal, and, after holding the body over the fire for a few minutes, eagerly devour it in its uncleaned state, and frequently eat so voraciously as to be in a condition of inactivity and torpor for several hours afterwards."[1] The same is true everywhere. The order seems to be, roasting or parching whole, roasting after preparation or in ovens, boiling by means of stones, boiling in pots, the use of stoves of some kind.

The Polynesians were aboriginally most delightful cooks and lived in blissful ignorance of the frying-pan. A pit small or large was dug, and heated stones put in the bottom. Upon these was carefully spread a layer of leaves and a layer of bread-fruit, properly dressed, and more leaves and more hot stones, and then earth and leaves. In half an hour the operation was finished and the food cooked to a turn. There are many now living in the States who remember the days of "roasting ears," when green maize was similarly baked in the ear and husk, a custom doubtless borrowed from the aboriginal cooks. The Polynesians frequently prepared a communal oven. A pit twenty or thirty feet in circumference was dug out, the bottom filled with stones, logs of firewood piled on them, and stones on the top of the wood, as in an open-air lime-kiln. The wood was then burned and the hot stones on top raked to the sides of the pit. Hundreds of ripe breadfruit were piled in the centre of the pit, leaves spread over them, hot stones piled on top like an arched oven, and a foot or more of earth piled on that. In a day or two a hole might be made in the side and the

[1] P. H. Eagle, *Ride Across the Frontier of Victoria;* also Brough Smith.

owners draw out the food until it was all consumed. The fruit would remain good for several weeks after the oven was opened.[1]

The Polynesians had no pottery or potstone ware, but they had a custom of putting "a quantity of arrowroot powder with the expressed milk from the kernel of the cocoanut into a large wooden tray or dish, and having mixed them well together, they threw in a number of red-hot stones, which being moved about by their white sticks, heated the whole mass nearly to boiling, and occasioned it to assume a thick, broken-jellied appearance."[2]

The Andamanese, having pottery, may cook in a variety of ways. Food, in its skin or shell, may be roasted whole in, or on, or over an open fire, that is baked like an ash cake, seared or spitted and roasted. It may be carefully wrapped in leaves and cooked among burning logs or in the midst of hot stones. Finally, there is no end to the ways in which the earthen pot and the food may become acquainted. The result of this variety is that these savages never eat half-cooked food. They have invented a method of preserving food through cooking so ingenious that it is worth describing here. Having procured and cleaned a length of bamboo, they heat it over a fire that the juices contained in it may be gradually absorbed. When this is satisfactorily accomplished, half-cooked pieces of pork, turtle, or any other food, are packed tightly into it, and the vessel is again by degrees put over the fire, in order to heat it slowly, lest the rapid expansion of the meat should cause a crack. When steam issues forth, the bamboo is taken off the fire, and after the opening has been closed by leaves, is set aside with its contents until a meal is required, when it is replaced on a fire, for it is a peculiarity of these savages to eat this food in an almost boiling state. As soon as the meat has been once more thoroughly baked, the bamboo is split with an adze, or other implement, and all take a share in the

[1] Ellis, *Polynes. Res.*, London, 1859, vol. i. p. 40.

[2] *Ibid.*, Bohn, vol. i. p. 49.

feast. Meat thus prepared will keep good for several days.[1]

The Campas of Peru roast, smoke, and boil. They use salt, of which the Lorenzos are ignorant. When they cannot consume by night all the game that they have taken during the day they smoke it by hanging it above the fire.[2]

Smoke as a preservative of food is a very early invention. No sight is more common in a savage hut than that of a frame suspended over the fire in the centre of the cabin for holding fish or meat to be dried out and smoked for future use. It will be readily seen that this was a potent factor in the increase of longevity, not only securing provisions for time of famine, but eliminating a portion of the noxious creatures that prey on subsistence and shorten life. On every American farm and plantation might be seen a few years ago a building called the "smoke house." Here were hung hams and shoulders and sides of hog's flesh to be stored and occasionally smoked to keep off the flies.

The Indians of the Plains and other parts of temperate America had learned also that fat is a preserver of fresh meat. Just as the country housewife nowadays keeps her sausages fresh indefinitely through the winter by boiling them to kill the germs, packing them in stone jars and pouring melted fat over them, the Indian dried his buffalo meat, ground it to meal by pounding, packed it away in skins and poured over the mass melted buffalo tallow or bears' grease. The germ theory of decay was totally unknown, but the power of heat and hot fat to preserve flesh were quite well understood.

Before Benjamin Franklin invented stoves, the whole world, from the beginning of time, warmed itself at the open fire. The Eskimo used his lamp for drying clothes and heating his igloo quite as much as he did for illumination. The West Coast people of America, who dwelt in communal houses, arranged their clans about a central fire, whose

[1] Man, *Andaman Islanders*, London, 1883, Trübner, *sub voce*, FOOD.

[2] Ollivier Ordinarie, *Rev. d'Ethnog.*, 1887, vol. vi. p. 271.

smoke enfolded them in a common misery before it escaped through the openings of the roof. The Algonkian tribes, the Iroquois, the various stocks along the slopes of the Rocky Mountains, not only ate the food from a common pot in gentile groups, but warmed their bodies in consanguine groups. The proverbial fireside was one of social importance. Around it the clans ate, and warmed themselves, and slept, and wrought and told their legends. The fireside, as a place of common gathering and genial friendship, lies at the bottom of more refined ideas to be especially mentioned.

Indeed, two absolutely different sets of inventions spring forth from this humble differentiation of ideas. The cooking-stove and the parlour-stove, the range and the heating apparatus, both started out from piles of smouldering embers in ancient smoky wigwams, and both grew up to their modern stature under the influence of great predominating needs. Hunger and cold created two kinds of stoves, the former dwelling in the kitchen, the other in the drawing-room, the office, the shop, the assembly-room, as of old.

The latter has limitations of climate, the former has no limitations, for "hunger has no ears." The former has existed always. There have ever been kitchens of some kind. The latter, prior to the nineteenth century, got little further than the charcoal brazier, which is really a refined but dangerous substitute for the wigwam fires of the Sioux Indians or the wooden lamps in the Korak's polog. One can scarcely realise the savage discomforts for artificial warmth amidst which even kings and nobility dwelt in not remote times. The Orientals resorted to their ovens when they could no longer endure the cold ; but Europe was not acquainted with the kang.

The next use of fire among savages to which attention is called is for the purpose of illumination. In the most primitive society men do not like to be left wholly in the dark, so the natives of Middle America imprison the fireflies and compel them to illuminate their rude habitations. The Eskimo during the long and dreary winter light up their subterranean homes with lamps of soapstone, having wicks made of moss resting in a mass of fat.

The Indians of British Columbia and Oregon coast used the candle-fish for their primitive torch. The creature contains so much fat that it will burn with a wick of cedar bark drawn though it.

The aborigines of Eastern United States, as well as the first settlers, made use of the fat pine-knot. It is common to read of the students of those days conning their lessons by this smoky, primeval light. The burning of natural objects, in short, is the most primitive and simple fashion of illumination. The searching out of the most valuable substances in each area constituted the development of invention along this line.

From these simple expedients to the electric light the investigation lies along an interesting path, growing brighter and brighter, on which shines first the torch and the signal fire, then the open lamp, then the closed lamp, then the rushlight and the candle, all of them fed on vegetable and animal fats. Later, camphene lamps, olefiant gas, petroleum products, and, last of all, the electric light, bring the night nearer and nearer to the brilliancy of the day.

For artificial light the Hawaiians burned the kernels of roasted kukui nuts strung on slender strips of palm or bamboo. As the nuts burned, the remains were knocked off as soon as the next nut was ignited. The oil was also expressed from the nut and burned with a wick in stone cups. Animal fat was used as well for this purpose, and for a wick a dried rush or a fuse of kapa was suitable.[1]

"We procured cocoanut oil, and when it grew dark, breaking a cocoanut in two, took one end, and winding a little cotton-wool round the thin stalk of the leaflet of the tree, fixed it erect in the kernel of the nut. This we filled

[1] Brigham, *Catalogue Bishop Museum*, Honolulu, 1892, vol. ii. p. 35.

with oil, and thus our lamp, excepting the small piece of cotton wick, was the product of the cocoanut tree."[1]

Torches, made by women, of resin wrapped in a large leaf (*Crinum lorifolium*), is used by the Andamanese when fishing, or travelling, or dancing by night.[2]

The damar resin is made into long torches throughout the Malay area and wrapped in splints of bamboo. A deliciously-scented white resin exudes from the hyawa tree (*Iria heptaphylla*) in British Guiana. The rough masses of this, which are very inflammable, are often collected and stored by the Indians for the purpose of lighting fires. Sometimes it is broken up into small pieces which are put into hollow sticks to be used as torches.

As a servant, the savage man utilised fire to do the work of mechanical tools, to assist him in conquering the beasts of the field, in levelling the discouraging forest, and in overcoming his enemies. Imagine the sons of men without fire, the sport of every wind. No wonder that so many mythic tales begin with the time when a race of fireless men dwelt wretchedly on the planet. The first men did not study economy, but waste; for they had to burn their way into a right to stand upon the earth.

The use of fire in hardening wood is alluded to again and again by travellers. The well-known digging-stick employed in collecting molluscs on the shore, and roots and vermin from the earth, which antedates all pickaxes, hoes, and spades, was thus tempered.

The Tanana Indians of Southern Alaska, belong to the Athapascan stock. They have no better wood than willow, frequently for bows; but, by dint of heating and rubbing many times over, they succeed in giving to the weapon considerable elasticity.

The fishing arrows of the Omahas were made without heads. The end of the shaft was cut to a point, then about four inches of the end of each shaft was held close to the

[1] Ellis, *Polynesian Researches*, vol. ii. p. 252.
[2] Man, *op. cit.* p. 186.

fire, and it was turned round and round till it was hardened by the heat.[1]

Other Sioux tribes use arrows without points for certain game and fish. The shafts are merely sharpened at the ends, which are also hardened in the fire.

The Pueblo agricultural tribes employ often the fire-hardened stick in the raising of their crops.

The Eskimo and Coast Indians were conversant with the use of fire in bending wood, both by heating the green sapling and by boiling the wood to be bent. In the chapter on woodworking, the use of fire in this regard will be more fully explained under boat-building. The dishes of the Eskimo and Indians, as well, are made of thin pieces of whalebone or spruce-wood heated until they may be bent into form. Many hundreds of dishes, especially on the Yukon river, are thus shaped, the ends being sewed together with spruce-root or whalebone strips.

The natives of Bowditch Island sometimes burned the trunk of a tree to make it fall, but as the fire occasionally ran up the heart of the tree as well, they usually cut away at the trunk with their shell hatchets day after day until it fell. It took from ten to thirty days to level a tree. Another plan was to dig down and cut the roots. In hollowing out a block of wood, they did the work by burning.[2]

The New Caledonians felled trees by means of a slow fire close to the ground, taking four days for the operation. For hollowing a canoe they cut a hole in the surface of the log with a stone axe, kindled a small fire, and burned down and along, carefully dropping water all around, to confine the blaze to a given spot.[3]

Fire opened the door of the primitive races to the use of metals. It is not necessary, as stated before, to believe that this mutual acquaintance of man and fire and metals is of very late occurrence. It is not true, probably, that men had to pass altogether through the rude and the polished

[1] Dorsey, *Third An. Rep. Bur. Ethnol.*, p. 301.

[2] Turner, *Samoa*, London, 1884, p. 270. [3] *Ibid.*, p. 343.

stone age before they learned that some stones cracked in the fire, some were scarcely affected by it, and others were rendered soft and tractable thereby.

The copper art, not the copper age, in the north central states of the Union, just now being carefully studied, teaches that a metallic age may co-exist with one of rude and polished stone combined. This copper is found in outlying masses and in shallow pits. The rocks surrounding the almost pure masses were removed by means of fire and water. The metal is said to have been hammered cold. But there is nothing unreasonable in the suggestion that the open wood fire was invoked in the process of making drills, needles, chisels, plates, ornamental discs, &c.[1]

It is in Africa, however, where the ingenious savage of to-day best understands the handling of metal by means of fire. Good ore from Nature's hand furnishes the material cause of the art. Then follow anvils, sledges of stone, charcoal, bellows of skins, tongs, and a host of cold chisels, punches, swedging apparatus, constituting quite an array of tools. It may be hinted that the Africans were taught these arts by wiser men. But the fact remains that peoples no higher in culture than North American Indians are practising a rude metallurgy by many ingenious processes now peculiar to themselves.

The negroid tribes inhabiting East Africa use weapons and tools made of iron, which they manufacture themselves. Among the Kaffir tribes native smiths are numerous, but their knowledge of metallurgic art is very primitive. Two round boulders of greenstone serve for an anvil, on which the red-hot iron is beaten with a rude hammer, whilst another Kaffir minds the charcoal fire, which is always made in a small hole in the ground. Two goat-skins are carefully sewn up, and meet in a hollowed-out bullock horn, one end of which is turned toward the fire. By alternately pressing the one goat-skin down with the hand closed, and pulling

[1] See Reynolds, *Pop. Sc. Month.*, vol. xxxi. pp. 519–531; *Am. Anthropologist*, Washington, vol. i. pp. 341–352.

the other up with the hand open, the air is forced into the coals, and sufficient heat is developed for the work. All the assegais of the Kaffirs, and the arrowheads of the tribes of the north, are made by native smiths, and most of them by smelting the iron directly from the ore. ·The natives also understand wire drawing. For this purpose they use small plates of iron, into which they bore holes.[1]

Livingstone's description of the South African iron ore illustrates well the help afforded by Nature to lower races in their arts. The material is obtained from the specular iron ore, and from the black oxide, the latter, being well roasted in the laboratory of Nature, contains a large proportion of the metal. It occurs in rounded lumps, and when found in river beds, it is easily detected by the oxide on the surface, and is dug with pointed sticks. Livingstone gives the report of an English gunmaker on this metal.[2]

The smelting furnace of the Dyoor is made of clay, about four feet high, with a conical base for charcoal, and a goblet-shaped top for the granulated ore. There are four perforations at the base through which pass the tewels, by means of which a strong current of air is supplied to the base of the furnace. In front of one of these is a pit for the accumulation of the beads of crude metal. The shaft is lighted below, the air is forced through, the ore is melted, and after about forty hours particles of molten iron begin to ooze and collect in the pit at the bottom. This crude metal, by means of stone hammers and repeated heatings in a forge, is cleansed and made fit for use.[3]

The Bongo furnace is a little more elaborate. It has three compartments, the middle one for the reception of ore and charcoal in alternate layers, the upper and the lower ones for pure coal. The chambers are separated from each other by ring-like incrustations on the inner wall. The bellows of the Bongo are formed of two trumpet-

[1] Griesbach, *J. Anthrop. Inst.*, London, 1872, p. cliv.

[2] *Travels, &c., in S. Africa*, New York, 1858, p. 695.

[3] Schweinfurth, *Artes Africanae*, London, 1875, pl. ii. figs. 11, 12.

shaped earthen vessels, covered on their outer end with leather, and opening into a third one. All the negro tribes of Africa use such a bellows, with immaterial variations. The valve is unknown, a very imperfect substitute being secured by piercing holes in the handle at the centre of the hide, and using the hands to let the air in and confine it.[1]

An interesting metallurgic art has sprung up in two regions of North America, namely, in Arizona and New Mexico, and in British Columbia and Alaska. It is not denied that since the possession of better tools, including files and emery paper, much better work is turned out. And, earlier still, Russians on the north, and Mexicans on the south, have influenced the craft. Still there is a residuum of doubt whether the northern Indians, as well as the Navajos of the south, may or may not have had a superstructure of their own on which to build. At any rate, in both regions there are now quaint silver- and coppersmiths who practise a curious mixture of savage and civilised handiwork in metal.

Matthews says the appliances and processes of the smith are much the same among the Navajo as among the Pueblo artisan. But the latter lives in a spacious house, and may have a permanent forge just the right height. The wandering Navajo constructs a temporary forge on the ground. Their tools and materials are few and simple: a forge, a bellows, an anvil, crucible, moulds, tongs, scissors, pliers, files, awls, cold chisels, matrix and die for moulding buttons, wooden implements used in grinding buttons, wooden stake, basin, charcoal, tools and materials for soldering (blow-pipe, braid of cotton rags soaked in grease, wire, and borax), materials for polishing (sand-paper, emery-paper, powdered sandstone, sand, ashes, and solid stone), and materials for whitening (a native mineral substance—almogen—salt, and water).

A forge built for Dr. Matthews by an Indian was 23

[1] Schweinfurth, *Artes Africanae*, London, 1875, pl. v. pp. 4, 5, 6.

THE SHOP OF A NAVAJO SILVERSMITH. (*After Matthews. Bur. Am. Ethnol.*)

inches long, 16 inches broad, 5 inches in height to the edge of the fireplace, and the latter, which was bowl-shaped, was 8 inches in diameter, and 3 inches deep. The forge was made of mud. Before this was completed a wooden nozzle was laid in for the bellows, where it was to remain, with one end about 6 inches from the fireplace, and the other end projecting about the same distance from the frame. Then he stuck into the nozzle a round piece of wood, which reached from the nozzle to the fireplace, and when the mudwork was finished the stick was withdrawn, leaving an uninflammable tweer. The nozzle of the bellows was tied to the protruding end of the wooden tube. The whole task of constructing did not occupy more than an hour. The bellows was a bag of goat-skin tied at one end to its nozzle, and nailed at the other to a disc of wood, in which is the valve. Handles and other accessories of the bellows were carved from wood. The nozzle was made of four pieces of wood tied together, and rounded on the outside so as to form a cylinder with a quadrangular hole in the centre about one inch square.[1] The bellows were worked by a horizontal motion of the arm.

For an anvil they employ any suitable piece of iron, but hard stones are still used sometimes for the same purpose. Crucibles are made of clay baked hard. After being in the fire two or three times they swell and become very porous. Some smiths use for crucibles fragments of Pueblo pottery. The moulds for casting their ingots are easily cut in sand-stone with a home-made chisel. Each mould is cut approximately in the shape of the article which is to be wrought out of the ingot, and is greased with suet before the metal is poured in. Tongs are made like sugar-tongs, and often nippers or scissors are used for tongs. Ordinary scissors are used in cutting the metal after it is wrought into thin

[1] This method of securing a bore, even in a crooked pipestem, is wide-spread. The Eskimo split a piece of wood, cut opposing gutters in the two halves, which are then joined again, and held firm by a lashing of raw-hide.

plates. Iron pliers, hammers, and files are purchased from the whites. The latter serve not only their legitimate functions, but are also used for punches and gravers. Metallic hemispheres for beads and buttons are made in a concave matrix by means of a round pointed bolt serving for a die. These matrices and dies are made by the Indians. On one bar of iron there may be many matrices of different sizes ; only one die fitting the smallest cavity is required to work the metal in all. For levelling the edges of the metallic hemispheres for buttons and beads, a small, roundish cavity is cut in the end of a cylinder of wood, of such size that it will hold the hemisphere tightly, but allow the uneven edges to project. The hemisphere is placed in this, and then rubbed on a flat piece of sandstone until the edges are worn level with the end of the wooden cylinder. For making charcoal they build a large fire of dry juniper, and when it has ceased to flame they smother it well with earth. If the fire is kindled at sunset the charcoal is ready for use next morning. The smith makes his own blowpipe, usually by beating a piece of thick brass wire into a flat strip, and then bending this into a tube about a foot long, slightly tapering, and curved at one end. They blow an intermitting current with undistended cheeks. The flame used in soldering is derived from a thick braid of cotton rags soaked in grease. For polishing they use powdered sandstone, sand, or ashes — all without water. For blanching, almogen (hydrous sulphate of alumina) is dissolved in water with salt. The silver, slightly heated, is boiled in this solution, and soon becomes white.

The Navajo silversmiths crouch on the ground while working. The whole process of producing their ware is minutely worked out by Matthews.[1]

But the splendid victory of man over the earth was achieved literally with the firebrand. The memory of conflagrations seems to haunt the dreams of bears and jackals

[1] *Cf.* Matthews, *Second An. Rep. Bur. Ethnol.*, Washington, 1883, pp. 171–178, pl. xvi.–xx.

and tigers, and all ferocious beasts. The naked African has only to kindle a little flame and lie down to sleep among ravenous lions. The wolf, the cougar, the wild cat, were long ago taught the hopelessness of resisting its fury. Great hunting excursions were made successful by setting the grass on fire. Venomous serpents and insects and bitter enemies of man, visible and invisible, had to yield to the brand.

The Shooli negroes hunt larger mammals thus by a co-operative method. Each man supplies a certain length of netting, and these are fastened together, it may be, to form a continuous barrier over a mile long. This is set up in the high grass along a space that has been burned over. Before each section of net a man is concealed. Several thousands of acres to the windward are fired, compelling the animals to run towards the nets, where they are killed by the men in hiding.[1] The hunting season commences when the grass is fit to burn. But should a person set fire to grass belonging to another proprietor, he would be condemned.

In the old camping days on the American prairies constant watchfulness and ingenuity were demanded in preventing the ravages of fire. The aborigines were well aware of this, and racked their brains to prevent the danger. They made fire in pits, cut trenches that the fire might not cross, and even saved their lives by burning a space before the great conflagration could reach them, and over this the flames could not pass.[2]

Not only in the pursuit of wild beasts, but in the art of war, fire held primarily a conspicuous place. The devices used will be described in the Chapter on War. The firebrand and its various processes in conquering and subduing the beasts inimical to man, and in compelling the earth to yield submissively to his dictation would form a study by itself. Even omitting the desolations of war caused thereby, one is

[1] Baker, *Ismaïlia*, New York, 1875, p. 457, with figure.

[2] *Cf.* Christy, "Why are the Prairies Treeless?" *Proc. Roy. Geog. Soc.*, London, 1892, pp. 78–99.

led to place this deadly weapon at the head of the list of useful inventions. Agriculture may be said to have been born of fire. There are many regions of the earth where this wasteful method still obtains. The forests are fired; the young growth is removed; the crops are planted; the blasted trunks are left to decay gradually. By and by the fertility of the ground is exhausted, and the prodigal husbandman makes another fiery onslaught upon the original forest. Fields are abandoned to grow up in such crops as seem them best. This was the primitive agriculture. In many places the modern farmer finds that he has been anticipated by men of whom there is no record.

The employment of fire as a mechanical means of determining definite portions of time must not be overlooked. Dr. Walter Hough has collected a number of examples under this head, among them the following [1]:—

The Polynesians skewer a number of the nuts of the candle-nut tree (*Aleurites triloba*) on a long palm-leaf midrib and light the upper one. Each kernel consumes in about ten minutes to a charred mass, which must be removed by an attendant when the next one below is ignited. The Marquesans tie bits of tapa at intervals along the torch.[2]

In China, the prescribed time during which the royal procession at the coronation of the emperor must move through the distance between the palace and the temple is regulated by a functionary who burns a joss-stick of a fixed length. At present in China, gong heung, or time-incense, consisting of five sticks of pressed wood-dust, made long or short according to the season, is burnt during the night, which is divided into five watches.

In Western China the water is raised by immense wheels belonging to the village, or to individuals who sell it to the peasants. The price is calculated by the quantity that flows from the wheel while a given length of joss-stick burns.[3]

[1] *Am. Anthropologist*, vol. vi. p. 209.

[2] See also *supra*, note on the Hawaiians by Bishop.

[3] See Rockhill, *The Land of the Llamas*, p. 42.

Chinese messengers who have a short time to sleep awake themselves by means of a lighted bit of joss-stick between the toes. In Korea the palace clock is an oiled paper lantern enclosing a rope of hemp soaked in nitre. Each hour is divided into four parts by cords tied to the rope. Time is announced by a lantern having transparent slides. The Koreans also reckon time after the manner of Wouter van Twiller—by the number of pipes smoked.

The European expressions, "marked candles," "King Alfred's," "auction by candle," "courting by candle," &c., are well known, and tell of the prolonged survival of a very ancient type of clock.

A series of human activities, connected for the most part with religion, have been ever associated with fire on the ceremonial side of life. In the simplest forms of Christian worship, from which almost all symbolism is eliminated, one constantly hears the word used in a figurative sense, in allusion to ancient altars. From this poetic allusion backward through symbolism, through sacrifices by fire, to fire-worship is a tolerably straight road.

The exorcism of horses and other animals among American tribes is well known. Matthews describes and illustrates with graphic power the introduction of fire into the Navajo medicine ceremonies. "The building of the great stack of juniper and cedar, twelve feet high and sixty paces in circumference, went on simultaneously with the sand painting.

"At the moment the music began the great central fire was lighted, and the conflagration spread so rapidly through the entire pile that in a few moments it was enveloped in great flames. A storm of sparks flew upward to the height of a hundred feet or more, and the descending ashes fell in the corral like a light shower of snow. The heat was soon so intense that in the remotest parts of the enclosure it was necessary for one to screen his face when he looked towards the fire.

"When the fire gave out its most intense heat a warning

whistle was heard in the outer darkness, and a dozen forms, lithe, lean, dressed only in the narrow white breech-cloth and moccasins, and daubed with white earth until they seemed a group of living marbles, came bounding through the entrance, yelping like wolves, and slowly moving around the fire. . . . When they had encircled the fire twice they began to thrust their wands towards it. One would dash wildly toward the fire and retreat; another would lie as close to the ground as a frightened lizard, and endeavour to wriggle himself up to the fire; others sought to catch on their wands the sparks flying in the air. One approached the flaming mass, suddenly threw himself on his back with his head to the fire, and swiftly thrust his wand into the flames." The end proposed in each case was to burn the bunch of down from the end of the wand, and with a show of dexterity to restore it.[1]

"Formerly," says Lacouperie, "the ancient kings [of China] had no houses. In winter they lived in caves which they had excavated, and in summer in nests which they had framed. They knew not yet the transforming power of fire, but ate the fruits of plants and trees, and the flesh of birds and beasts, drinking the blood and swallowing the hair and feathers (as well). They knew not yet the use of flax and silk, but clothed themselves with feathers and skins.

"The later sages arose and men (learned) to take advantage of the benefits of fire. They moulded the metal and fashioned the clay, so as to rear towers with structures on them, and houses with windows and doors. They toasted, grilled, boiled, and roasted. They produced must and sauces. They dealt with the flax and silk so as to form linen and silken fabrics."[2]

The simplest invention with regard to the preservation of fire grew to most important offices and ceremonies among

[1] Matthews, *Fifth An. Rep. Bur. Ethnol.*, p. 432, pl. xii. The reader should try to see the wonderful illustrations of this paper.

[2] *The Silk Goddess*, London, 1891, Nutt, p. 17; compare J. Legge, *The Li ki*, p. 369.

the cultured nations of antiquity. The Greek prytaneum and the temple of Vesta at Rome were the culmination of a series of cults, which began with the central tent fire of primitive peoples. M. Elie Reclus, in his article "Fire" in the *Encyclopædia Britannica*, works out this idea, but Professor Frazer more recently gives clearly the line of progress in the development of the Prytaneum, the temple of Vesta, the Vestals and Perpetual fires. "The Prytaneum, a round building with a pointed, umbrella-shaped roof, was originally the house of the king, chief, or headman (*prytanis*) of an independent village or town, and it contained a fire that was kept constantly burning. When a colony was sent out, the fire for the chief's house (*prytaneum*) in the new village was taken from that in the chief's house of the old village." "The Italian temple of Vesta, like the Greek prytaneum, was a round building. Tradition preserved the memory of the time when its walls were made of wattled osiers, and the roof was of thatch. The inmost shrine continued down to even late times to be formed of the same simple materials. Thus, looking back into the dim past, we descry the chiefs of the old Græco-Italian clans dwelling in round huts of wattled osiers with peaked roofs of thatch. And through the open door of the hut we see a fire burning on the hearth." "The gathering of sticks and putting them on the fire probably fell on those 'maids-of-all-work' in early households—the wife and daughters. Afterwards the fire in the hut, which royalty had relinquished to religion, was tended by maidens, who represented the daughters of the king."

"The creation of new fires, *i.e.*, the formal extinction and rekindling of fires at fixed periods, like the custom of maintaining perpetual fires, probably owed their origin, not to any profound theory of the relation of the life of man to the courses of the heavens, but to the elementary difficulty of lighting the kitchen fire by rubbing two sticks against each other." [1]

[1] J. G. Frazer, *J. of Philol.*, London, 1885, vol. xiv. pp. 145-172. The whole paper should be read.

Mr. Frazer is correct in the main, but it is not difficult to make fire with two sticks. The more probable motive in new fire was the finding of fire kindled by natural causes. The tribes on the Western Soudan extinguish every fire when they find a tree ignited by lightning. The Andamanese keep perpetual fire sacredly, because they have never learned how to kindle it.

CHAPTER IV.

STONE-WORKING.

"In seeking to ascertain the method by which the stone implements and weapons of antiquity were fabricated, we cannot, in all probability, follow a better guide than that which is afforded us by the manner in which instruments of a similar character are produced at the present day." —JOHN EVANS.

THE modern savage and his ancient representatives revealed in the study of archæology were good lithologists. They knew in each region what stone was best for their purposes in every emergency. They found out where this material abounded under the best conditions to be worked. They planned methods and invented apparatus for mining and quarrying it. They transported the material for long distances, half shaped it in their quarries to reduce the weight, made treaties with hostile tribes to secure the right to visit the coveted spot, and bartered the choicest of their own productions with fortunate possessors of the coveted material. All of these statements are known by archæologists to be true, and abundant examples may be cited to substantiate them.

But the savage man's knowledge of lithology did not stop at his acquaintance with materials. The qualities of substances were known to him, both as to working and as to using. He could tell you how each kind of mineral ought to be worked, and how it would do its work after it was put into shape. An examination of his workshops demonstrates that he understood cleavage and granular

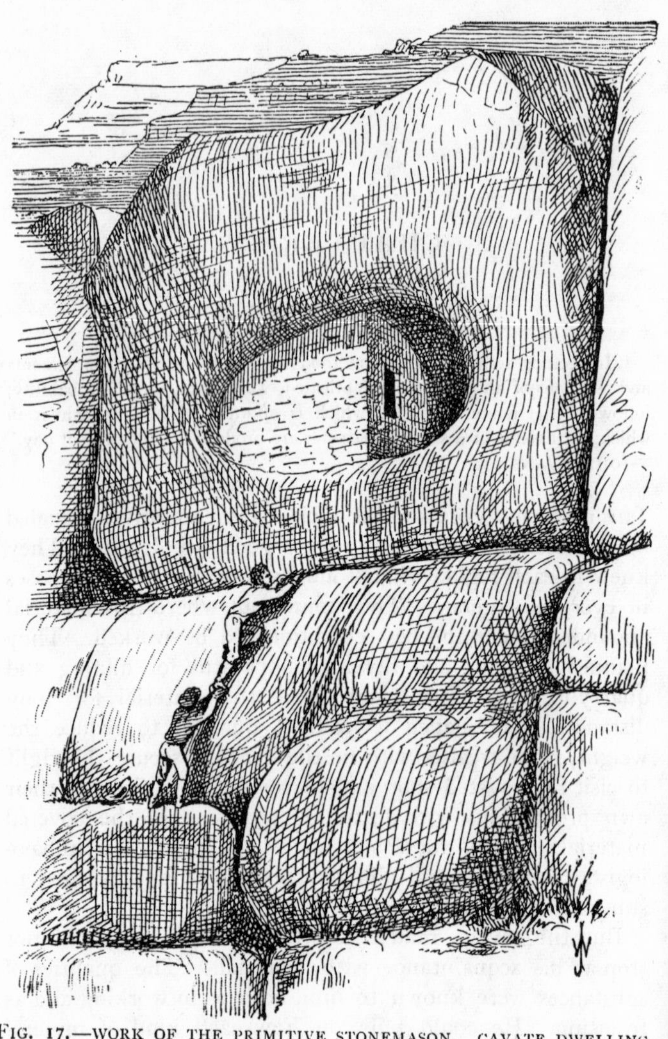

FIG. 17.—WORK OF THE PRIMITIVE STONEMASON. CAVATE DWELLING, NORTH MEXICO. (*After Jackson.*)

structure, and the idiosyncrasies of each stone. Many hundreds of rejected pieces about the quarries have been struck only one or two or three blows and then thrown away ; as though the ancient stone-worker communed with himself thus : "I have struck this one blow, and that is enough to prove that this rock or cobblestone is too soft, too friable, or without proper cleavage." After a second stroke, it would occur to him that there were flaws or hard lumps in the way. Perhaps all went well around one side of the pebble ; but on turning the stone over and flaking, he would be assured by long experience that the blade or other object he sought to make would be much too thick. This would result in abandoning the experiment. A careful study of this subject, by Holmes especially, leads to the inevitable conclusion that the prehistoric savage had nothing more to learn about the physical properties of minerals that were necessary for him to use in his avocations. So true is this that the most skilful flint-knappers of Brandon and elsewhere are not able to reproduce some of the more beautiful forms that are common in museums. The Smithsonian Institution has had a number of skilled workmen spend a great deal of time on the making of a leaf-shaped blade, but they have never succeeded in the effort.

The very earliest men suspected to have lived upon this earth by the French archæologists were the people of Thenay. But they and their ancestors must have walked along the shores of time far enough and long enough to study all minerals and all rocks, and to select the best one in all the earth for chipping and flaking. These flints of Thenay were found by the Abbé Bourgeois in what he believed to be undisturbed Tertiary formation, and so able an archæologist as Gabriel de Mortillet sustains him in the conclusion. Indeed, the distinguished archæologist has placed these Thenay flints at the head of the class in the St. Germain museum. This implies, as was just said, that man existed in the indescribably long ago Tertiary, an that he had even then graduated in mineralogy. There are other ways of explaining these interesting pieces. They may not after all be of human workmanship. They may not have been found in place in the undisturbed Tertiary. The Abbé was not a geologist, and sharper eyes than his have been deceived in such matters. Finally, these flints may not be the finished products of extremely ancient art work, but the rejected material of later flint chippers, just as one finds spalls of marble and blocked-out rejecta about the sites of ancient and modern stone-cutters' yards. Within the few past years investigations have been conducted in the United States, chiefly by the gentlemen connected with the Bureau of Ethnology of the Smithsonian Institution, which make it clear that millions of roughly chipped stones formerly thought to be ancient, on account of their form, are only the refuse left by men who were aiming to make blades. This leads to the conclusion that in other lands it is quite possible that many surface finds are "rejects," not implements. The question of antiquity is a geological one.

The geologic surroundings of savages had a pronounced effect upon their implements, masonry, sculpture, utensils, and weapons, limiting the forms and sizes, and determining to a considerable extent the kinds employed in the various districts, independently of biologic and other conditions.

The geology of the tide-water country in the United States is altogether unlike that of the highland, and the rocks available to the aborigines in the two regions were not only different in distribution, but peculiar in the shapes they received and in other features that affect the character of the utensils made and employed.

The workable stones, such as argillite quartz, quartzite, rhyolite, jasper, and flint were much sought by the aborigines of the lowland. Fragmental material was to be obtained almost everywhere upon the surface, but choice varieties were confined to limited areas and often to distant regions; and, where surface exposures were not sufficient

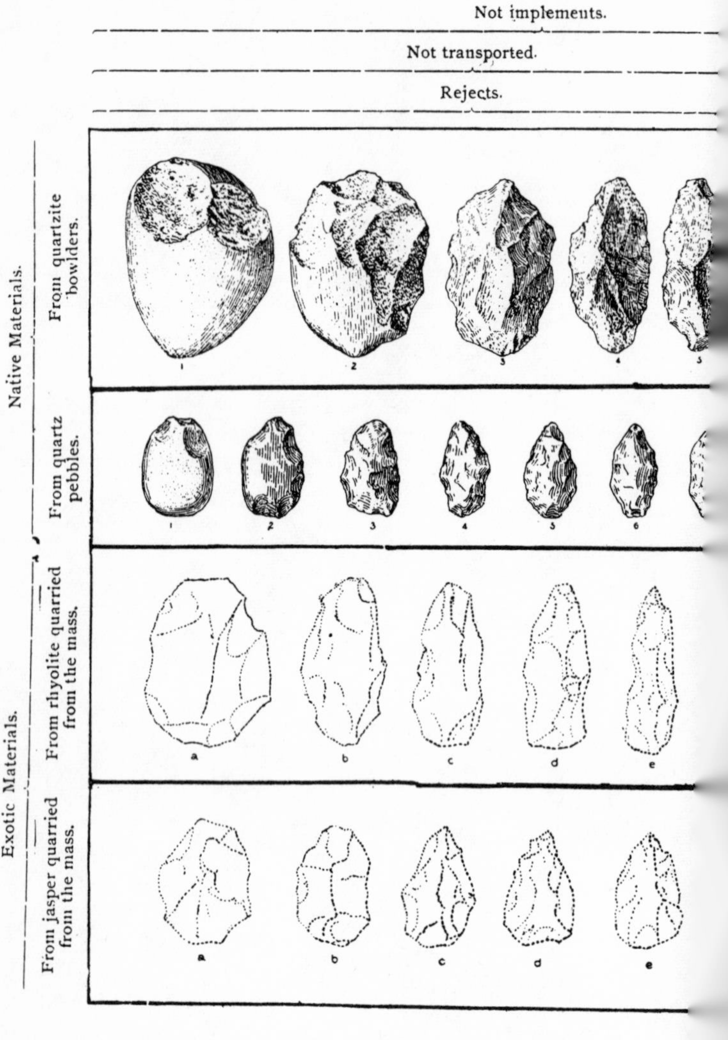

Not implements.

Not transported.

Rejects.

Native Materials.

From quartzite bowlders.

From quartz pebbles.

Exotic Materials.

From rhyolite quarried from the mass.

From jasper quarried from the mass.

FIG. 18.—THE ELABORATION OF CHIPPED STONE IMPLEMENTS.

(*After W. H. Holmes, Am. Anthropologist*, vi. pl. 2.)

ts. Implements.

Transported.

he forms. Specialized forms.

8 9 | 10 11 12 13 14 15 16 17 18

0 11 | 12 13 14 15 16 18 19 20

h i | j k l m n o p q

n | i j k l m n o p q

to supply the demand, quarrying was resorted to, and the work of securing, transporting, and trading or exchanging the stone must have become an important factor in the lives of the people. The masses of rock were uncovered, broken up, and tested; the best pieces were selected and reduced to forms approximating the implements to be made, and in this shape they were carried to the lowland.

In the lowland all varieties of hard stones are fragmental, and the species are intermingled in various ways. The fragments of rock are not merely broken, angular pieces, but rounded masses and bits known as boulders, cobbles, and pebbles, and comprise chiefly such tough flinty, homogeneous stones as are available in the arts of primitive man. Nature, in her own way, selected from the highland along the stream courses the very choicest bits of crumbled rocks, reduced them in hundreds of cataract mills and in the breakers on the shores of ancient seas to rounded forms, and deposited them in the lowlands in great heaps and beds. Nature has not provided any other form of the several tough varieties of stone so perfectly suited to the purposes of the implement flaker as the boulder or pebble.

Nature selected the rocks used by the tide-water peoples and distributed them in groups, varying with the original location, with hardness, with toughness, with shape and with size. The effect of these conditions of distribution upon the stone art of the various districts was necessarily very pronounced. One community located near deposits of large boulders used them, and the tools shaped therefrom are large on the average, and *vice versâ*.[1]

The most important result of Holmes's investigation is the emphasis laid on the fact that thousands of spalls and wrought stones lying about, which have been called paleolithic implements, because they are so rude, are only the rejecta of the quarrymen and of the blocking-out process. By actual experiments, Mr. Holmes and others have shown

[1] Consult Holmes, "Distribution of Stone Implements, &c.," *Am. Anthropologist*, Washington, 1893, vol. vi. pp. 1–14, 2 pl.

that for one boulder that yields a good implement at least ten are thrown away after one, two, or more blows that reveal the weak spot.

In repeating the processes of the ancient cutlers and armourers, Holmes found among the "rejects," or "wasters," three well-marked steps of progression. The first class of rejects were discarded after one, or at most two, blows, revealing bad material. The second class have one face chipped over, the other still showing the unwrought mineral ; but this half-working brings to light a hump or fault on the chipped side. The third class are wrought on both sides, but on the dressing down of the second side some weakness revealed itself, and the "turtle-back" was discarded. Unless geological evidence is forthcoming to prove that a given piece is ancient, Holmes does not regard the form of the artefact to be good evidence either of antiquity or of the utility of the specimen. It is simply rubbish.

It is worthy of note in this connection that, in many of the European paleolithic sites, the most beautiful leaf-shaped implements in the world are found associated with so-called paleoliths. And this is exactly what Mr. Holmes finds to be true in his quarry sites—boulders, cracked boulders, turtle-backs, and broken leaf-shaped pieces, fractured at the last moment, all in one confused mass. But near by, on an adjoining camp-site, knives, spear-heads, arrow-points, and the rest abound. "Nearly all rude, bulky implements of chipped stone, and all failures or rejects of manufacture, are, as a matter of course, found upon or near the sites from which the raw materials were derived."

Again, the percentage of failures—turtle-backs and other refuse of manufacture—decreases rapidly with the distance from the source of supply of the raw material, extending little beyond it.[1] In one instance Holmes was able to trace up a rhyolite quarry from a camp site a hundred miles away by

[1] Holmes, "Distribution of Stone Implements," *Am. Anthropologist*, Washington, 1893, vol. vi. pp. 1–14. Also H. C. Mercer, *Pop. Sc. Monthly*, New York, 1893, September.

simply following up the implements and wasted chips of this material along the lines of their greatest abundance.

The earliest period of human industry is called the "Stone Age," because, in digging about among the graves and remains of the past, archæologists find relics made of stone always lower down, or in older beds, than relics made of metal; and it is conceivable that there might have been a time when men were so rude as to use naught but apparatus of stone in their industries. But there is no evidence of such a period. The careful study of modern savages proves that for every stone tool showing evidence of human workmanship there were many more constructed of wood, bone, shell, hide, &c. And these very pieces of stone themselves were accompanied with, and attached to, other and perishable material. Even the hammer-stone was an apparatus for making something else which could not be used until it was so attached. The stonecutter's age began very early in the history of workmanship, but even then he was only one of many craftsmen. It is convenient, however, if we keep this explanation in mind, to speak of a "stone age," in which was developed the search for material and the study of its qualities with a view to working it into the general scheme of mechanical appliances in vogue.

The aboriginal stone-working art may be subdivided in several ways. Men have been wont to speak of a palæolithic and a neolithic age of the world, or status of culture. In the former, the products are said to have been only rudely chipped and flaked; in the latter, they were battered and ground, or polished.

These two terms have become firmly embedded in the vocabulary of archæology, and when properly used are convenient. But they have not always been judiciously employed. I here take for granted that men have practised every art rudely at first, and have learned to work at it more cleverly later on. This is true of every human occupation.

But there is another truth that must be also tenaciously

held in mind. Arts, industries, occupations, are all determined by the proximity of material. It is easier to peck a granular stone than it is to chip a flinty one. Polishing one stone on another is a simpler art, and easier to learn, than making a delicately flaked dagger of flint, and it does not require half the knack. It is even conceivable that a region might be so favourably situated that savage men could more easily develop an age of metal therein than one of polished or of chipped stone. There are vast tracts of earth where there is not a mineral having conchoidal fracture. Here were discovered tribes of aborigines in the polished, or at least the hammered, stone age. But they were cannibals, and their language and social system both showed that they were low in the scale of culture. The conclusion is not that these savages had passed through a chipped stone age before they were found, but that they did the easiest thing within the capabilities of their environment. The Polynesian Islanders, the natives of the West Indies, the Tlingit Indians are examples of this class.

In certain portions of Africa, in Canada, and perhaps in Michigan, the metal age is as old as the stone age, and from some areas all evidence of the stone age is absent. The wiser method of looking at this matter is to hold that each art has had its elaboration in the home of its proper materials, beginning with simple and almost infantile processes, and working along through greater complexity to their perfection. That there has been degradation and dissolution of skill and industry is equally true. But, when men lost their knack in polishing stone they did not take to chipping or flaking. They simply made a more sorry job of battering and polishing. This process may be seen in any city where new methods are introduced. A few non-progressive persons will hold on to an old art, and it will decline in their hands.

In studying out the evolution of invention as regards the stoneworker's art, therefore, little help will come from the terms "palæolithic" and "neolithic." These are excellent in their place in the classification of a definite series of art products like that of Western Europe, where flint was first blocked out, then chipped or ground to an edge merely, then polished over the whole surface.

From the point of view here assumed, whence we may look down upon the workman, and his work, and his reward, and the demands of a more and more exacting public, it is better to disregard this historic speculation, and turn the eye upon the various artisans and tools engaged. The aboriginal stoneworkers, or stoneworkings, may be thus classified :—

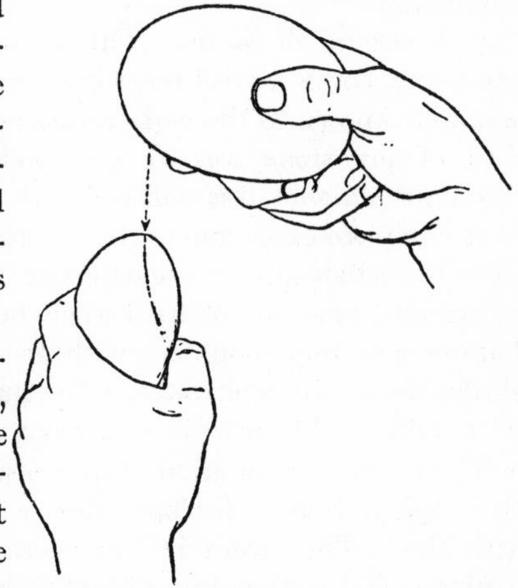

FIG. 19.—The first blow in Stone Implement Making. (*After Holmes.*)

1. The stone-knappers : makers of spalls, and artefacts with large facets. Their implements were at first other stones, then stone hammers specially selected and formed, and after that knapping hammers of metal. The art consists in breaking stone with a blow.

2. Stone-chippers and flakers : makers of chipped products. Their tools were small hammers of stone, but more especially pointed pieces of bone or antler, which were used as pitching tools, or for pressure.

3. Hammerers of stone : makers of mortars, pestles, axes, sculptures of all kinds. Their apparatus was the stone hammer, ancestor of modern bush-hammers.

4. Stonecutters, *par excellence :* workers in soft materials at first, such as soapstone and the less compact volcanic rocks. The tools were chisel-like, or gravers, and were

worked with the hands rather than struck with mallets. The modern carvers are their descendants.

5. Sawers of stone and other hard materials. Their tools are not well understood, since, strange to tell, few white men have ever reported observations on the subject.

6. Borers of stone. Their tools were drills of soft material chiefly, used with sand, and, in boring soft stone, harder stone points.

7. Polishers of stone. Other stones with or without sand, and corals or ochres, were the means employed. It was well known to the early artificers of this class that the dust of any stone was its best polisher. After the same fashion the diamond is polished with diamond dust.

A complete catalogue of the workmen of this class would have to include also the quarriers of stone, with rude shovels of scapula, crowbars of hard wood burnt at the end, sledge hammers of huge boulders, with or without hafting, and a skilful use of fire and water. The human beast of burden, with wallet of skin or woven basketry, would complete the list.

The recent examination of immense aboriginal quarries in the United States discloses this whole series of ancient activities. They were in full operation always when the explorers and settlers first visited each region.

Above all, it should be remembered that two or more of these artisans, and indeed all of them, in the early culture periods, might be one and the same person. The Indian quarriers about the city of Washington, doubtless, who dug into the hill at Piney Branch and extracted the boulders, were the same men who tested and trimmed them, who flaked and chipped them, and who wore the products out in the chase or in war. The differentiation of crafts came later. In the West Indies, surely, the same artists quarried the volcanic rocks and worked out the stone collars and mammiform stones.[1]

[1] Mason, "Antiquities of Porto Rico," *Smithson. Rep.*, 1876, pp. 372–393, many figures. Also *Smithson. Rep.*, 1884. The Guesde Collection.

In our day there would be division of labour in such matters. In any stonecutter's yard may be seen at work the spall-maker and stone-breaker, with great hammer or with mallet and steel chisels. Indeed, he will on occasion break one stone with another.

In the flint-knapper's humble shops, about Brandon, the chipper of gun flints holds the office of the old-time dagger-makers near by.

On all public buildings the busher, pounding away on a block of granite with several plates of steel fastened together, is the lineal descendant of the very ancient wielder of the stone hammer in battering and pecking stone.

In the steam marble works strips of soft iron sawing backwards and forward separate the material through the co-operation of sand and water as in primeval times.

It is not decidedly known when the ancestors of gem-cutters and borers began to use emery or corundum powder. But the lapidary's wheel and the mechanical devices for polishing industrial stone look backward easily to the first savages rubbing one stone against another of the same sort or of harder material. The same stone that does the rubbing is itself rubbed, and in many operations the sand or emery is reduced to a fine flour that makes the best of polish.

For the more civilised successors of the savage quarrymen the reader will have to make a journey to Peru, Mexico, Easter Island, Southern India, or ascend the Nile and examine the beds from which were cut, with very simple appliances, the great blocks for Egyptian sphinxes and obelisks, or he must visit the quarries of Baalbec or Persepolis, or the later-worked beds of Italy and Greece.

What with steam excavators, and drills, and dynamite, and travelling cranes, the modern quarryman has gone very far past the early processes. Yet he cannot even guess how his predecessors removed and set up the great monuments of the past.

The knapping of stone as an art is not practised in

precisely the same fashion throughout all savage areas, owing to national traits and cleverness, to material and to the end proposed. The Eastern Indian of the United States doubtless held a boulder of quartz or a mass of rhyolite or argillite in his left hand, and with another suitable piece, or with a specially hard specimen which he kept for the purpose, held in the right hand, struck a smart blow upon the former to determine its quality. At this point the material piece was either accepted or thrown away. If the workman with his right hand kept on striking sharp, hard, ringing blows, carrying away conchoidal flakes from first one side then the other, until the stone assumed the correct form for dressing down.

The problem ever before the workman's mind was to prevent the occurrence of a hump or monticule in the middle of the object. Should such a hump be left, the subsequent processes for making a beautiful leaf-shaped blade could not go on.

It is contended among archæologists of one school, however, that these pieces with thick centres, called "turtle-backs" in the United States and Chelléan or Moustérian implements in France, were designedly blocked out, that they are really one type in an evolutionary series, that the maker of chipped implements must needs have gone through this form of work before he invented processes by which he could avoid the hump in the middle and secure a laminated blade. The American archæologists, who have laboured long to repeat the processes of the aborigines in stone work, find themselves unavoidably making "turtle-backs" when they are really trying to create the leaf-shaped blade. If that be so, then such pieces found in earlier geological horizons are really palæolithic, and there can be no objection to such an opinion.

In the country of obsidian and of the finest calcareous flint the object of the knapper in earliest times was not only to secure leaf-shaped or almond-shaped implements. Long razor-like blades were in great demand for scarifying, shaving, sacrificing, and for domestic purposes. The California Indians used a "cold-chisel" or pitching-tool of antler struck with a hammer of wood or stone for such results.

No historic reference is found descriptive of the way in which the ancient Mexicans and the savages of Western Europe struck off long and even blades of obsidian and flint. It is possible, by carefully studying the texture of these materials, to do this work with a blow. Some references are to be found to the use of great and steady pressure. But the evidence is clear enough that, with the proper

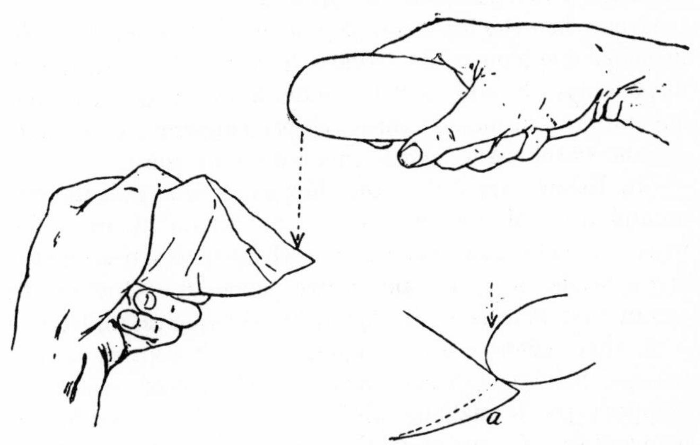

FIG. 2C.—Removing Flakes with Stone Hammer. (*After Holmes.*)

knack, either the ancient Dane or Frenchman might with a single blow of his stone hammer remove a flint blade nearly a foot in length. The Mexican knapper was equally clever, as the abundant relics of his handicraft testify.

The tools and processes of the stone-chipper or flaker are more varied. The author has seen both Indians and white men pound a small chip of jasper into excellent shape for an arrow-head with a small pebble of quartz alone. For the scraper blades and coarser knives and smaller weapons there is no doubt that this process sufficed. To effect this, take a thin chip of any conchoidal stone between the left thumb and forefinger. With an elongated pebble of hard stone strike a series of quick, light, elastic blows along one margin of the chip, barely touching it. The nearer one comes to missing the edge the better. The blow is better struck downward and slightly under. Turning the chip over, the other margin may be similarly trimmed, and, by reversing it end for end, the processes may be repeated on the margins from different sides. Any one possessing a large series of American arrow-heads will observe that the margins of many of them have been trimmed by the end to end reversion when the chip is revolved.

Even when the little chipping hammer has not been the principal implement, its services have not been altogether dispensed with. There is no doubt, however, that the chief apparatus in the manufacture of chipped implements throughout the world has involved some kind of pressure.

The Eskimo are still in the chipped stone age all the way around from Mackenzie river to the Yukon mouth. The most delightful raw material for whale lances, deer lances, arrow-heads, scrapers, and knives abounds. One of the commonest objects to be found in ethnological collections from that region is the chipping tool. It consists of two parts, a handle of walrus ivory exactly carved to fit the chipper's palm, enabling him or her to have the firmest grip and to exert the greatest pressure. All of these handles are finished with the utmost care and highly polished.

At the working end of this ivory grip is a groove dug out about two inches long, half an inch deep, and less than a quarter of an inch in width.

Into this groove is fitted a strip of very hard antler or bone, extending, say, one inch beyond the end of the handle. The two parts are firmly seized together by a band of sinew or fine raw-hide string. The apparatus is ready now for action.

Before describing its process, it ought to be said that the two parts have each a *raison d'être*. A piece of antler or bone could not be found large enough for the grip or handle.

And that material is entirely too slippery and hard for the chipping point. Wood is too soft, ivory is too hard. A material is needed which is tough enough to break stone and yet soft enough to allow the stone to sink into its substance a little way to get a hold. Hard bone or antler are of all things the best, as every Indian had found out before Columbus discovered America. Expert Indians will do the finest chipping with a steel point. But this can be made very sharp at the end, and does not slip. White men who make arrow-heads prefer the point of steel.

The method of using the chipper among the Eskimo is to

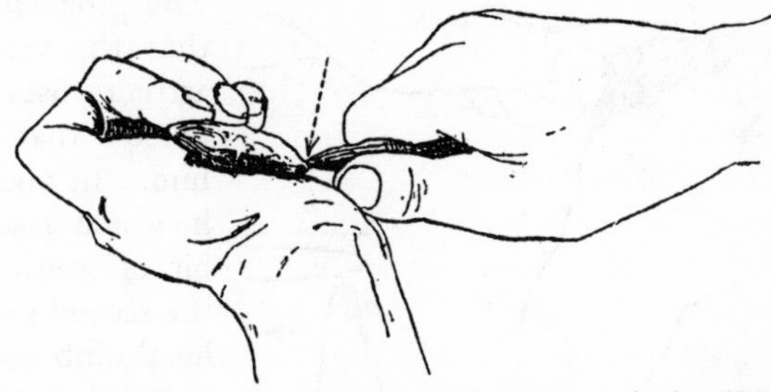

FIG. 21.—Chipping Stone by downward pressure. (*After Holmes.*)

lay a glove or piece of hide in the left hand, to place on this the bit of stone to be wrought, and then to hold it in place by means of the thumb. Thus prepared, the workman or workwoman (for both sexes chip stone in Alaska) grasps the chipper in the right hand and presses downward along the edge of the stone, making thirty or forty efforts per minute, feeling the way along, gauging the width of the chips deftly, until the point or butt is reached. The piece is turned over or reversed, as in the process last described. The especial feature of the Eskimo work is this downward pressure, which to a civilised artisan would seem to be working in the dark.[1]

In the United States National Museum is a collection of the fibulæ of the deer, pieces of very hard bone about a foot

[1] Figured and described in Murdoch, *Ninth An. Rep. Bur. Ethnol.*, Washington, 1892, p. 288, figs. 279–281.

long and pointed at one end. They are the chipping tools of the Shoshonean tribes of the Great Interior Basin of the United States. These Indians made their arrow-heads, spear-heads, and knives of jasper chiefly. Their method of procedure was to grasp the chipper near the working end, so that the other end might be firmly braced in the forearm. The bone in drying becomes extremely tough and strong, having the two qualities requisite in a chipper, strength and bight. Testimony is conflicting as to whether the Shoshonean presses upward or downward. The probability is that the versatile artisan uses the process that suits him. In one case he would grasp his bit of stone with the second joint of his thumb and the first joint of his forefinger, and push off his little flakes upward.

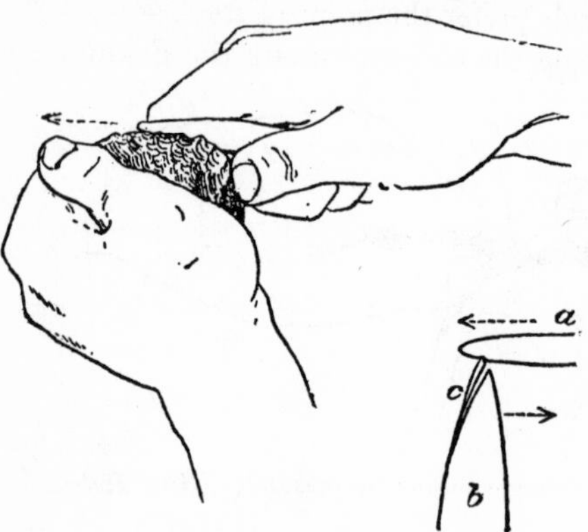

FIG. 22.—Flaking Stone by outward pressure. (*After Holmes.*)

This process may be repeated by any one who will grind the end of a tooth-brush handle to a point and follow the directions. By the other process the Shoshonean imitates the Eskimo, lays the piece of stone upon a bit of leather in his palm, and presses upon the edge in a downward direction.

Mr. Cushing informed the writer that the long and beautifully crenated surfaces of choice daggers and other blades were produced by placing little bits of soft gum along the midrib at regular intervals, and then using pressure. The writer has for years sought for an Indian who could do this fine dagger work, but he has failed. This is indeed one of the lost arts. The English gun-flint makers are able to take a core of flint and divide it into a series of laminæ with marvellous skill; but no amount of reward has been able to tempt one of them to produce a leaf-shaped blade. Mr. Edward Lovett, of the Bank of Scotland, went to great pains for the author to secure the services of a knapper to do this work, but the specimens turned out to be utter failures. The deft hands that were once so numerous have lost their cunning, and there may never stand on earth another who can imitate what they wrought.[1]

"The Andaman Islanders employ chips of quartz for lancets and razors. No piece is used more than once, and several may be required for each operation. Those having a sharp, blade-like edge are reserved for shaving, while others with a fine point are kept for tattooing or scarifying. Flaking is regarded as one of the duties of women, and is done by them. For making chips two pieces of quartz are needed. One is first heated and allowed to cool; it is then held firmly and struck at right angles with the other stone. The smallest flakes are obtained in this way and not by pressure."[2]

The rejected flakes and cores are thrown upon the refuse heap to prevent their cutting the feet of children, and Dr. Man thinks that this accounts for supposed rude stone implements found in Kjökkenmöddings.

As intimated previously, there are vast regions of the earth, once inhabited by savages, where there does not exist a bit of quartz, or jasper, or flint, or obsidian, or any other sort of stone capable of being flaked. But there are volcanic as well as sedimentary rocks, which may be pounded or battered into shape. These same materials

[1] Evans, *Ancient Stone Implements*, pp. 34, 35. Murdoch, *Ninth An. Rep. Bur. Ethnol.*, pp. 287–89. Mason, *Smithson. Rep.*, 1886, part i., pl. xxi., figs. 92, 96. The largest cache of chipped stone implements ever discovered was found in a mound near Chilicothe, Ohio, by W. K. Moorehead. They are seven thousand in number, and weigh three tons and a half. See Moorehead, *Primitive Man in Ohio*, New York, Putnam, fig. 35.

[2] Man, *Andaman Islanders*, London, 1883, Trübner, p. 160.

abound all over the world, even where there is abundance of flakable stone.[1] If the striking of one stone against another is as easy as breaking one stone with another, then the rudiments of stone pecking, as before suggested, are as old as those of chipping stone, and the hammer is as ancient as the flaker.

The practical method of working among savages is to select the toughest bit of stone accessible for a hammer. Holding this in the right hand, between the thumb and two middle fingers, and placing the forefinger on the top, the workman administers quick, sharp blows at the rate of three hundred or more per minute. The hammer has an excursion of three inches, more or less. At short intervals the worker brushes away the loosened material with a little broom of stiff fibres and begins again. At first, only common boulders were selected for hammers, but it did not take the ingenious ones long to discover that little pits on either side enabled the thumb and middle finger to relax their hold a little just as the blow was struck, in order to avoid injury by the concussion. So the conventional hammer stone was invented. Mr. McGuire says that after an hour's work with a tool without the "finger-pits," his arm grew very sore. The pecking of stone, among modern savages, reveals all the steps through which the art passed in its early evolution. In the Acorn region of California, the women take a common boulder for the nether millstone. Around the margin of the upper side they place a hopper of basketry, sometimes luting it fast with the pine-tree gum, sometimes holding it down with their feet, while they do the grinding with their hands.

For a pestle these primitive millers use an elongated piece of hard porous rock, the corners of which have been

[1] Recently Mr. Joseph D. McGuire, of Washington, has devoted two years to the study of the stone hammer and its various uses, with important results. J. D. McGuire, "The Stone Hammer," &c., *Am. Anthropologist*, Washington, vol. iii.; also by the same author "The Aboriginal Lapidary," *id.*, vol. v. pp. 165–176.

battered away by pecking. With this apparatus the milling begins. The constant beating on the nether stone with the pestle gradually excavates a bowl-shaped cavity, a shapeless mortar, as simple as it can be. There are numerous examples of apparatus of this elementary sort, in which the mortar is carved on the surface of a great stationary rock near some camp, and thither resorted one family after another to prepare their flour.

But families, or, more properly speaking, clans and tribes in that grade of culture move from place to place. It would be necessary to have among the household effects a portable mortar. In the eastern portions of the United States great numbers of mortars are found that are extremely rude in shape. A piece of granite or sandstone, not over six inches thick and one foot across, was battered into an exceedingly rude outline on the outside, and hollowed out a few inches on the inside. The pestles are better shaped, being oftentimes quite cylindrical. A few examples are very heavy, and we are informed that they were suspended from the limb of a tree and kept in motion with the two hands. It is just possible that these rude shallow mortars had basketry hoppers. Even in the Ohio valley, where some kinds of stone pecking were done with great skill, the mortars are still rude. But along the Pacific Coast of America, from Alaska to the Mexican States, the mortars were carved out with exquisite care. The Tlingit Indians of South-eastern Alaska were especially clever with the stone hammer. A block of sandstone or serpentine or porous volcanic rock was hammered into a symmetrical bowl-shaped form on the outside, and hollowed out so as to leave a wall one inch or less in thickness. Upon the exterior of these, carved projections were left of considerable beauty.

The pestles for these mortars were of the same materials. The fundamental form was bell-shaped, but in many examples the top is cut in the likeness of one of the totemic animals of the tribe. Other forms were used, but

FIG. 23.—AN ABORIGINAL STAIRWAY; CAVE DWELLING, NEW MEXICO. (*After W. H. Jackson.*)

every one of them was produced simply by pounding one stone with another. Upon the surfaces of both mortars and pestles hundreds of little pitted marks are left, showing where the blows were struck. On the coast of California, about the Santa Barbara Islands, were great manufactories of mortars and pestles. Huge blocks of sandstone were quarried, pounded into form and then pecked out into the symmetrical shape of a great bowl. These were finished up on the outside and the inside with much care, and they were articles of trade throughout the neighbouring region.

Further south, in Mexico and throughout Tropical America, mortars and pestles give place to metates and mullers. But these, where they are not mere slabs of gritty stone, slightly modified by pounding, are carefully pecked into shape with the hammer stone. Even those that are made with legs, and have their borders adorned with sculpture, are not exceptions. They are the product of the stone hammer alone, and were wrought out by incessantly pounding one piece of rock with another. Quite similar to these table-form metates are the stone chairs or stools seen frequently among archæological collections from Middle America, and they were wrought out in the same shops. Their furrowed ornamentations are the work of stone picks, of hard material which could be easily re-pointed, now and then as occasion demanded. A more universal product of the stone hammer is the stone axe and celt and sledge. There are many edge tools in all European collections that went from the knapper's hands to the grindstone, and were gradually transformed from chipped implements into polished implements in the process of being worn out. This was necessary where excellent flint occurred, and better material was scarce. But in both Americas, and almost everywhere out of Europe, and in many parts of Europe, axes and celts were made by means of the stone hammer. This process of pecking-out axes is hinted at in the older writers, and Loskiel affirms that it required a lifetime frequently to fabricate a single piece. To determine the truth of these assertions, Mr. J. D. McGuire, of Ellicott City, Maryland, conducted a careful series of experiments at the Smithsonian Institution for the author. The results are given in a series of papers published by Mr. McGuire. Any ordinary grooved axe or celt could be produced in less than fifty hours of continuous work. A grooved axe of jadeite was wrought from the rough spall in eighty-six hours. But for the occurrence of a flaw in the material the axe would have been brought to an edge in one hundred hours.

The examination of many hundreds of specimens reveals the process of the aborigines, which seemed to be somewhat like that of the sculptor. The workman looking over many hundreds of boulders or spalls sees one in which his imagination detects the outline of the celt or the axe desired. His motive is the same as that which has reigned in the minds of artisans from the beginning of industry, namely, to achieve the greatest result in the expenditure of the least effort. It is interesting to note how little labour will transform a pebble into a grooved axe, when the workman knows what he is about. His skilled eye detects at a glance the very best piece for a specific result, and after he has pecked away a few hours the metamorphosis is complete. Upon the surface of most objects of this class the marks of the hammer are left, frequently from an ornamental motive, and there is also oftentimes a large surface that Nature wrought. But the enormous amount of work done by the stone hammer does not appear in the museum, because the effects of the blows have been obliterated by polishing.

Among the celts and axes seen in collections there are many that are too highly ornamented ever to have been much used. Objects of this kind are common especially from the West Indies. But they were made by the self-same process that produced the commonest axe, namely, by striking one stone against another hundreds of thousands of times. The old proverb, "A continual dropping wears away a stone," must surely have been written of the stone hammer.

The stone hammer itself has had a curious history in this series of operations. Many hundreds of them are to be seen in the museums of the world. Schliemann mentions the finding of thousands of them without suspecting their function.[1] The simplest hammers are merely battered masses of stone, while the best of them are almost lenticular in shape, with pits on the sides and evidences of work over the entire rim.

Artists of antiquity had recourse to the stone hammer as well as did the artisans.

There does not seem to be a habitable part of the world where the aborigines have not left their marks upon boulders, cliffs, standing stones, walls of caves, by means of this implement. The simplest process is the production of shallow lines in intaglio, portraying beasts and human beings and illegible figures. In places where men have been wont to congregate the surface of the rock has been gone over several times, until it is an inextricable confusion of etchings. The scribe or artist, as the case might be, simply took a pointed hard stone in his hand, and by means of a succession of blows traced out a shallow writing or picture. Mr. Pickwick's famous inscription of Mr. Stumps's autograph was no doubt wrought after the same fashion. Innumerable savage carvings were cut in soapstone and other soft material by scratching, etching, cutting, and the like. But in all granular material there has been but one implement and one method, the easiest and the simplest, the implement being the stone hammer and the method that of a rapid succession of blows.

Next in line of evolution in this class of work is a kind of low or flat relief produced by pounding away the intervening material and leaving the figure outstanding. It is seen in many rude Indian carvings, but in its perfection in the Maya

[1] Schliemann, quoted by McGuire, *Am. Anthropologist*, Washington, 1893, vi. p. 314; vii. p. 358.

inscriptions of Central America. Here the artist has in mind a certain symbol which has been painted on pottery or other material, and seeks to reproduce the same image in flat relief by pecking away the unnecessary stone. This kind of work was done by Mr. McGuire with great rapidity upon a block of tough lava, now in the United States National Museum. With stone hammers he first made the surface flat. Then having marked thereon with a bit of charcoal the Maya symbol, he, by means of pecking tools

FIG. 24.—EASTER ISLAND IDOL HALF EXECUTED. (*After W. H. Thomson, U. S. Nat. Museum.*)

chipped to a point, removed the surplus stone. The constant wearing of the point demanded frequent renewal, but this was easily accomplished. The result was that in less than thirty hours he worked out of a rough block the rabbit symbol in flat relief. Archæologists in their writings have frequently remarked upon the lack of square-cut incisions in American sculpture. This phenomenon is sufficiently explained by the instrument.

From this form of carving, common enough in both hemispheres, to high relief was only a matter of normal growth. The processes were the same. There has not been found a sculpture in Mexico or Central America or in South America that was cut in any other way. The statuary was wrought after the same fashion. In the Old World it is not possible to speak with the same precision. Some of the more ancient sculptures of Egypt represent men standing on platforms and pounding with stone hammers. This tool is capable of such work. No one knows who has not tried what results it will achieve. It is not necessary to introduce steel or adamant, as one born out of due time, to do the work of that faithful implement which stood by our race from the beginning. All the primitive sculptures of the world were wrought with the stone hammer alone. The early sculptures of Egypt, Babylonia, and India were not beyond its powers. Indeed, it would not be difficult to prove that the stone hammer was more capable of effecting these works than were the first efforts in bronze and iron.

The modern stonecutter is a familiar object, with his white cap and apron, his tools of steel, which he holds lightly in his left hand and his wooden mallet wielded in his right. His prototype in savagery is not difficult to find.

In many portions of the United States and without doubt elsewhere in the world, there are quarries of soft slate, serpentine, alabaster, steatite, and other materials. The aborigines were wont to quarry these substances and actually to carve them into form. The eastern Indians, finding a protruding mass of steatite, set themselves to cutting out a block large enough for a cooking-pot or pan, leaving a great deal for waste. These blocks were removed by means of long, wedge-shaped picks or chisels of quartz. Deep gashes were cut in from the top and sides until the piece could safely be removed with levers or wedges or heavy stone mauls. The pot was cut into shape with the same edge tool of quartz and not by pecking. The stonecutter did not hold his chisel in one hand and a mallet in the other, but seized the chisel in both hands and used it on the stone after the manner of an adze. In very many of the examples that were broken in the working these scarfs are visible in parallel rows, showing that the implement was about an inch wide at the edge, and could take off a chip from one to three inches in length. When the stonecutter's work was done on the vessels, the marks of his tools were obliterated by scraping and polishing. For scraping there was no end of sharp chips of flinty rocks, and the grainy sandstone would complete the smoothing process.

The most beautiful stone-cutting done by modern savages is that of the Haida Indians of Queen Charlotte Islands, on the Pacific coast of British Columbia. They quarry a black slate, which is very soft at first, and this is really whittled into shape with knives having metal blades. There seems to be a difference of opinion as to the existence of any antique forms of this ware, owing to the difficulty of working such materials with stone knives. Formerly the same designs were wrought of wood in the greatest abundance.

Hearne remarks of the Indians in North-west Canada : "Their household furniture chiefly consists of stone kettles and wooden troughs of various sizes ; also dishes, scoops, and spoons, made of the buffalo or musk-ox horns. Their kettles are formed of a pepper-and-salt coloured stone ; and though the texture appears to be very coarse, and as porous as a drip-stone, yet they are perfectly tight, and will sound as clear as a China bowl. Some of those kettles are so large as to be capable of containing five or six gallons ; and though it is impossible these poor people can perform this arduous work with any other tools than harder stones, yet they are by far superior to any that I had ever seen in Hudson's Bay, every one of them being ornamented with neat mouldings round the rim, and some of the large ones with a kind of flute-work at each corner. In shape they were a long square, something wider at the top than bottom, like a knife-tray, and strong handles of the solid stone were left at each end to lift them up."[1]

The Eskimo still saw such hard stones as the pectolite,

[1] Hearne, *Journey*, &c., London, 1795, Strahan, p. 168.

which is placed among the jadoid materials for texture and temper. There are two ways of effecting this sawing. One piece of stone may be cut with a thin spall of the same or harder material, or the work may be done by means of sand and a soft stone on a splint of wood to carry it. The Eskimo also know how to bore a row of holes as quarrymen do, and split the material thus.

The modern sand-saw, of soft iron, helped out by plenty of water and grit, is out of the question in savagery. But the principle was discovered by them that a very soft substance could cut a very hard one by means of sharp granules of a denser material. The application of water in the work has a great deal to do with the rapidity of the cutting. To keep the scarf clear of worn-out material a fresh supply beneath the carrier is quite necessary. Mr. McGuire gets by far his best results with sheets of cold-hammered copper. The opinion that raw-hide and sand, either wet or dry, will cut stone needs confirmation. Experiments with such materials failed utterly in the Smithsonian experiment.

Savages knew how to bore holes with stone, and how to perforate stone The perforator of stone was a long piece of carefully-chipped jasper or other hard material, in shape of an ordinary nail. For a grip a broader head was left at the butt-end. This could be held in the hand and worked by a reciprocating motion exactly after the manner of the awl or reamer. For the thousand and one uses to which an awl could be put, however, other substances, such as bone, horn, antler, and ivory were used. The stone reamer of very hard rock was excellent on soft rock.

The simplest composite drill now in use among savages consists of a shaft of wood, into the lower end of which is inserted a point of stone, held in place by a seizing of raw-hide or sinews. This may be twirled between the palms of the hand. But the most effective method of boring stone is by means of the pump-drill, or the bow-drill, or the strap-drill, using not a hard point of stone or other material, but one of wood or bone, with plenty of sand. The process is well described by Mr. McGuire. Dr. Rau was partially successful in boring diorite. Mr. McGuire has made the pipe, the ceremonial axe, and the double conical perforation. He has, by means of the most primitive appliances and a little emery, reproduced in quartz crystal the bore made with the solid drill, and that with the delicate core standing in the middle. If with sand alone, and in the use of the bow-drill or pump-drill or strap-drill any savage artisan could perforate beads of jasper, chalcedony, or jade, that is now among the lost arts.

"The Brazilians use ornaments of imperfectly crystallised quartz, from four to eight inches long, and about one inch in diameter. Hard as they are they contrive to drill a hole at each end, using for that purpose the pointed leaf-shoot of the large wild plaintain, with sand and water." [1]

This quotation from Sir John Lubbock has been oft repeated, and the writer, after all he has been saying, is not going to deny the account. Certain it is that beads of very hard material and very small bore are common products of savage manufacture. But it must be admitted that, without emery, the author cannot do it, and all experiments to repeat the Brazilian processes with sand and water have so far failed at the Smithsonian Institution.

Polishing and grinding stone were among the earliest, as they are now one of the latest, arts. The preparation of gems, the making of plate glass and ground glass, the grinding down and polishing of ornamental stones, are some of our most active modern industries. Many patents are still taken out for improvements in the art, and many hundreds of skilled labourers make their living in this way. Now the first grinders of stone are usually relegated to the neolithic period of history; with what reason it is difficult to say.

Probably no discovery is older than the fact that friction would wear away wood or bone, or even stone. Practically the modern savage uses stone to grind other substances, and

[1] Lubbock, *Preh. Times*, New York, 1878, p. 561.

also grinds one stone with another. The North American Indian woman keeps near her side a bit of whetstone, on which she sharpens her bone needles and bodkins and crimping tools. Her companion is similarly provided with whetstones for repairing the edge of the implements in his crafts. He also uses sandstone, on which he has chipped a little groove to rub down the shafts of his arrows, and other woodwork. Stone is his sandpaper, scraper, plane, spoke-shave, and polisher in one. The potter, after building up her vessel by coiling, rubs away every mark of the fingers from the surfaces with grinding and polishing stones. The woodworker, having chipped out a dish, or a sculpture with adzes and chisels of stone, obliterates the scars by means of a grinding stone. The cook, noticing that in the grinding of food both upper and nether millstone give to each other a polished surface that must needs be removed with a stone hammer, undesignedly invented the process of all subsequent gem-cutting and stone-polishing.

The polishing of one stone upon another is, however, the characteristic art of neolithic times. In the course of grinding the edge of an axe or chisel a polish would be communicated. But many hundreds of these implements are polished beautifully over their entire surface. Some of them have the brilliancy of a mirror. They are not only ground, but they have been polished by rubbing down with buckskin and fine powder of some kind.

Just a word should be said in this connection also about the polishing that comes from use. Upon the surfaces of many of the hoe blades coming from the Mississippi valley, there is a lustre or nacre that was not designedly added. It would seem that the mere act of using had enabled the implement to take on a vitreous glaze. Furthermore, pipes, battle-axes, adzes, axes, chisels that have been used a long time, acquire a gloss and brilliancy that cannot be conferred at once. The greasy hand of the savage, seldom washed, also communicates a beautiful surface to shell, ivory, and stone implements.

Among the stone implements in New Zealand collections may be seen the *hoanga*, or grinding stone, oval in shape, formed of coarse sandstone, with a hollow oval groove in its upper surface. In these, implements were ground down with the aid of water.

As the natives performed the tedious process of shaping their implements, at which they spent most of their time, they sang a song of first voice, second voice, and chorus. The first voice asks what the grinding of the stone is for. The second replies that it is to shape the tool, to sharpen it, and describes the flying of the chips, the splitting of the stone, &c. The chorus encourages the workman, urging him to continue his work, with an appeal to the goddess of axe-sharpening.[1]

In the service of this goddess the Polynesians were indeed faithful. Their material was extremely hard. None in the world employed for implements or weapons was more refractory. It was wrought into useful and grotesque forms, and received a surfacing that was truly remarkable. In our day examples command fabulous prices both on account of the material and the skilled labour bestowed upon them.

In speaking of the lapidary art among savages it is necessary to include all those materials which have been required to take the place and do the work of stone. This is quite essential in studying the arts of those island areas where stones for chipping or for grinding are absent. For instance, there were very skilful natives among the Bahamas, but they made their chisels and adze blades of shell. In a similar plight were many of the Pacific islanders, but the Tridacna, the Margarita, and other molluscs, were at hand with their kind offices. "Among the materials upon which primitive man set his eye, shells of molluscs are not to be overlooked. Everywhere within the range of these creatures their soft parts afforded ready and most nutritious food, and the hard parts were the servants of innumerable wants. The early men not only

[1] *Cf.* A. Shand, *J. Polynes. Soc.*, Wellington, 1892, vol. i. p. 82.

used shells for dishes and tools and art purposes, but wrought in them; and the shells in turn wrought on the men, suggesting forms and uses, and touching the fancy by their beauty of colour, and even their power of music."[1]

The shell could be chipped, or sawed, or ground, or polished, or perforated like a stone. The same is true of ivory, whales' teeth, fossil tusks, antler, and the hard shells of some seeds, and even nuggets of copper and iron. All of these may be classed with the material of the lapidary, upon which he exercised his ingenuity and secured his patents. They taxed his patience and evoked his faculties, and were turned into the currents of the world's great industries in very ancient times before the historian had learned to write. . . . .

[1] Holmes, "Art of Shell in the Ancient Americans," *Second An. Rep. Bur. Ethnol.*, Washington, 1883, pp. 179-305, pl. xxi.-lxxvii.

CHAPTER V.

THE POTTER'S ART.

"This earthen jar
A touch can make, a touch can mar;
. . . . .
To-morrow the hot furnace flame
Will search the heart and try the frame,
And stamp with honour or with shame,
These vessels made of clay."

LONGFELLOW, *Keramos.*

PORCELAIN is the glorification of pottery, of the processes of making and decorating it, and of the purposes for which it is created. If, then, we are able to find the origins of modern pottery in primeval times, and the survival of those times in the ware of our own day, there is shown an unbroken genealogy between sun-dried vessels of the first ceramists and the most delicate work of Worcester, Sèvres, Meissen, Höchst, and Berlin. It might be refreshing to the reader, before making on foot this tedious journey to the humble dwellings of the first potters, to look over the works of Brogniart, Jacquemart, Birch, Bowes, Garnier, and the South Kensington handbooks. No harm, but much inspiration, would also be experienced by spending a day or two among the brickmakers, the terra-cotta works, the old-fashioned fabricators of cheap stone ware, the potteries located here and there in all lands, or, if occasion permits, to take in the finer works of Staffordshire, Sèvres, Meissen, Berlin, or of Trenton and Cincinnati. There will be no need in our visit to the earliest ceramists to speak of earthenware and stoneware and porcelain. There will be plain ware and lustred ware, but glazes and enamels that are not purely accidental will be unknown. Clay there will be in abundance, but the artisans will not be quite so particular about the hydrated silicate of alumina and pure white sand and chalk and felspar and calcined bones and potash. Neither will they require such costly machinery for grinding the clay and the flinty ingredients until a paste as fine as flour dough can be made. The washings and settling, the mixing and weighing, will not be so scrupulously done. You will see no potter's wheel, nor any machinery like unto it, nor any device that at first sight will remind one of it. Yet, the ends achieved thereby will be reached. The primitive artisans will both mould and model, though in quaint ways. The modern potter invented neither moulding nor modelling. Engobe or slip will be applied to the surface of the vessels, and dainty figures will be painted thereon, though it must be freely admitted that the paint-shop does not contain so many nor such lasting colours. No matter, all our art is the lineal descendant of theirs through many vicissitudes which constitute the charming story of the ceramic industry.

In the very earliest graves and camp-sites no fragments of pottery occur. If our first parents were makers thereof, we should know it, because this most brittle of human works is also among the most enduring. Fire-making devices were invented before pottery, because all of it was effected by means of fire, if we except sun-dried bricks and lamp-stoves. The bow and the arrow, the spear and the fish-hook, are older. They are found in older graves. Can it be that this art came in with the grinding of food? At any rate, it long antedated Homer, for the potter's wheel is mentioned by him (Il. xviii. 600). The simpler hand epoch antedates all books and writings, and there are many, many tribes of uncivilised peoples on the earth making beautiful ware, who do not read at all. The

Lake-dwellers had pottery, and so had the Mound-Builders, and the people of very ancient Troy. In Peru beautiful specimens come from the oldest graves, and over the cañons of Colorado, and especially of its tributaries, hundreds of complete vessels, and millions of fragments, are scattered similar to that made near by to-day.

We need not stop to inquire about the first person in the world who fabricated a clay vessel, nor try to conjure up the manner in which the invention was made. Clay is the most docile of all materials. It has its limitations, but compared with stone, bone, horn, wood, hide, fibre, and so forth, how easy it is to work—so pliable, and yet so superior to all the above-named substances in the fire. The first man who trod in clay must have noticed that he had made a pan impervious to water. The earliest cooks put hot stones into tight baskets to boil their food. Soapstone pots did tolerably well if the walls were left thick enough. But, just as soon as people had fire, became sedentary, ate farinaceous food, the pot came to be born. And in cold regions, the use of fire would, as we shall see, compel the invention of pottery.

In the last and simplest analysis, sun-dried adobe or bricks are the most primitive things made of clay. They are masses of rude paste worked up by hand, not at first in moulds, and dried in the sun. In all rainless regions of the globe they exist. In Babylon, in Egypt, in Peru, in Mexico, it is the same story. Given the material and the arid climate, and the thing is done, by that universal law, in human affairs as in nature, of following the lines of least resistance. This may not be the oldest treatment of the material since climate is a ruling factor, but it is the least complicated method of handling it.

The next simplest process is to be found in vogue in our day among certain Eskimo tribes on the tundras about the peninsula of Alaska. These cunning people, when most spread out, occupied the northern shores of America from Southern Labrador all the way around to Kadiak Island in

Alaska. Almost everywhere they utilised fire only in the lamp-stove. Forests being absent, and even drift-wood being scarce, their only resource has been to burn the blubber or fat of the seal, whale, walrus, and other animals that abounded in that area. There was no lack of fuel. Of the mosses and vegetable fibres that came in their way they fabricated the wicks. For a lamp they took a slab of soapstone about two inches thick, straight along one margin, and curved on the other. This was excavated to form a shallow dish, in which the blubber was put, and the wick. The Eskimo knew both the firesticks and the flint and pyrites method of exciting fire, so it was never difficult to make a blaze. Now, there are in the west, regions where no soapstone exists of which to make lamp-stoves, so the ever quick-witted housewives knead clay with blood and hair, and form it into a thick shallow dish or bowl with the hand, and after drying it only a little, proceed to make thereof a true lamp-stove. The constant use of this simple device hardens it by burning, so that there is no need of firing the ware at all. Nothing save a sun-dried brick could be simpler. The first real potter seems in this way to have been a fabricator of lamps and stoves. Now and then rings are incised around these objects, commencing already in the most simple manner the process of decoration. No rims, nor handles, nor legs, nor bases, nor paint, nor modelled ornaments occur. We are behind the history of the art.

True pottery, hardened in the fire, if we are to trust the testimony of the living and the dead, is and was confined within certain boundaries. Within these, since clay is almost universally distributed, the fictile art was generally practised, though the Australians and the Polynesian race have always been ignorant of it. They cook in open fires and pits, and drink from gourds. No doubt at first the art was stimulated by the absence of other material. In South-western California, where the potstone was abundant, great numbers of globular ollas are yet found, some of them capable of holding several gallons, and scraped down very thin. The light and tough boiling basket also no doubt deterred many other migratory tribes from this method of cooking. The geographic distribution of the art will be found amenable to those natural and mental laws whose co-operation we are tracing.

At the present moment there is no other spot on the earth where the primitive potter can be studied to such advantage as in the south-western portion of the United States, in the drainage of the Upper Rio Grande and the Colorado river. There, at the present time, are tribes belonging to the Keresan, the Tewan, the Zuñian, the Shoshonean, and the Athapascan stock. Whatever their ancestors may have done in other habitats for vessels, here they have all learned to make pottery and to build adobe walls. The Navajo may perhaps be excepted from the wall-builders, dwelling in hogans with his cousin, the Apache, and both of them migrants from the Mackenzie drainage. To this workshop let us go and sit at the feet of the primitive artisan.

Mr. Cushing, who has spent many years of his life among these Pueblos, says, "There is no other section of the United States where the potter's art was so extensively practised, where it reached such a degree of perfection, as within the limits of these ancient Pueblos. . . . In these regions water not only occurs in small quantities, but it is attainable only at points separated by great distances, hence to the Pueblos the first necessity of life is the transportation and preservation of water. The skins and paunches of animals could be used in the effort to meet this want with but small success, as the heat and the aridity of the atmosphere would in a short time render water thus kept unfit for use, and the membranes once empty would be liable to destruction by drying." [1]

In the early times the Zuñi used large sections of reed, and now a common sight is a water-carrier employing a

[1] Cushing, *Fourth An. Rep. Bur. Ethnol.*, Washington, 1886, p. 482.

gourd or a basketry bottle, lined like the ark of Moses, within and without with pitch.[1]

As to the supply of material, Holmes says : "Nature was lavish in her supply of the material needed. Suitable clay could be found in nearly every valley, both in well-exposed strata and in the sediment of streams. I have noticed that after the passage of a sudden storm over the mesa country, and the rapid disappearance of the transient flood, the pools of the arroyos would retain a sediment of clay two or three inches thick, having a consistency perfectly suited to the hand of the potter. It would not be difficult, however, to find the native clay among the sedimentary formations of the neighbourhood. Usually the material has been very fine grained, and, when used without coarse tempering, the vessels have an extremely even and often a conchoidal fracture."[2] This clay from the arroyos, or from its natural beds, the women of the Pueblo gather and transport on their backs to their workshops under the open sky, either on the mesa or on the housetops. It is further washed carefully to exclude foreign bodies, and to render it pliable to the artist's hand.

We read a great deal about the employment of what the French call *dégraissant* in working clay, that the walls of the vessel are rendered stronger and less liable to crack thereby. Be that as it may, the Pueblo woman mixes sand or pounded potsherds with her clay. In the ware of the Mound Builders pulverised shells, old pottery, mica, and other tempering materials were employed. The pottery of the southern half of Africa is formed of clay found near ant-hills, in which case the mixing is done by the little creatures. The compounding of these materials is not a haphazard affair. Too much or too little of the tempering material would be disastrous. The shells must be carefully burned previously, or the fragments would be calcined in the firing, and afterwards slake. Cushing says that the

[1] *Fourth An. Rep. Bur. Ethnol.*, p. 491, fig. 520.

[2] Holmes, *Fourth An. Rep. Bur. Ethnol.*, Washington, 1886, p. 267.

quality of the clay is not uniform throughout the Pueblo region, and to this cause he attributes the fact that in some places "fragments of the greatest variety in colour, shape, size, and finish of ware occur in abundance." In other spots where the architecture of the houses is equally well executed, potsherds are coarse, irregular in curvature, badly decayed, and exceptionally scarce.[1] This is another confirmation of the principle of mutual relation between the inventive faculty and the natural resources upon which it operates.

The potter is now ready to construct her vase, and in its edification employs one or all of three methods, *modelling*, *moulding*, and *coiling*. We should waste time in discussing which is oldest among these three processes. They will be discussed in the sequence named because that is the order of their importance in the region under consideration. Just as the Eskimo woman takes a lump of prepared clay and with her fingers models her stove-lamp, the Pueblo woman and savage women nearly all over the world model vessels from the lump, model also the rims and bases and handles and raised decorations of ware made in the other methods mentioned. This very earliest and rudest act of clay-working remains and is glorified in the sculptor's art.

Moulding pottery is a common method at present, and it must have been practised most extensively in former and in ancient times. All over the United States bits of ware are picked up on whose surfaces are deep furrows and nodes whose true structure is declared by making casts of these markings in plaster, clay, or wax. The deeper portions saved from attrition are clearly thrown up, and reveal the presence of basketry, textile, or netting. In many cases, to be mentioned later, these impressions are ornaments produced by means of textile substances. But the ninety and nine were made in nets or baskets or bags. In such examples the markings are on the outside.[2] It is just as easy, how-

[1] Cushing, *Fourth An. Rep. Bur. Ethnol.*, p. 493. [2] *Ibid.*, p. 484, fig. 501.

ever, to work the other way and build the vessel on the outside of the mould, in some cases a gourd, as in Arkansas pottery, in others a basket, as in Pueblo examples. Further south, in Mexico and in Ancient Peru, modelling and moulding were more commonly practised than on the Rio Grande. The Mound Builders also had certain forms which they achieved by these processes. The Guadalajara potter of our own day understands perfectly this art of moulding. He seems to be the cleverest artist in the world, producing portraits and animal groups with marvellous exactness by means of spatulas of hard wood, a brush or two, a bit of tin and stones for rubbing smooth. He does not model *en bloc* as one of our artists does, but moulds and models his man first and then dresses him by laying on the parts. He appears like one who has taken a lesson or two in modern sculpture and is trying also to hold on to his old traditions,—with marvellous cleverness, however, for he never saw a throwing wheel or a studio.

It is very doubtful whether true casting was ever practised by savage peoples. And some writers forgetting that the primitive ceramist could supplement moulding by modelling, seeing on the inside of vessels the imprint of natural or artificial objects have assumed that the clay was plastered all over the inside or outside of the mould and that the latter was removed always by burning. There may have been instances of this, but the savage woman thought too much of her net or basket or gourd to destroy it so ruthlessly.

The third process of building up pottery was far the most common, at least in the Western Continent. It is a kind of potter's wheel of a slow velocity, only the hand travels round and round instead of the clay. "The ancient Pueblo potter rolled out long, slender fillets, or ropes of clay, varying in width and thickness to suit the size and character of the vessel to be constructed. They were usually, perhaps, from one fourth to one half an inch in thickness. The potter began by taking the end of a single fillet between the fingers and proceeded to coil it up on itself, gradually forming a disc. At first the fillets overlapped only a little, but as the disc grew large and was rounded upward to form the body of the vessel the imbrication became more pronounced. The fillet was placed obliquely, and was exposed on the exterior side to probably one half of its width. Strip after strip of the clay was added, the ends being carefully joined, so that the continuity might not be broken until the vessel was completed. The rim generally consisted of a broad strip, thickened a little at the lip and somewhat re-

FIG. 25.—Coiled Pottery, showing suggestion from coiled basketry. (*After F. H. Cushing.*)

curved. The exterior imbricated edges were carefully preserved, while those on the inner surface were totally obliterated, first by pressure, and finally by smoothing down with an implement or with the fingers, imprints of the latter being frequently visible. So thoroughly were the fillets pressed down and welded together that the vessels seldom fracture more readily along the lines of junction than in other directions." [1]

The suggestion of this peculiar mode of building up a dish was doubtless given by coiled basketry. In the Chapter on Textiles in this work the whole process will be minutely described, and as both arts are confined to the sphere of

[1] Holmes, *Fourth An. Rep. Bur. Ethnol.*, p. 274, figs. 217, 218.

woman's work, there will be no difficulty in seeing how she could coil in one case as well as in the other. Indeed, the Havasupai Indians put a thin lining of clay inside the basket trays by means of which they winnow their grass seeds and parch them with hot stones. Cushing thinks that they actually started their coiled ware on the bottom of a basket. Either the outside or the inside of the bowl would serve the purpose.[1]

A precisely similar process of coiled work is described by Man among the Nicobarese, and Atkinson among the New Caledonians, both of which peoples use some device to facilitate the turning of the work. The Nicobarese start the coiling on a lump of clay moulded in the bottom of a dish supported on a pad-ring, and the New Caledonians make theirs on the outside of a roundish pebble. On this they put a small dab as a beginning; round this the coils of clay are wound and the pot built up. As the under side of the stone is roundish it becomes a natural primitive potter's wheel.[2]

The former is preferable, because after the walls have been built up for some distance, the same basket serves as a support and primitive potter's wheel. Mr. Cushing remarks on this ingenious discovery that the fabrication of large vessels thereafter was no longer effected by the spiral method exclusively. "A lump of clay, hollowed out, was shaped how rudely so ever on the bottom of the basket [moulding] or in the hand [modelling], then pressed in a hemispherical basket bowl and stroked until pressed outward to conform with the shape, and to project a little above the edges of its temporary mould [moulding], whence it was built up spirally until the desired form had been attained [coiling], after which it was smoothed by scraping."[3]

[1] Cushing, *Fourth An. Rep. Bur. Ethnol.*, p. 497, fig. 524, and p. 500, fig. 529.

[2] E. H. Man, "Nicobar Pottery," *J. Anthrop. Inst.*, vol. xxiii. pp. 21–27, 1 pl. J. J. Atkinson, "New Caledonian Pottery," *id.*, p. 90.

[3] Cushing, *Fourth An. Rep. Bur. Ethnol.*, p. 499, figs. 526, 527, 528, 529, 530.

Any one who is clever may follow this description by using either artist's clay or paste from the pottery. In the building up of the vessel the workman has certain decorative motives in mind of which mention will be made presently, and a portion of these may be realised in the treatment of the coil. It should be previously fixed in mind, however, that in most cases all traces of the coils are to be obliterated either when the clay is soft, by means of little paddles of gourd or shell or pottery or wood, or they are subsequently rubbed away with fine grained polishing stones.

In addition to all this smoothing and scraping and rubbing the potter of the Pueblos was acquainted with slip, which was really very fine clay thinned with water and applied as a wash, previously to decoration with colour.

At this point, indeed, the fineness, the form, and the finish of ware becomes differentiated by the functions which it has to perform, whether it will have to go to the spring and become a vehicle, to stand in the house and be a receptacle of ever-needed water, whether it be the storehouse for grain and foods, a vessel of dishonour in the cooking pot, a dish for serving food, or a vessel of honour in the work of art, or the sacred meal bowl. Abundance of fragments all over the world attest that the fire was often applied to the rude piece, with no other decorations than a few scratches with a pointed stick. But the primitive potter, as we know her, was not satisfied with this for noble uses.

In the first place, the student, looking at a very great collection like that in Washington, notices that untarnish ware ranges in colour from nearly white and grey and light creamy yellow to red and brown and jet black. This is owing, in the first place, to the selection of the clay, and to the natural colour imparted by clays in the different localities. The burning also has to do with the body colour, but more of that anon. On the authority of Holmes, "reds and browns result from the presence of iron, which may have been oxidised in burning, or the red oxides may have been used in rare cases as colouring matter in kneading the clay. The surface is often lighter than the mass ; a condition probably resulting from colouring matter in the clay, which is destroyed on the surface and remains unchanged within. In the south the colours of the paste are often slightly reddish or yellowish in hue. It is notable that a small percentage of the ware of all localities is red." [1]

Brick-red is the rule in Peru and common among the sacred pottery of Mexico. That of the mounds is a dirty brown or dark grey, with a sprinkling of red, and each area has its characteristic colour. Both the texture and the shade of fragments discovered in various places have their say in deciding concerning the status of the makers. Indeed, among American archæologists and ethnologists who have been making the map of the Northern Continent according to tribes, the limits of certain textures of pottery fragments have been useful in determining boundaries.

The forms of primitive pottery are an ever-pleasing surprise to the archæologist and the technographer. Of course, if a bowl or dish or pot or jar or vase be moulded inside or outside of a basket or gourd, the shape is pre-ordained up to the point where the work has to be constricted or gathered in. There is the dividing line where the artist has to withdraw the mould and proceed alone. In modelling the rest of the jar or in building by coiling she has got to imitate natural objects or those fabricated from other materials. Of these there are abundance in the endless shapes of gourds and shells and horn and wood and bark and basketry." [2]

The causes of modified forms, after the primitive idea has been adopted, have been tabulated by Holmes :—

"Incapacity of the material to assume or retain form.
"Incapacity of the artisan.
"Changes in the methods or processes of manufacture.
"Changes of environment.
"Changes of use.

[1] *Op. cit.* p. 269.
[2] Compare Holmes, *Fourth An. Rep.*, *&c.*, figs. 210–216 ; 466–472.

"Lack of use.
"Influence of new or exotic forms.
"To enhance usefulness.
"To please the fancy for the beautiful or the grotesque."[1]

FIG. 26.—Vase with flat top, from Tusayan or Moki, Arizona. (*After Holmes.*)

FIG. 27.—The same seen from above.

[1] Holmes, *Fourth An. Rep. Bur. Ethnol.*, pp. 450–457.

Most of these are easily understood. If, however, the reader is unacquainted with the poor resources of the primitive potter, he will scarcely realise what a toilsome journey it is across the top of a jar after it begins to narrow until the rim is reached.

For some of the older pieces of flat jars taken from the ruined Pueblos fabulous prices have been paid, on account of their rarity and the great difficulty in building them up. Indeed, there are some things clay cannot do. The woodworker, or the basket-maker, or Nature herself pronounces a word, gives expression to a formal thought, in presence of the potter; the latter repeats it as best she can. After all she is a novice, and her imitations display her limits. The composition and resolution of her desires, the patterns before her, the limitations of her material and the effects of environment, account for all ceramic forms.

The ornamentation of primitive pottery was effected by engraving the surface, by adding parts and by colouring. According to Holmes, the sources of decorative motives were :—

"1. Suggestions of features of natural utensils and objects.

"2. Suggestions of features of artificial utensils or objects.

"(*a*) Functional, as handles, legs, bands, perforations, &c.

"(*b*) Structural, as the coil, seam, stitch, plait, twist, &c.

"3. Suggestions from accidents in construction, as marks of the fingers, of implements, of moulds, &c.

"4. Suggestions of ideographic features or pictorial delineations." [1]

These primary suggestions are afterward modified by the natural law of least effort and survival according to the methods governing change of form. The æsthetic sense begins to assert itself before the vase is finished. Very

[1] Holmes, *Fourth An. Rep. Bur. Ethnol.*, p. 453. Illustrations of these, *id.*, p. 454, *seq.*, figs. 475–479.

pleasing effects are produced in Pueblo pottery on the unchanged surface of the coiled ware by pinching the coils regularly in accordance with some pre-ordained plan, by the finger-nails, the finger-tips, or with pointed sticks, after the fashion of the good housewife decorating the edge of a pie. In Holmes's work upon the pottery of the ancient villages of Tusayan is figured a vase, belonging to the Hemenway collection, which seems to me to be certainly the most beautiful specimen of rude aboriginal pottery in the world.[1] The coils are indented and left plain in such a manner as to cover the whole surface with triangles in light and dark shading. No colour has been applied; no tool but a woman's delicate fingers has touched the gracious surface. It is a brilliant recitation in the old-time ceramics, an example of fingering which all lovers of art behold with pleasure, and conviction that in the stone age men and women were under the spell of the beautiful.

It will be noted that the feminine gender is used throughout in speaking of aboriginal potters. This is because every piece of such ware is the work of woman's hands. She quarried the clay, and like a patient beast of burden bore it home on her back. She washed it and kneaded it and rolled it into fillets. These she wound carefully and symmetrically until the vessel was built up. She further decorated and burned it and wore it out in household drudgery. The art at first was woman's.

The Caribs are very skilful potters. The manner of their working is precisely like that of the Pueblo people of the United States, only the Caribs commence the work by laying out a flat circular sheet of clay on a small piece of board; the rest of the material is rolled out between the palms of the hands into long cylindrical pieces as thick as a man's thumb. One of these rolls is laid round the edge of the foundation so as to stand up like the rim of a tray. This is made solid, smoothed up, and other rolls added until the whole is complete. Colours and glazing are done

[1] *Fourth An. Rep. Bur. Ethnol.*, p. 297, fig. 253.

with vegetable dyes and certain barks, burnt and ground and mixed with the clay, give to the ware a black colour.

The Dinka tribes make pottery off-hand. Handles are wanting in nearly all the pottery of Central Africa ; the exterior is usually marked with incised lines, which afford a rough surface to the bearer's hands.[1] The negroid inhabitants of New Guinea are governed by the same rules. The water-jars are globular and as symmetrical in form as the pretty little ollas that come from Chiriqui. The women carry them on the shoulder supported by one hand, and not on the head as do their African congeners.

Among the Andamanese, both men and women make pottery at present. A clay, unmixed with other substances, is used, found only in a few places, and there the work is carried on. The tools employed are a pointed stick, an Arca shell, and a kneading board. The clay is first cleaned of stones and other foreign bodies, washed and kneaded to a proper consistency. Rolls about fifteen inches long and one half-inch thick are then made, and the pot is built up by coiling one of these after another, the inequalities of surface being removed by the shell, and the surface ornamented with wavy, checked, or striped designs by means of the pointed stick. The pot is then dried very carefully in the sun and over the fire. When it is sufficiently hardened, it is baked thoroughly by placing burning pieces of wood both inside and around the vessel. It is then allowed to cool and is considered ready for use. With good management a pot is ordinarily fit for use by the close of the day on which it is made. They are quite uniform in shape, and vary in size from that of a cocoanut shell to a capacity of two gallons or more.[2] When needed for travelling, pots are fitted in a light wicker frame of bamboo like a conserved ginger jar.[3]

[1] Schweinfurth, *Artes Africanae*, London, 1875, Sampson Low, tab. 1, fig. 6.

[2] Man, *Andaman Islanders*, London, 1883, Trübner, p. 154.

[3] *Ibid.*, p. 179.

To any one who has traced his name in the sand or pressed a seal on wax it will occur how the first ceramists engraved their ware. Scarification of the whole surface was and is practised universally. Modigliani figures such a vase from Nias;[1] and one cannot go amiss for examples in any part of the world. Indeed, the same process has become a craze both in pottery and wall decoration in civilised countries. Even these patternless scratches are frequently quite decorative, and they lead quickly to geometric patterns, endless in variation and instructive in their evolution.

A study of aboriginal textile processes is absolutely essential to a correct understanding of the leading strings which the primitive designer followed in working her way out of plainness into most refined combinations. There was no difficulty in accomplishing this, since women also everywhere invented the textile art. Diagonal and diaper weaving in basketry or matting were revelations to the imagination of the potter. Her first movement away from straight lines and square corners, which she practised as a weaver, was toward triangles and herring-bone and an endless variety of geometric forms on the softer material. In the former, she was bound by the rigidity of her filaments; in the latter, her hand had no such limitations.

A little further on, the study of form and colour on painted ware will enable the reader to follow this transfer certainly, even into more modern and higher art life. The shifting of textile patterns to pottery was effected in two ways, by freehand tracing with a pointed instrument and by stamping, that is, by printing designs carved on wood, by pressing natural objects on the surface, or by pressing textile work on the clay.

Tracing or etching on pottery with any sort of pointed tool seems to have been a necessity with the oldest of potters. It is difficult to find fragments that have not been

[1] *Un viaggio in Nias*, Milano.

so treated. Work of this sort is freehand, however, and in seeking to reproduce the geometric lines on textiles the novice errs, comes short, and finally gets lost or bewildered, doing the thing that is easiest. Many of the designs of this class on ancient ware are only the shreds and distorted remnants of older patterns that once had meaning.

Printing began with the impressions of the nails and the markings on the finger-tips. This was quickly followed, however, by deftly laying bits of string or some other textile object or surfaces of natural objects upon the soft ware regularly. And this suggested the making of stamps. If cut in relief or constructed by attaching bits of wood, &c., to a plain surface, the design would be intaglio. If the pattern were simply cut into the block or stamp the figures would be in relief like the image on a pound of butter. All of these methods were employed in savagery.

Quite in advance of this marking on the surface was the creation of decorations in relief by modelling. Nothing could be simpler than the first efforts. The pristine modellers in clay were like children playing with putty or wax, or making mud pastry. They simply took a little surplus material and, working it into familiar form, luted it on the side of the vessel, where it remained permanently fixed.[1]

These simple elements were ready at once to modify their forms by the methods already laid down, to combine in patterns innumerable, and to become the alphabet of a language which has been spoken by nearly all the races of men. It is really a species of overlaying. The textile worker found out a dozen ways of adding feathers, shells, and other pretty objects by sewing, and the potter with a little wet clay repeats the process.

The Pueblo potter, however, was not fond of overlaying. Her work is in this regard severely plain, since she gets her embellishment through surface etching and colour. To this list must be added the plain Chiriqui ware, all Eastern

[1] Holmes, *Fourth An. Rep. Bur. Ethnol.*, p. 283, figs. 233–238.

American, and much that comes from Africa, the Malay region, Fiji, and even the older wares of China, Japan, and Corea.

The Mound Builders of the Mississippi Valley, the Mexicans, the Central Americans, and West Indians, the South Americans especially, developed methods and types of added ornament that are quite remarkable. They not only combine bosses and fillets, and coils and scrolls in harmonious grouping, but they ventured out into sculpture, commencing with the crude and grotesque, and ending with genuine portraiture of men and animals. These are so faithfully executed at times that the naturalist has no difficulty in naming the species. Vegetable forms are often copied in the body of the ware itself, but rarely are vegetable forms luted on as decoration.

Animal forms are combined, and monstrosities produced, but there is little dramatic grouping to be seen, though there is, especially in Central American ware, abundance of animal life in action. The monkey is an especial favourite for this class of modelled decoration on vases.

By examining a large series of pieces from any region the student is able to mark the decay of gross forms, corresponding to what Max Müller has termed phonetic decay in speech. Upon this vase, near its mouth, sits in lifelike pose a frog, a monkey, or a bird. In the next, the image is not quite so plain, and as the eye ranges along the series the creature seems to be sinking into the material until a little boss or two on the surface for eyes or ears or crest, marks the spot where the full image ought to be, a kind of shorthand, standing first for letters and then for whole words. And this may raise the question which cannot be settled here, namely, whether the plastic art was not deeply concerned in that evolution through which pictography became phonography in the history of written speech. After appreciating the exuberance of fancy in this Middle American ware, one would have to wander long through collections of the finest ceramic products in the world to find a conception worked out that has not its barbaric prototype in those. Not that the former is to be mentioned in the same day with the latter. No such doctrine is promulgated in this book. The lily is more beautiful than the lily seed, yet the former is all in the latter potentially.

After all, the most gorgeous decoration in pottery is by means of colour. Savages generally did not paint their ware, nor do they at the present moment. The Peruvian was economical in the use of the brush. The Mound Builder, the Eastern Indian, the Mexican, the ancient peoples of the old world, aborigines of the negroid area in Australasia, sparingly used this style of ornament. The Egyptians, indeed, made a pseudo-porcelain, in which quite a variety of plain colours were used, but their most common ware was in monochrome. The Assyro-Babylonian and the Hebraco-Phenician branches of the Semitic family adhered closely to the natural colours of the paste and the slip, with sparing use of black. Even the Greeks, the Etruscans, and the Romans used only body colour, to which in the climax of art a plain glaze and black pigment were added. With these simplest of all resources, the Greek potters created products that are the astonishment of the world. Highly coloured pottery did not make its appearance in Europe until the Mohammedans asserted their sway. What was doing in the Celestial Empire when the most cultivated peoples of Ancient Europe were still working in terra cotta, it is difficult to say. Very coarse earthen and stone ware come from the Far East. But the Chinese invented porcelain. Even in the old blue ware, most antique of all, there is the greatest wealth of animal and vegetal forms, of scenery and grotesquerie.

It is a pleasure to turn aside from these earliest Eastern Hemisphere efforts at colour to study for a while the ancient and modern potters in Central America, and in the Pueblo country. Enough has been said of the manner in which the artist brought about the smooth vase ready for the colour. Nothing more need be said concerning the slip or wash of fine, thin clay laid on the surface, before or after the last finishing touches of the engraver and modeller.

One will see quite frequently in collections of Southwestern pottery and in other regions doubtless, little compound vessels with two or more cavities, looking like

FIG. 28.—Painted Decorations on Vase from Moki Pueblo, Arizona. (*After Holmes.*)

a number of small cups that got stuck together in the making. They are now and then labelled incense cups, cosmetic cups (which they are frequently), or even condiment cups. But they are the paint cups of the potters, holding white, black, red, brown, yellow mineral or vegetal colours, all mixed ready to be laid on.

By the side of these curious paint "tubes" will be seen

half a dozen brushes made of very finely shredded textile fibre, or of hair daintily seized to the rib of a leaf or the quill of a bird. The portfolio of the artist is the whole pictorial world around her. But her first hesitating efforts in colour on her ware will be the imitation of the works of her own hands. Later on, the natural world, the realm of fancy, and the mythologic host will furnish her daughters with motives. Lastly, the descendants of these will yield themselves to that unseen and unsuspected but irresistible current of art evolution, in which their mysterious designs will connote nothing in the world to those who employ them and the entire metamorphosis from naturalism to convention will be completed. Mr. Cushing, who lived several years with the Zuñi for the sake of studying them thoroughly, makes some ingenious suggestions concerning the development of colouring that ought not to be omitted.

Decoration in colour began when the smooth surface was reached. As long as the dish was left with the corrugations on, there was no motive to use paint. Vessels are for cooking, for serving food, for carrying water, and for storage. For eating and drinking vessels the interior surface at least would better be smooth. The Zuñi eating bowls were painted inside, not because that portion is more in view, but because bowls were made smooth on the inside even when they were left corrugated on the outside.[1]

This style of decoration once coupled with a kind of ware, or with a definite part of a vessel, retained its association permanently. Furthermore, every student of basketry knows that the coiled basket bowl has a right side and a wrong side, whereon the ends of filaments are fastened off. The half-painted pot or bowl is a close imitation of the basket in this regard.

It cannot have escaped the eye of ingenious and vigilant potters in early times that clays of various kinds when

[1] Cushing, *Fourth An. Rep. Bur. Ethnol.*, p. 498, fig. 525. This is an excellent illustration.

burned change in colour, and produce a great variety of shades. This fact was quite sufficient to explain the use of clay-washes as paint, and as a permanent decorative

FIG. 29.—Detail on Moki Vase.

agent on the surface of ware. Among the more advanced tribes, vegetable paints are also employed. But the most

FIG. 30.—Detail on Moki Vase.

primitive decoration of this sort was undoubtedly of mineral origin.

In the South-western States of the Union occur in abundance whole pieces and fragments of plain, rough grey ware; of whitish ware decorated in black; of red ware, either plain or adorned with geometric patterns in black and white.

The grey or brown colour was produced when a corrugated jar was smoothed down with stones and burned without slip. There would be an exception to this where a ferruginous clay abounded, and was used in building up the vessel.

The tempering material of sand or broken pottery or *dégraissant* necessary to prevent the cracking of the vessel in drying, left the surface rough. This led to the use of slip, and a vessel thus prepared and burned had a creamy, pure white, red-brown, or other colour, according to the clay used.

Black was the next pigment discovered. Perhaps, Mr. Cushing suggests, because the mineral blacks used in staining splints for basketry would naturally be tried on pottery, and those that would remain became standard. One slip would also colour another. At any rate, ancient remains in the South-west show that white and black varieties came first, then red and black, and later the red, with white and black decoration. It was easy to employ the red clay for the first wash, the blue clay, which burned white, for the white pigment, and any of the black pigments for that colour. Or the process might be interchanged. But there are no examples of black ware with decorations in red or white. The designs were applied to the surface of the ware by means of brushes made of the fibre of the yucca, finely shredded. The lines are in some Pueblo ware, and in much of the lower Central American States, drawn with as much care as one of our own artists would bestow.

In all the painted ware the designs have been laid upon surfaces that had been specially prepared to receive them. The parts selected were generally those exposed to view, but there are older reasons than that. The one suggested by Mr. Cushing had to do with earlier forms. Granting, also, that natural objects and basketry furnished the first suggestion of form, they would likewise have to do with the earliest attempt at ornamentation. Once the fashion was set, there was nothing else to do subsequently but to follow it.

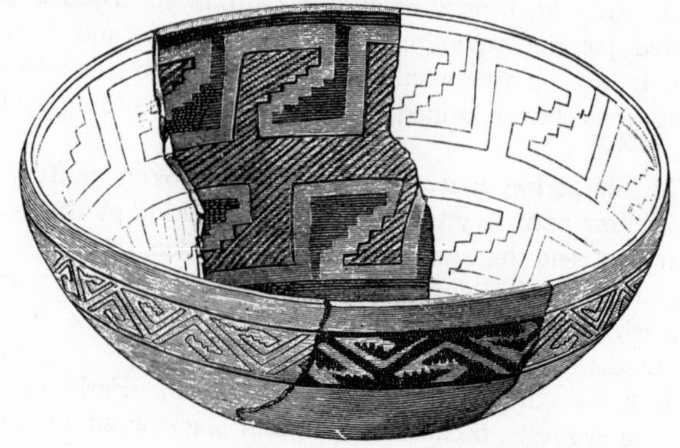

FIG. 31.—Pottery Bowl from Rio San Juan, S.W. United States. (*After Holmes.*)

"Generally the neck furnishes the space for one zone of devices, and the body that for another, while the shoulder, where it is wide or particularly accented, suggests the intro-

FIG. 32.—Painted Design on Bowl from Rio San Juan.

duction of a third. In vessels of irregular form the figures take such positions as happen to have been suggested by the available spaces, by the demands of superstition, or the dictates of fancy pure and simple." [1]

[1] Holmes, *Fourth An. Rep. Bur. Ethnol.*, p. 302.

Now, the very same motives have actuated potters in our own day, or at least in historic times. Whether the parts of the vessel to be decorated were best selected by savages, civilisation follows suit, and has made little changes therein. The same is true of colours in clays. It is to be seriously questioned whether the most experienced potters could go to our south-west country and find better materials for ware than those sought out by the aborigines, and often carried to great distances.

There is no better way of showing the first suggestions and the modifications of decorative motives than to study a few designs evidently transferred from one art to another. Holmes shows the commonest form of two-colour pattern. Every embroiderer knows that a child could soon be taught to imitate it in bead or sampler-work. In wampum belts, in basketry, in primitive embroidery these geometric patterns are ever obtruding themselves.[1]

In the basketry bowl, made in the same coiled combination by the Pima Indians, living on the Gulf of California, just where California and Mexico come together, the meander is perfectly worked out, not in straight lines, but in true radii emanating from one pole of the sphere, and true parallels following the coils around.[2]

Now let us compare with this a bowl found on the Rio San Juan, not far distant from the corner-stone of the four political divisions of Colorado, New Mexico, Arizona, and Utah. "A narrow zone of ornament based upon the meander encircles the exterior margin of the rim, and a broad, carefully drawn design, consisting of two parallel meanders, occupies the interior. The meandered fillets of the interior are in white, and the bordering stripes and the upper and lower rows of triangular interspaces are in solid black, while the median band and its connected triangles are obliquely striped. The oblique portions of the meander are indented or stepped." The drawing is a reconstruction

[1] *Fourth An. Rep. Bur. Ethnol.*, p. 487, figs. 507, 508, 509.
[2] *Sixth An. Rep. Bur. Ethnol.*, p. 220, fig. 323.

from fragments, but the extended pattern will show what the design was on the bowl. This pattern would be very easily reproduced in basketry by a Mohave woman, and the steps or indentations that she would reproduce have been carefully copied by the painter.[1] In the paper from which this design is taken, the author has produced an extended series of portraits of the most precious specimens of Pueblo art, and in the accompanying figures unrolls the designs by a sort of Mercator's projection, so that the elements may be more carefully studied. Nothing is more apparent than the rounding of corners and the little liberties which gradually lead the painter on to the point where she may venture in freehand to produce vegetal and animal designs. Geometric and textile motives are seen sparingly also on the ware of Mexico, Central America, Peru, and in the Mississippi Valley. The aborigines of these regions, however, were painters on bark and skin and other surfaces. The painted robe of the Plains Indians, the Mexican and Maya Codices, and works of that kind, lent their pedagogic aid in giving bias to the decorator's mind. It can easily be imagined that the Indians of our North-west coast, were they to become suddenly potters would transfer all their wood and slate carving skill to ceramics. Their luxuriant variety and beauty of basketry work would also have their influence. They would combine the geometric work of the Pueblos with the rounded forms of the Mounds and of Chiriqui. The Pueblo stocks do not seem ever to have painted on any integument or paper. They made their coloured figures in the sand by drawing furrows on the natural surface and filling these with sands of different colours. Conventional figures for beasts and birds and insects were painted on pottery, but we must go further south for the native home of animal forms on ceramic ware.[2]

[1] Holmes, *Fourth An. Rep. Bur. Ethnol.*, p. 317. Compare figs. 290 + 291; 292 + 293; 294 + 295 + 296; 302 + 303 + 304 + 305; 310 + 311; 318 + 319; 321 + 322 + 323; 324 + 325 + 326; 327 + 328 + 329; 332 + 333 + 334; 336 + 337; 338 + 339; 340 + 341; 342 + 343.

[2] Cushing, *Fourth An. Rep. Bur. Ethnol.*, p. 515, fig. 551; p. 519, figs. 559, 560.

The Mound Builders' pottery departs from the Pueblo pottery chiefly in the modelling of animal forms. It is at the same time so inferior to the ware from the middle American region that one might regard it as the rudimentary state of the latter. Furthermore, those who have studied the matter discover local peculiarities even in this area, dividing it into an upper Mississippi type, a middle Mississippi type, and a lower Mississippi type. The material of this ware is often mixed with the rich alluvial soil of the region, and tempered with broken or pulverised shells. These natural supplies gave a special characteristic to the work of Mississippi pottery.

Two types of body colour exist, the dark and the light, these effects being due partly to the clays used and partly to the manipulation. The fundamental forms are bowls, pots, bottles, or jars, and vessels imitating natural or artificial objects. In the first three types there is a similarity with Pueblo work, but in the specialised forms the Mississippi ware is unique.[1] Holmes figures an animal-shaped dark vase from Arkansas, combining a number of marks peculiar to this area. The whole was moulded, the figure lays no claim to portraiture, the feet are luted on, the surface was engraved, after baking, with rings and involuted lines. In this same fashion and very much after the order of the stone pipes, but inferior in execution, occur imitations of fish in the freshwater streams, of the wild animals, of birds, and of the human form, besides grotesque objects in endless variety. Their surfaces are ornamented by trailing, incising, or excavation in the soft clay ; by stamps and impressions ; by engraving on the dried or on the burnt surface, and by painting in white, red, brown, and black.

A very great number of the animal and human forms hint at the derivation of the first suggestion from a carving rather than from a weaving people. The most noticeable feature in the wooden dishes of the T'lingit and the Haida stocks is the form of a seal, bear, bird, or other creature

[1] Holmes, *Fourth An. Rep. Bur. Ethnol.*, p. 404, fig. 416.

common to this forest region, lying on the back, and the stomach or body forming the vessel, while the head and the tail or the head of another animal form the handles. At times the animal is erect, and the opening is in the back. In very many cases the head and tail are merely carved in low relief at the ends. The work on the old ware is extremely well done. A very popular delineation in this woodware is the head of an animal reaching above the edge of the dish and forming a handle on one side. In the great museums of Europe and America is to be seen an abundance of this North-west Coast woodware.[1]

Along this same line of development the Mound Builder made a bottle of the human form, thinking at the same time of a gourd in the matter of pouring out the liquid. The standard human effigy vase from this region is shown in Holmes.[2] A woman sitting on her heels with her hands on her knees, with very good profile, is the subject. Holmes says of this vase, which is ten inches high, that it is well modelled, a good deal of attention having been given to the details of anatomy. The back is very much humped, and the vertebræ are represented by a series of knobs. The knees, calves, ankles, and the various parts of the feet are indicated with an approach to accuracy. Balfour shrewdly observes, with reference to the grotesque in all this savage art, "However rude these representations may be, the intention is realistic, and the greater or less resemblance to nature is only a question of skill. But want of skill may of itself tend to alter the character of such designs. Imperfect realism readily degenerates into the grotesque, and this may partly account for the great prevalence of fanciful representations of objects among so many savage peoples." [3]

The Mound Builder was capable of better work in pre-

[1] For a series of grotesque handles see Holmes, *Fourth An. Rep. Bur. Ethnol.*, p. 388, figs. 383, 384; for vessels, figs. 385 to 391; 415–419; 445–447.

[2] Holmes, *Fourth An. Rep. Bur. Ethnol.*, p. 425, figs. 452 and 453 b.

[3] Henry Balfour, *Midland Naturalist*, 1890, vol. xiii. p. 10.

senting the human form in clay. An example of this is seen in a head five inches high, figured by Holmes. The walls are from one-eighth to one-fourth of an inch thick. "This vase was modelled in plain clay, and allowed to harden before the devices were engraved. After this a thick film of fine yellowish-grey was applied to the face. The remainder of the face, including the lips, received a thick coat of dark red paint. The whole surface was then highly polished."[1]

By far the most elegantly decorated ware in America is found in the central regions, from the isthmus northward to Nicaragua. Both geometric and animal forms are delineated in such manner as to show that the makers were already in the metallic age. The fabrics and the ceramic ware both manifest the influence of that curious metal-work which characterises the region. But, in what Mr. Holmes calls the "Lost Colour Group," a new feature in laying on ornament is introduced. "The paste is fine grained, and usually of a light yellow grey tint throughout. The surface was finished either in a light-coloured slip or in a strong red pigment. In some cases the light tint was used exclusively, and again the red covered the entire surface, but more frequently the two were used together, occupying distinct areas of the same vessel, and forming the ground work of decorative patterns in other colours. They were polished down with great care, giving a glistening surface upon which the markings of the tool can still be seen. The bright red colour is only a ground tint, and is not used in any case in the delineation of design. The patterns were worked out in a pigment or fluid now totally lost, but which has left traces of its former existence through its effect on the ground colours. In the beginning of the decoration a thin black colour, probably of vegetal character, was carried over the area to be treated, and upon this the figures were traced in the lost colour. When this colour, or taking-out medium, disappeared, it carried the black tint beneath, exposing the light grey and

[1] Holmes, *Fourth An. Rep. Bur. Ethnol.*, p. 407, fig. 420.

red tints of the ground, and leaving the interstices in black. These interstitial characters are often mistaken for the true design." [1]

There is one instructive body of evidence that must not be overlooked before the potter's art is dismissed. True it is that not much remains of the textile industry of antiquity. Here and there a few charred examples, and in desert or rainless countries, like Peru and Egypt, the relics of more advanced weaving and embroidery, that is all.

But it occurred to the potters of antiquity to adorn the surfaces of their ware with textile patterns. The threads made deep furrows on the clay while it was soft. The burning fixed the impression. Centuries of exposure have removed from these fragments the painted decorations and the engraved designs. But dust has crept into the meandering furrows, just as dust enveloped the Mesopotamian cities. By washing the fragment and pressing on the surface soft wax or putty or artist's clay, the most delicate filaments of the ancient textile stand out revealed.

Curiously enough, the potsherds of North America, with few exceptions, bear the impress of fabrics made to this day by the living aborigines. More than this, the Lake dwellings of Switzerland reveal among the charred objects not only the plain weaving with which all are familiar, but the twined style, common all over North America and Africa. These, and even the very knots in the nets, are similarly demonstrated on pottery, as well as a kind of embroidery on bark, such as the Polynesians employ in making stamps for tapa cloth. [2]

[1] Holmes, *Sixth An. Rep. Bur. Ethnol.*, 1888, p. 113.

[2] Sellers, *Pop. Sc. Monthly*, New York, vol. xi. p. 573. Holmes, "Prehistoric Textile Fabrics," *Third An. Rep. Bur. Ethnol.*, Washington, 1884, pp. 393–426, pl. xxxix., figs. 60–115.

PRIMITIVE GLEANERS.—UTE WOMEN GATHERING WILD SEEDS.
(*After Powell.*)

CHAPTER VI.

PRIMITIVE USES OF PLANTS.

> "Inventor, Pallas, of the fatt'ning oil,
> Thou founder of the plough and ploughman's toil;
> Come all ye gods and goddesses that wear
> The rural honours and increase the year,
> You who supply the ground with seeds of grain,
> And you who swell those seeds with kindly rain."
>
> VIRGIL, *Georgics*, i.

PRIMITIVE peoples approached the vegetal kingdom for *food*, for *fibre*, and for *woods* or *timber*.

In the matter of aliment, no one can doubt that the earliest peoples helped themselves from the bounty of nature, and their inventions, if they made any, related merely to searching for food and carrying it home.

In progressing beyond this natural harvest and consumption, they adopted the order of selection and preparation previously laid down in this volume, proceeding always from the simple to the more complex in every respect. As an example of a great variety of plants utilised by a savage race whose wants were not excessive, Ellis, in his *Polynesian Researches*, mentions the apape, used by the natives in building their canoes, and the faifai, employed for the same purpose; the aito, or toa (*Casuarina equasitifolia*), called also ironwood, for making weapons; the reva (*Galaxa sparta*), quite as valuable; tiairi, or candle-nut tree (*Alurites triloba*); tamanu, or ati (*Callophyllum inophyllum*), an insect-proof wood out of which furniture and keels for canoes are wrought; purau (*Hibiscus tiliaceus*), furnishing wood for paddles, boards for native vessels, and rafters for ordinary dwellings; auti (*Morus papyrifera*), or Chinese paper mulberry, for manufacturing the cloth worn in the islands; mate (*Ficus prolixa*), the berries supplying a scarlet dye and the bark furnishing fibre for the large and durable salmon nets; romaha (*Urtica argentea*), from the bark of which are made strong elastic fishing lines and smaller nets; bread-fruit (*Artocarpus*, many species), the chief food tree of the islands; taro (*Colocasia esculenta*), the most serviceable article of food; the uhi, or yam (*Dioscorea alata*), a valuable food root; the umaia, or sweet potato (*Convolvulus batatus*), also grown for food; pia, or arrowroot (*Chailea tacca*), grows spontaneously, and is sometimes cultivated; the haari, cocoanut (*Coccos nucifera*), used for spears, wall plates, rafters, pillars, kitchen knives, parts of canoes, fences, bagging, fuel, house screens, mats, baskets, clothing, hats, food, apparatus of worship; maia, plantain and banana (*Musa paradisiaca* and *M. sapientum*), thirty varieties cultivated; vi, or Brazilian plum (*Spondias dulcis*), an abundant and excellent fruit; ahia, or jambo (*Eugenia Mallaccensis*), the most juicy fruit in the island; mape, or rata (*Tuscarpus edulis*), used in times of scarcity as a substitute for bread-fruit; ti-root (*Dracanæ terminalis*), baked and eaten; to, or sugar-cane (*Saccharum officinarum*), often carried on long journeys to appease hunger and thirst as well. To this list, Ellis says, many more were added, but this will be "sufficient to show the abundance, diversity, nutritiveness, delicacy, and richness of the provision spontaneously furnished to gratify the palate and supply the necessities of the inhabitants." [1] The Kew Gardens could multiply this list ten times easily.

De Candolle discusses the origin of cultivated plants under the following heads:—

1. Plants cultivated for their subterranean parts.
2. Plants cultivated for their stems or leaves.

[1] Ellis, *Polynes. Res.*, London, 1859, Bohn, vol. i., chaps. ii. and iii.

3. Plants cultivated for their flowers.
4. Plants cultivated for their fruits.
5. Plants cultivated for their seeds.

In noting the unequal distribution of cultivable plants, he enumerates the inducements held out by Nature to cultivation.

1. A region with plants worth the trouble of cultivating.
2. With not too rigorous a climate.
3. Having only moderate duration of drought.
4. Furnishing security to the cultivator.
5. Yet pressing the necessity of cultivation by the absence of game, fish, and indigenous nutritious plants, such as chestnuts, dates, bananas, breadfruit and the like.[1]

As an example of a plant worthy of cultivation, a sago palm tree of good size will produce enough food to keep a man a whole year. Wallace says it is truly an extraordinary sight to witness a whole tree-trunk converted into food with so little labour and preparation. The great Sago district is in the island Ceram. When sago is to be made, a full-grown tree is cut down, the leaves cleared away, and a broad strip of the bark taken off the upper side of the trunk. The pith is dug out with a club of hard wood having a sharp piece of quartz embedded in one end, leaving a skin not more than half an inch thick. This material is carried away in baskets, made of the sheathing bases of the leaves, to the washing machine, of which the large sheathing bases of the leaves form the troughs and the fibrous covering from the leaf stalks of the young cocoanut the strainer. Water is poured on the pith, which is pressed against the strainer until the starch is dissolved. The water charged with the starch passes on to a trough, with a depression in the centre, where the sediment is deposited and the water trickles off by a shallow outlet.[2]

Quite as remarkable are the banana, the breadfruit, the

[1] *Cf.* De Candolle, *Origin of Cultivated Plants*, New York, 1885.

[2] Wallace, *Malay Archipelago*, New York, 1869, p. 383, *seq.* with pictures of apparatus.

date, the yam, and perhaps others, besides all the great cereals.

Vegetable food species, says Payne, may be arranged in regard to the way in which they most quickly and amply repay the labour which the process involves, into three groups: (1) Plants bearing succulent fruits, stems, or leaves; (2) plants with succulent roots; (3) culmiferous or cereal grasses, supplemented by edible nuts and other materials for the miller. They fall into the same order when it is sought to group them with reference (1) to the order in which man appears to have been led to appropriate them as natural bases of subsistence; (2) to the amount of labour necessary to adapt them for human consumption; (3) to the amount of labour necessary to convert them from a natural to an artificial basis; (4) to the value possessed by each relatively to their bulk and capacity for storage; and (5) to the degree in which their cultivation contributes to advancement.[1]

The harvester, the carrier, the miller, the baker, the cook, the purveyor, are all to be seen in embryo in the occupations of the first men and the first women in their *rôle* of going to the plant kingdom for foodstuffs. Very often one woman will be found in her daily cares performing all of these functions.

At first, the planter, the farmer, or the gardener, did not appear. The order of her coming (for this art primarily belonged to women) seems to have been first as a field botanist, familiarising herself with nutritious plants; second, as a weeder, removing from proximity with the useful plant others that impeded its growth and were of no use; third, as a planter, putting the root or seed where it would grow and take care of itself. An example or two taken from actual life will illustrate at once how this primeval gleaner came at last to dress and keep the earth.[2]

"The Panamint (Shoshonean) Indians of Death's Valley,

[1] Payne, *Hist. of America*, New York, 1892, vol. i. p. 333.

[2] Consult Ling Roth, "Origin of Agriculture," *J. Anthrop. Inst.*, London, 1886, vol. xvi. pp. 102–136.

California, eat large quantities of the nuts of *Pinus monophylla*. In early autumn, after the seeds have matured, but before the scales have opened, the cones are beaten from the trees, gathered into baskets and spread out on a smooth piece of ground exposed to the heat of the sun. The scales soon become dry and crack apart, and the seeds are shaken out by blows of a stick. They are then gathered into baskets and most of them are cached in dry places among the rocks for use during the year. To prepare them for food the nuts are put into a basket with some live coals and shaken or stirred until they are gradually roasted."[1]

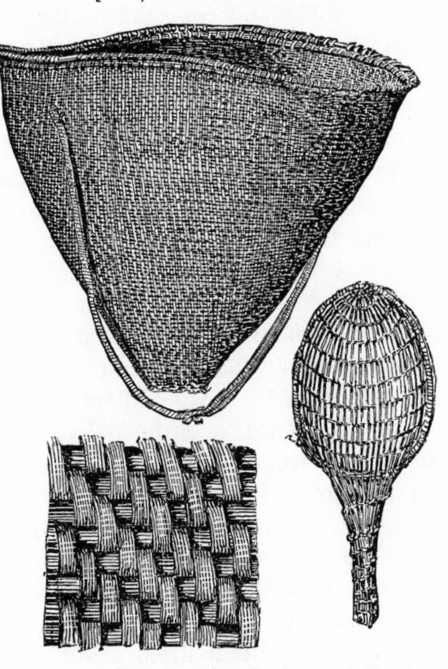

FIG. 34.—Ute Carrying Basket and Reed Gatherer's Fan. (*U. S. Nat. Museum.*)

The seeds of many other plants, indeed of almost all that are not poisonous, are also used. The method of procuring them is different with each plant, but the process of roasting is always the same. The common sand grass of the desert (*Oryzopsis membranacea*) produces an abundance of seed and is generally used. In gathering it the squaw carries in one hand a small basket and in the other a paddle made of wicker-work resembling a tennis racket. With this she beats the grass panicles over the basket.

[1] Coville, *Am. Anthropologist*, Washington, 1892, vol. v. p. 352.

One of the prickly pears (*Opuntia basilaris*) used by these Panamint Indians has taxed the ingenuity of the harvester, and her methods of overcoming the difficulties is very ingenious. "In May and early June the flat, fleshy joints of the season's growth as well as the buds, blossoms, and immature fruit are fully distended with sweet sap. They are broken off with sticks and collected in large baskets. Each joint having been rubbed with grass to remove the fine, barbed prickles, is exposed to the heat of the sun. Instead of the drying process, another, more elaborate, is sometimes adopted. A hole about ten inches in depth and three feet in diameter, is dug in the ground and lined with stones. Upon this a fire is built and other stones thrown in. When they are thoroughly heated—the ashes, coals and all—one layer of stones is scraped away, and some fresh or moistened grass spread in the hole. Next a layer of cactus joints is added, then more hot stones, and so on, until the pile is well rounded. The whole is then covered with a mat, and lastly with moist earth. After about twelve hours of steaming the pile is opened, and the nä'-vo is salted and eaten. A portion is dried and preserved."

There grow in the desert several large crucifers. The process of preparation is as follows : The leaves and young stems are gathered and thrown into boiling water for a few minutes, then taken out, washed in cold water and squeezed. The operation of washing is repeated five or six times, and the leaves are finally dried ready to be used as boiled cabbage. The washing removes the bitter taste and substances that produce nausea and diarrhœa.

The ripe pods of the mesquit bean (*Prosopis juliflora*) contain very hard, bony seed, but the spaces between them, the body and septa of the pod, are stored with a small amount of nutritious matter, consisting principally of sugar. These are gathered and often cached until spring by the California Indians. The pods are ground in a wooden mortar and the flour sifted out.

The common reed (*Phragmites vulgaris*) furnishes the

Panamint Indians with sugar. In early summer, when the plants have attained nearly full size they are cut and dried in the sun. When perfectly brittle the whole plant is ground and the finer portions separated by sifting. This moist, sticky flour is moulded by the hand into a thick gum-like mass. It is then set near a fire and roasted until it swells and browns slightly, and in this toffy-like state it is eaten.

In April when the flower buds of the *Yucca brevifolia* are swelling, they are in a proper condition for food. The buds are terminal on the branches, and are protected by a close rosette of serrated, stiff, pointed, almost dagger-like leaves. The fibre of these leaves is so tough and the situation of the buds on the stem such that it would be difficult to cut one out even with an axe or hatchet. The Indians substitute dexterity for instruments. The four or five uppermost mature leaves are drawn together over the apex of the bud and grasped by the hand ; then by a quick, sidewise bend the head is broken off. The leaves and tips are discarded, leaving an egg-shaped, solid, juicy mass. This is roasted on hot coals and eaten at once or kept until cold.

In some parts of the Charleston Mountains, Nevada, the Piutes use a small mescal (*Agave utahensis*). Where the plant abounds are numerous mescal pits, in which it has been cooked in previous years. These mescal pits, as they are called, are circular depressions in the ground six to twenty feet in circumference, and sloping evenly to a centre one three feet in depth. They are deeply lined with coarse gravel. A great fire was built in a pit, and kept up until the stones were very hot. The fire was raked out, the mescal plants placed in and covered with grass. After two days of steaming the pile was opened, and the mescal was ready for eating.

Mr. Coville, the Botanist of the Department of Agriculture in Washington, in his function devotes much energy to searching for the native American plants that might yet be brought under cultivation. In this work it is necessary to know what species were of use to the aborigines, and the result is most astonishing. The list would include hundreds of species.[1]

Wherever savages have been visited in their native simplicity, they seem to have found out just how to garner the products of plants in the best manner. The Calispel Indians gather cranberries with rakes; the Ojibwa woman paddles her canoe among the wild rice, and with a proper wand beats the seeds into a coarse mat spread on the bottom. The gatherers of certain water-plants know the feeling of the desired roots with their feet, and use these members to secure them.

After all has been said about other devices, the digging stick is the beginning of agricultural implements, the progenitor of the hoe, the spade, the plough. It would be difficult to find a tribe so low down as not to know its use. A patent-office examiner would declare that nothing could be simpler than to rub a stick on a rock to give it a point. Hardening this point in fire came later. At any rate, the California tribes dig clams, the Oregon tribes dig roots, the Australian tribes dig vermin, the Andamanese dig yams, and all over Africa root-food and animal food are forced out of the ground with the digging stick. As for taro, the Polynesian both plants it and harvests it with the same implement.

Ellis says: "The chief and almost only implement used was the *ô*, a stick sharpened at the point, and used in loosening and turning up the earth. Formerly they hardened the end with which they penetrated the soil by charring it in the fire. No use is made of the foot in thrusting the spade into the soil, but the person digging assumes a crouching attitude, pierces the ground, and breaks up the earth by the strength of the hands and arms. The

[1] Consult Sturtevant, "Origin of Garden Vegetables," *Am. Naturalist*, vol. xix.; Newberry, "Plants used by North American Indians," *Pop. Sc. Monthly*, vol. xxxii. See also Goodale, "Useful Plants of the Future," *Proc. A. A. A. S.*, 1892, pp. 1–38, and "Official Guide to Kew Gardens."

making and repairing of fences occupies much of the time of those engaged in the cultivation of the soil." [1]

This implement is not only useful in taking from the soil

FIG. 35.—Ancient Roman Plough. Bronze figure.

the crops that Nature has planted and raised, but it is the beginner of artificial planting.

An exceedingly ingenious device in the nature of a

[1] Ellis, *Polynes. Researches*, London, 1859, vol. i. p. 137.

harvesting implement is the kelp-knife of the Oregon Indians. The stalks of kelp are harvested and tied together to form the long lines used in deep-sea fishing. To cut loose the stalk from its root a slip-knot of cedar withe containing a knife-blade is passed over the kelp-bud at the top of the water. A stone sinker carries the apparatus down the stem to the bottom, and the harvester, holding the kelp stem with one hand, quickly pulls a line which he had previously attached to the cutter, and severs the stalk close to the root. When wet these stems are very strong, but when dry they snap like pipe-stems.

The farming corporations in new countries now burn off the prairie, and by means of the steam-plough are able to cultivate thousands of acres under one management. They follow this up with wheel, spring-tooth, and crushing harrows, and plant their crops with buggy drills. All this seems easy enough, for it is naught but riding. Many who read this will remember their grandfather's farm, where the processes were far more simple and the work more laborious. Some have travelled in Mexico and Sicily and Tunis, where the ploughs are as primitive as they were in the days of the prophets. But this is nowhere near the beginning of agriculture. Let us see. A company of Cocopa or Mohave or Pima women set forth to a rich and favoured spot on the side of a cañon or rocky steep. They are guarded by a sufficient number of men from capture or molestation. Each woman has a little bag of gourd seed, and when the company reach their destination she proceeds to plant the seeds one by one in a rich cranny or crevice where the roots may have opportunity to hold, the sun may shine in, and the vines with their fruit may swing down as from a trellis. The planters then go home and take no further notice of their vines until they return in the autumn to gather the gourds. This is the testimony of E. Palmer, who spent many years as a collector among the American aborigines. Seed-time and harvest : no preparation of the soil, no tending of the young plants ; ingathering, that is all.

The Polynesians propagated the aute (*Morus papyrifera*), or Chinese paper mulberry, from scions, just as in England the osiers are raised. The arum was multiplied by transplanting the small tubers which grow round the principal root, or by setting out the top or crown of the roots that had been used for food.[1]

The breadfruit tree was propagated from shoots springing out of the roots. The uhi or yam (*Dioscorea alata*) was cultivated with much care. On the sides of hills and banks terraces were formed of mixed rich earth and decayed leaves. The roots were kept in baskets till they began to sprout. The eyes or sprouts were then cut off with pieces of the root attached to them, and these spread out and left to dry. The rest of the yam was eaten. The native agriculturists had a theory that it was better to plant the eye with only a thin shaving of the root than with the whole tuber or a thick slice attached as we do. When the slips were sufficiently dry they were put into the ground sprouts uppermost, and a small portion of leaves and mould laid over them. When the roots began to swell the farmers kept hilling them up, as we call it, until they were sufficiently mature to harvest. The umara or sweet potato (*Convolvulus batatus* or *chrysorizus*) received the following treatment. A mound of rich black mould is raised, nine or ten feet in diameter and three feet high. In the top of this mound they inserted a small bunch of vines, which germinated and produced abundantly. Arrowroots were planted whole, and a number of tuberous roots formed at the extremities of the fibres.

On his first voyage Columbus found the natives cultivating what is called a yam (*Dioscorea alata*). "They powder and knead them, and make them into bread; then they plant the same branch in another part, which again sends out four or five of the same roots."[2]

The "Campas of Peru burn and cut down the lofty

[1] Ellis, *Polynes. Res.*, London, 1859, Bohn, vol. i. pp. 34, 344.

[2] *J. of Christopher Columbus*, Hakluyt, London, 1893, vol. lxxxvi. . 112.

forests around their cabins (*panguchis*), and in the area thus opened they plant or sow bananas, coca, maize, yuccas, beans, *mangonas*, and *uncuchas*, solanaceous plants analogous to potatoes. The cotton plants spring up spontaneously about their houses, as nettles do with us."[1]

The reader has already been reminded of the beneficent co-operation of fire in the subduing of the earth. One who has lived in Canada, or the Western States of the Union, or

FIG. 36.—Survival of ancient form of Plough in Spanish agriculture.

in the colonial possessions of the Great Powers of Europe, knows how powerless man would even now be without this generous aid.

"The chief use of the hatchets among the Delaware Indians of New Jersey," says Kalm, "was to make good fields for maize plantations. If the ground was covered with woods, they cut off the bark all round the trees with

[1] Ollivier Ordinaire, *Rev. d'Ethnog.*, 1887, vol. vi. p. 271. Payne, *Hist. of America*, Oxford, 1892, vol. i. p. 365 *seq*.

their hatchets at the time when they lose their sap. The trees thus girdled died, and the ground was a little turned up with crooked or sharp branches."[1] This attack upon the tree is absolutely efficacious, even in old forests where the amount of undergrowth that savages could get together would not be sufficient to burn the life out of the giants.

In the line of inventing the plough, Livingstone calls attention to a double-handled hoe used by the women in Portuguese Africa. It consists of two limbs of a tree joined by a short piece of the trunk, into the grain of which the shank of the hoe is driven. The implement is worked by a dragging motion, and in the excellent figure given there are knobs cut on the extreme ends of the handles which might well serve to attach traces for these primitive draught animals.[2]

Naturally the question would arise, why all regions are not alike ready to receive the hoe and the plough. In this search de Candolle will be of great service. He says, "The various causes which favour or obstruct the beginnings of agriculture, explain why certain regions have been for thousands of years peopled by husbandmen, while others are still inhabited by nomadic tribes. . . . Into Australia, Patagonia, and even in the south of Africa the plants of the north temperate zone could not penetrate by reason of the distance, and those of the intertropical zone were excluded by great drought or by absence of high temperature. At the same time the indigenous species were very poor. It is not the mere want of intelligence or of security which prevented the inhabitants from cultivating them."[3]

The preparation of the ground opens the way to the simplest fashions of tillage. The Zuñi head-man mounts the top of his house every morning and gives his orders to each member of the clan about what to do in the peach

[1] Kalm, *Travels, &c.*, London, 1771, vol. ii. p. 38.

[2] Livingstone, *Trav. &c., in S. Afr.*, New York, 1858, p. 442, figured.

[3] De Candolle, *Origin of Cultivated Plants*, New York, 1885, D. A. & Co., p. 3.

orchard, the pumpkin patch, the cornfield. The hoeing is indeed very scanty, but one thing cannot be intermitted; rapacious animals and birds are ever on the alert. There is no genus or species of scarecrow that they are not familiar with, so the Zuñi farmer has a hard time of it. His prayers, his vigilance, and his labours never cease until the crop is safely in his stomach.

The African farmers are equally alert. Most of the agriculture is woman's work; it is of the African woman that it might be written, "Woman's work is never done." More than in the Western Continent, the great beasts and small trample down and eat up the crops. The fence, therefore, derives its origin quite as much from a desire to protect plants as from a military motive. Almost all the pictures of African and Indo-Pacific home life include this feature, and the Pueblo peoples know how to stand poles on end and fasten them together by strong withes in twined weaving.

There are many things that sorely try the patience of the Dyaks as they watch with unflagging interest the growth of their crops. It is true the pigs and deer are excluded by means of the wooden fence, but nothing short of the most untiring vigilance can keep out the climbing and winged pests such as monkeys, squirrels, rats, and sparrows, some of which are sure to visit the farm as the paddy is ripening.[1]

The elaboration of mechanical devices and methods of ingathering began with mere plucking, just as animals do. The many inventions for climbing trees are really harvesting apparatus. The savage appliances for going to the great harvest-fields of Nature would be just as useful for the ingathering of cultivated crops. Knives of stone were the first sickles, and human backs were the earliest farm waggons. Very old forms of bronze sickles are also to be seen in European museums. Many of those who read these pages will have seen a scarcely improved one of iron in use in Norway and Sweden.

The same is true of threshing and winnowing, or the

[1] Ling Roth, *J. Anthrop. Inst.*, London, 1892, vol. ii. p. 23.

separation of the chaff from the grain. In America, fire is universally used for this purpose. A mass of gathered heads is placed in a conical basket and rubbed with the hands. Then the mixture of seed and chaff is put into a "roasting tray," which is a large flat basketry dish. Then red-hot stones are laid on the mass by means of wooden tongs. The seeds are parched to a turn and the chaff is burned. The first flail was the commonest kind of a stick, the first fanning mill was a forked stick co-operating with the wind. These were all primarily used with the wild seeds. The transition from the dropping of grain seeds naturally to the intentional planting of them was not difficult to learn. Neither did the threshing and winnowing of the crops from planted seeds need or receive any change of treatment.

Fig. 37.—Tree Climbing in India. (*After Bishop Hurst.*)

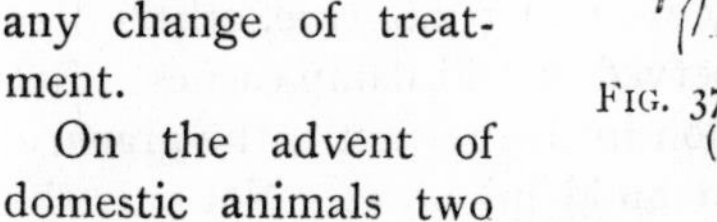

On the advent of domestic animals two styles of threshing came into vogue: the one was simply driving donkeys or other animals over the straw; the other was the use of the *mowrej* or *tribulum*, not yet gone out of use in North Africa and South-western Asia.

"The common mode of threshing in Palestine," according to Thomson, "is with the mowrej, drawn by a yoke of oxen, until the grain is shelled out and even the straw is ground into chaff. Bits of rough lava are fastened into the bottom of the mowrej and the driver sits or stands on it. The construction of the floor is very simple. A circular space, from thirty to fifty feet in diameter, is made level, and the ground smoothed off and beaten solid. The entire harvest is brought to them and there threshed and winnowed, and the different products carried to their respective places. The Egyptian mowrej has rollers which revolve on the grain. In the plains of Hamath I saw this machine much improved on by having circular saws attached to the rollers. On some floors here at Yebna, there was no machine of any kind; but boys rode or drove donkeys and horses or oxen round upon the grain. No one continued long in the same direction, but each changed every few minutes to keep the animals from becoming dizzy. The grain as it is threshed is heaped up in the centre of the floor, and when the wind blows the mixture is thrown up with shovel and fork to have the dust and chaff and straw blown away."[1]

Of scarcely less importance in the art of feeding on the vegetable kingdom is the storing of food. The preservation for future use demands a knowledge of the nature of the crop and the ever-present danger of destructive animals and decay and thieves called for eternal vigilance on the part of the owners.

In the Chapter on Fire reference was made to the drying and cooking of animal and vegetable food. Many seeds may be kept indefinitely by excluding moisture, others, like the succulent roots, are preserved in cold, damp caches. But the most important invention in this respect is the granary. The animals taught men to build granaries. Not only the ant has something to say to the sluggard. In California the woodpeckers bore holes in trees, and store in each one

[1] Thomson, *The Land and the Book*, New York, 1880, vol. ii. pp. 149 to 155. Five excellent illustrations. The reader will again notice that all the intricate invention in the modern thresher is not in the working part, which remains practically the same, but in the manual part which substitutes machinery for man or beast.

—COMPLETE OUTFIT OF A CALIFORNIA ACORN GLEANER AND MILLER, SHOWING BURDEN BASKET, GRANARY BASKET, MILL COMPLETE, HEADBAND FOR CARRIER, PICKING BASKET, AND DETAIL OF BASKET HOPPER.

an acorn for future use. The Indians, in times of scarcity, actually rob these acorn stores to supply themselves. Bees may be mentioned whose stores are also appropriated, and there are a thousand and one animal and vegetal gums and resins of which men help themselves. The natives of every land commence their life of true economy by invading the stores of their animal neighbours. Not content with killing and snaring the creatures they proceed also to administer on their effects.

The gleaning of the fields and of the waters really supposes the use of storehouses. William Strachey says, "Their corne and indeed their copper hatchetts, howses, beades, perle, and most things of value they hide in the grownd within the woods, and so keep them untill they have fit use for them." [1]

Granaries or public receptacles were by the people in Florida built of stone or earth, and roofed with palmetto leaves and clay. They served as depositories for maize, fruits, nuts, nutritious roots, dried fishes, alligators, deer, dogs, and other jerked meats. Hoards of corn, nuts, and meat are frequently mentioned in the early narratives as existent among the primitive peoples of this region.[2]

" The general storehouses of the Muskhogean tribes were circular in form—their walls constructed of stone and earth, and their roofs fashioned with the branches of trees, grass, clay, and palmetto leaves—were located in the neighbourhood of streams and in retired spots, where they were protected from the direct rays of the sun. They were built and furnished by the common labour of the tribe, and in them were stored corn, various fruits, and the flesh of fishes, deer, alligators, snakes, dogs and other animals, previously smoked and dried on a scaffold." [3]

Among the Nutka Indians of Vancouver Island, in

[1] *Hist. of Travaile into Va.*, London, 1849, Hakluyt Soc., p. 113.

[2] *Cf.* Jones, *Smithson. Rep.*, 1885, vol. i. p. 900.

[3] Jones, *So. Indians*, New York, 1873, p. 12, refers to pl. xxii., xxiii., xxiv. in *Brevis Narratis*, also p. 35.

addition to stores of oil, dried fish, and spawn, several varities of seaweed and lichens, as well as the camass and other roots were regularly laid up for winter; while berries, everywhere abundant, were gathered in great quantities in their season, and at least one variety preserved by pressing in bunches.[1]

The Hurons of Canada, according to Cartier, 1535, had good and large fields of corn, which they preserve in garrets at the tops of their houses. Champlain speaks to the same effect (1610).[2]

According to Bandelier, Mindeleff, and other explorers of the Pueblos, the rooms of the ground storey in ancient times were used for granaries and storage. Even the roofs were used for the temporary storage; "and in autumn, after the harvests have been gathered, the terraces and copings are often covered with drying peaches, and the peculiar long strips into which pumpkins and squashes have been cut to facilitate their desiccation for winter use." In another Pueblo a bin was built upon a ledge in a corner of a room by means of slabs and clay. "In many houses, both in Tusayan and Cibola, shelves are constructed for the more convenient storage of food." Another device for the storage of food is a pocket or bin built into the corner of a room.[3]

Schweinfurth says that granaries for corn, resting on posts, are in use in the whole of Central Africa as far as the tropical rains and white ants extend. In the example and figures of the author there is nothing left out which would be thought essential in modern storehouses. They are secure from dampness below and rain above, vermin and thief proof, and universally have a look-out on the top.[4]

[1] Bancroft, *Native Races*, vol. i. p. 187.

[2] Referred to with other references by Lucien Carr, *Ky. Geol. Survey*, vol. ii. p. 14.

[3] Victor Mindeleff, *Eighth An. Rep. Bur. Ethnol.*, Washington, 1891; consult index *sub voce* "Storage."

[4] Schweinfurth, *Artes Africanae*, London, 1871, pl. vi. fig. 1.

An ingenious granary is built by the Sehre negroes, neighbours of the Niam Niam, embracing the three ideal characteristics of such a structure, namely, protection against moth and rust and thieves,—in this case termites and vermin, dampness and thieves. A post is set in the ground and peeled, and on the top of this the receptacle for corn is made, vase shaped, from the clay taken from the mushroom-form structures of the *Termes mordax*. Over the clay structure a thatched roof is built in a very tasteful manner.[1]

Cisterns as granaries are of very ancient date. In limestone countries, no sooner did men begin to lay up provisions than they discovered ants and mice and rats. The cisterns of Palestine were proof against dampness as well as from vermin and plundering robbers. Indeed, the owner of the crops has taken refuge in them, and there also has he incarcerated his enemies. In the old sieges, lasting many years, the preservation of grain must have been a study. Indeed, there is no reason to doubt that the ability to preserve grain for a long time, as in Mexico, Peru, and Egypt was a factor in the early and more rapid development of their populations. At any rate, the oldest of historic nations elaborated early the granary.[2]

There was no such thing as a food-safe in the economy of the ancient Hawaiian house, and to preserve victuals from pigs, dogs, and rats, it was necessary to suspend the gourds that contained them beyond their reach. Usually a pole was fixed in the floor of the house, and to the top was fitted a notched cross-bar, from which a number of gourds might be hung.

Before the introduction of cats the Hawaiian used a bow and arrow to kill rats and mice. But they never employed this weapon in war.[3]

The Kyan, Kiñah, and Lanahan Dyaks stow their paddy

[1] Schweinfurth, *Artes Africanae*, London, 1875, pl. xx. fig. 4.

[2] Wilkinson, *Anct. Egyptians*, quoting the Bible, Diodorus, and other writers; index, *sub voce*, "Granaries."

[3] Brigham, *Cat. Bishop. Mus.*, Honolulu, 1892, vol. ii. pp. 23, 31.

in barns built for the purpose. The floor is six feet above the ground and the posts are encircled with wooden discs to keep off the rats.[1]

Compare this with the more primitive device of the Fijians, cut from a solid block of wood. In form it resembles a clumsy spindle, with shaft and whorl of one piece. By the part above the whorl the apparatus is suspended. At the lower end of this axis any number of hooks are cut, and from these all sorts of food are suspended. The wooden disc prevents vermin crawling down the shaft.

Plants were early used in healing disease. It is well understood by all ethnologists that disease is thought by savages to be the possession of the individual by evil spirits. The cure of maladies, therefore, would be akin to sorcery, and the doctor is exorcist, seer, conjurer, fortune-teller, and physician all in one. But, with this knowledge fully before us, we are bound to own that a deal of experimental medicine and surgery were early developed in spite of such wrong theories. When a Florida Indian doctor scarified the forehead of a patient with a shell and sucked therefrom the demon of disease, he was cupping and leeching his sick man and nothing short. When, again, he compelled the patient to inhale the smoke of tobacco or other medicinal herbs, he was fumigating him, and unwittingly discovering a little in bacteriology. These same doctors had found out purgatives and emetics and astringents to drive away with disgust the evil spirits he had in his mind ; but the disease departed quite as soon for him as for us, when he gave the proper medicine. Bathing, sun-baths, exercise, massage, and even faith-cure had a part in this practice. "The Inhabitants giue great credit vnto their speeche, which oftentimes they finde to bee true."[2]

The number of medicinal plants in the pharmacopœia ot any savage people is surprisingly large. Many of them

[1] Ling Roth, *J. Anthrop. Inst.*, London, 1892, vol. xxii. p. 30.

[2] Hariot, quoted by Jones, *So. Indians*, pp. 30–34. See Max Bartels' interesting work, *Medicin der Naturvölker*, Leipzig, 1893.

remain in use; but in the forests are many more, which they have taken once and abandoned for ever.

Not enough has been here said about the improvement of plants under domestication. The inventive genius of early man is shown most remarkably in this industry, if the reader will consider the fact that the very earliest historians and poets and myth-makers record the names of all the well-established fruits and vegetables. No grain is a better evidence of this than maize. Payne says, "That the maize as we now have it is practically the creation of human labour and ingenuity is proved not only by the insignificant size of the euchlœna grasses, the cognate wild species, but by its rapid deterioration if allowed to become feral; the first step of which is the familiar sodden corn of the American farmer."[1]

And this brings us, in the second place, to consider woods or timbers. Wood-workers at first began with the use of saplings, and as their descendants became more clever they attacked small trees and then larger ones. Where the sapling would not serve the purpose, there was always abundance of fallen timber, and the Arctic peoples relied upon drift wood. The savage woman was a collector of faggots, which she broke for her fire with her hands or with the great stone, ever near. Both she and her man early learned to sever the slender tree trunk by means of fire or a sharpened stone, whether to make a club, a tent-pole, or a handle.

The very first settlers in New England observed that "The Indian houses are made of poles, large end in the ground, in a circle and bound at the top with the bark of walnut-trees, covered with sedge sewed together with the splinter of bone of a Cranes legge. Leaving several places for doors which are covered with mats which may be rolled up and let down with pleasure."[2] These sewed mats will be referred to again under textiles. They are mentioned by

[1] Payne, *Hist. of America*, New York, 1892, vol. i. p. 337.

[2] Morton, *New English Canaan*, 1632, p. 19; also Bozman, *Maryland*, p. 70.

MAKING AND BAKING TORTILLAS IN YUCETAN, MEXICO.

John Smith, and are now made by the Indians of Washington State.

Concerning the natives of North-west Canada, Hearne relates, "Agreeably to the Indians' proposal, we remained ten days, during which time my companions were busily employed (at their intervals from hunting) in preparing small staves of birch-wood, about one and a quarter inch square and seven or eight feet long. These serve as tent-poles all the summer while on the barren ground, and as the fall advances are converted into snow-shoe frames for winter use. . . . All the wood-work was reduced to its proper size for the sake of making it light for carriage."[1]

None of these incidental avocations interfere with or retard the Indians in their journey, for they always take advantage of every opportunity which offers as they pass along, and when they see a tree fit for their purpose, cut it down, and either strip off the bark, if that be what they want, or split the trunk in pieces, and after hewing it roughly with their hatchet, carry it to the tent, where in the evenings or in the mornings before they set out they reduce it with their knives to the shape and size which is required.[2]

The prettiest wood-working is done by the South Sea Islanders upon their clubs, paddles, and ceremonial adzes. To dig up the tree of the proper size, cut it off, dress it down, and prepare it for the carver was the work of the digging-stick, the adze, the scraper, and the sharp-grained rubbing-stone. As soon as the wood was sufficiently seasoned it was further hardened and dried by means of fire. The surface was heated enough to destroy the grain of the wood. The engraver then with a shark's tooth etched upon the surface a lace-work of geometric patterns, varied now and then with a bit of true sculpture.

The Panamint Indians of California make their bows from the desert juniper (*Juniperus californica utahensis*). The Indian prefers a piece of wood from the trunk or a limb of a tree that has died and seasoned while standing. At low

[1] Hearne, *Journey, &c.*, London, 1795, p. 87. [2] *Ibid.*, *op. cit.*, p. 280.

altitudes in these desert mountains moist rot of dead wood never occurs. A mature tree subjected to the intensely drying heat of the region is in perfect condition for this use. The bow rarely exceeds three feet in length, and is strengthened by glueing to its back a cover composed of strips of deer sinew laid lengthwise along it. The string is made of twisted sinew, or sometimes of cord prepared from Indian hemp. Arrows are made from the stems of the reed (*Phragmites vulgaris*) and from willow shoots. The shaft is about three and one half feet long. Nearly mature but still green leaves are cut, their leaves removed, and the stems dried and

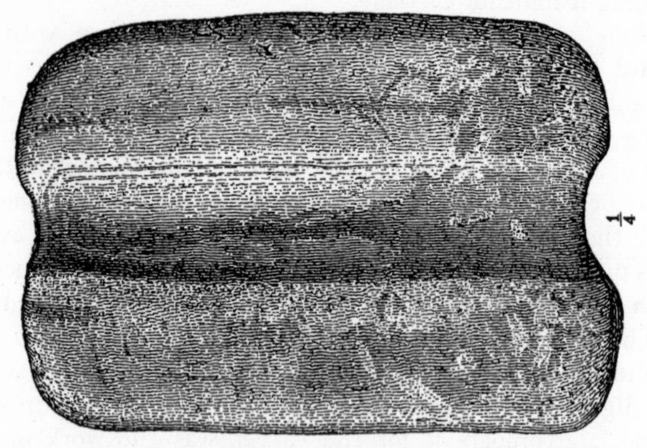

FIG. 40.—Grooved Sandstone for polishing Arrow Shafts. (*U. S. Nat. Museum.*)

straightened in the hands before a fire. In the straightening process use is often made of a small stone, across the face of which have been cut two grooves large enough to admit an arrow shaft. This stone is heated, and a portion of the crude arrow is laid in one of the grooves until it is hot. The cane is then straightened by holding it crosswise in the teeth and drawing the ends downward. By repeating this process throughout the whole length a marvellously straight arrow is produced. The head of the arrow is a pin of very hard wood, taken, I believe, from some species of *Atriplex*, or grease-wood. It is about five inches long, and tapers evenly to a blunt point. The base is inserted about three-fourths of an inch into the hollow of the reed, and rests against the uppermost joint. It is bound in place by a thin band of sinew. At each joint of the arrow-shaft is burned a ring of diagonal lines. The base of the shaft is notched and feathered with three half feathers, bound on with sinew, and slightly twisted to give the arrow a rotary motion. Willow arrow-shafts are "sand-papered by drawing the stick backward and forward in the angle between two flat stones held in the palm of the hand." [1]

The drums of the Dinka are made by stretching a cleansed goat-skin over the broader, open end of a scooped-out piece of the trunk of a tree (mostly of the tamarind), and in the fashion of our drums, tightened by means of cross-straps strung together by a second skin covering the lower, massive end. Kettle-drums of similar build and identical shape are much used in the East Indies. Nor was ancient Egypt destitute of them.[2]

The "dug-out" drum has a very wide distribution. In looking over a large collection of musical instruments the reader will find that the dug-out drum was first a sonorous log, then a hollowed piece, then with one head, afterwards provided with two heads of membrane, and finally with tightening apparatus, while the cylinder underwent adaptive modifications.

Besides the making of holes through wood, described in the Chapter on Tools, the ancient wood-workers needed often the help of a tube.

In many of the Eskimo and North-west Coast pipes, the long stems are frequently crooked, and the problem is to bore a curved or bent hole for the smoke. In some cases two holes are bored, one from the mouthpiece, another from the other end, these holes meeting about half way and

[1] Coville, *op. cit.*, p. 360.

[2] Schweinfurth, *Artes Africanae*, London, 1875, Sampson Low, p. 1 and fig. 1.

near the underside, making a very obtuse angle with each other. The meeting of these holes is generally so near the underside of the stem that a small piece of ivory or wood is there let in, and may be removed to clean the pipe. At the butt-end a third hole is bored down upon the main perforation to connect with the cavity in the bowl and the main perforation, and from this point outward is stopped with a plug. Another device of these cunning savages for perforating their crooked pipe-stems is to split the stem and cut a half-round gash or furrow in each, so that when the two halves are brought together the gashes form a continuous tube. In so doing the carver at the same time often brought the cuts so near the surface at one place in the underside of the stem that a little trap-door could be let in neatly, and removed for the purpose of cleaning the pipe. When it is remembered that the boring of savages is generally done from two sides, it will be seen that we have only to elongate the auger bit to get the result.

The sarbacan, or zarabatana, is made of two separate pieces of wood, in each of which is cut a semi-cylindrical groove, so that when they are placed in contact they form a long wooden rod pierced with a circular bore. The native use in this work the incisor teeth of rodents. The bore being carefully smoothed, the two halves are laid together and bound by means of a long, flat strip of jacita-wood wound specially around them. To the lower end is fastened a mouthpiece with a trumpet-shaped opening. Cement is then rubbed over the whole weapon, and it is ready for use.[1]

The manufacture of the blow-tube called pucuna is thus conducted in Guiana. The gigantic reed, *Arundinaria Schomburckii*, furnishes the material. A straight piece, varying from eight to fourteen feet in length, is cut between two widely-separated nodes, and is thoroughly dried, first by the fire and then in the sun. To obviate warping the straight, slender stem of a palm is bored throughout its

[1] *Cf.* Wood, *Unciv. Races*, Hartford, 1870, vol. ii. p. 583.

length with a long pointed stick, and within this tube the reed is inserted to keep it straight. Peccary teeth are sometimes fastened close together and parallel on the outside for sights. The darts for these tubes, and the curious manner of carrying and discharging them, will be described elsewhere.[1]

This blessed partnership between man and some special natural product is well explained by a note of Wallace concerning the bamboo: "Almost all tropical countries produce bamboos, and wherever they are found in abundance the natives apply them to a variety of uses. Their strength, lightness, smoothness, straightness, roundness, and hollowness, the facility and regularity with which they can be split, their many different sizes, the varying length of their joints, the ease with which they can be cut, and with which holes can be made through them, their hardness outside, their freedom from any pronounced taste or smell, their great abundance, and the rapidity of their growth and increase, are all qualities which render them useful for a hundred different purposes, to serve which other materials would require much more labour and preparation. The bamboo is one of the most wonderful and most beautiful productions of the tropics, and one of nature's most valuable gifts to uncivilised man."[2]

The midribs of the immense leaves of the sago palm in Ceram, and other Malayan islands, supply the place of bamboo. They are twelve to fifteen feet long, and often as thick in the lower part as a man's leg. They are light, consisting entirely of a firm pith covered with a hard, thin rind. Entire houses are built of these; they form excellent roofing poles for thatch; split and well supported they do for flooring; and when chosen of equal size and pegged together they make neat panels for houses. In

[1] *Cf.* im Thurn, *op. cit.*, p. 300; *cf.* Wood, *Unciv. Races*, Hartford, vol. ii. pp. 583–90.

[2] Wallace, *Malay Archipelago*, New York, 1869, p. 87; with many references to the plant and its uses.

place of boards they do not shrink, require no paint or varnish, and are not a quarter the expense. When carefully split and shaved smooth, they are formed into light boards, with pegs of the rind itself.[1]

The bark of trees was a standard material among savages, for cloth and textiles, as will be seen, for all kinds of vessels used in housekeeping, for roofs, and for boats. The cedar, elm, and birch tree were indispensable to the tribes of Canada.[2] In the tropical parts of South America the natives were very skilful in taking off enough bark to make a boat from a single piece, but the North American bark canoe was usually constructed of many pieces sewed together, and caulked with gum and pitch. The following detailed account is from Hearne :—

"Immediately after our arrival at Clowey, the Indians began to build their bark canoes, and embraced every convenient opportunity for that purpose ; but as warm and dry weather only is fit for this business, which was by no means the case at present, it was the 18th of May before the canoes belonging to my party could be completed. On the nineteenth we agreed to proceed on our journey; but Matonabbee's canoe meeting with some damage, which took near a whole day to repair, we were detained till the twentieth. Those vessels, though made of the same materials with the canoes of the Southern Indians, differ from them both in shape and construction ; they are also much smaller and lighter, and though very slight and simple in their construction, are nevertheless the best that could possibly be contrived for the use of those poor people, who are frequently obliged to carry them a hundred, and sometimes a hundred and fifty miles at a time, without having occasion to put them into the water. Indeed, the chief use of these canoes is to ferry over unfordable rivers, though sometimes, and in a few places, it must be acknowledged that they are of

[1] Wallace, *Malay Archipel.*, New York, 1869, p. 382.

[2] See Hind's *Canadian Red River, &c.*, London, 1860, vol. i. pp. 200, 203; vol. ii. p. 63, for tents covered with birch bark instead of skin

great service in killing deer, as they enable the Indians to cross rivers and the narrow parts of lakes; they are also useful in killing swans, geese, ducks, &c., in the moulting season. All the tools used by an Indian in building his canoe, as well as in making his snow shoes, and every other kind of wood-work, consist of a hatchet, a knife, a file, and an awl, in the use of which they are so dexterous that everything they make is executed with a neatness not to be excelled by the most expert mechanic, assisted with every tool he could wish."[1]

The lightest and most easily-made boats in Guiana are "woodskins," made of the bark of the locust (*Hymenæa courbaril*), or the purple heart (*Copaifera pubiflora*). A strip of bark of sufficient length is first carefully taken from a tree and cut to an oblong shape, the natural curve being accurately preserved. About two or three feet from each end a wedge-shaped piece, or gore, is cut from either side of the bark, the ends bent up until the edges of the gores meet, when they are sewed together with "bush rope." This process raises the bow and stern at an angle, while the body of the craft floats parallel to the water-line. Sticks of strong wood are sometimes fastened around the gunwale. Pieces of squared bark are laid on the floor to serve as seats for passengers or rests for goods, and the craft is ready. These canoes are so light as to be portable around falls or obstructions to navigation. When not in use they are sunk in the water to prevent splitting or warping under the action of the sun. Paddles are hewn out of solid block or out of the board, like natural buttresses of the paddle tree (*Aspidospermum excelsum*). These paddles differ in form from tribe to tribe.[2] On the Columbia river the Callispels, and on the Amoor the Giliaks, cut the gore so as to make the canoe bow and stern pointed under water.

To make a bark canoe the Dyak goes to the nearest stringy bark tree, chops a circle round it at its base, and

[1] Hearne, *Journey, &c.*, London, 1795, Strahan, p. 96.

[2] *Cf.* im Thurn, *Ind. of Br. Guiana*, London, 1883, p. 296.

another circle seven or eight feet from the ground ; he then makes a longitudinal cut on each side and strips off as much bark as is required. The ends are sewed up carefully and daubed up with clay, the sides being kept in position by cross pieces. The steering is performed by two greatly developed fixed paddles.[1]

The natives of Gippsland, Australia, make a boat of a single sheet of the *Eucalyptus sirberiana*, the ends being tied up. The interior people use for the same purpose the bark from the convex side of a crooked tree, and stop the ends with balls of mud. They are propelled by poles and by means of a circular piece of bark, six inches in diameter, which is used as well to bail out the canoe. Two men with six hundred pounds of flour will cross a lake in one of these frail craft.[2]

The lumbermen among savages were no mean craftsmen. They knew the quality of every kind of tree around them, and what its bark and timber were good for. Their art consisted in felling the trees, splitting them into the proper lumber, working this down to the desired object, and transporting either material. Their work was felling, riving, dressing, excavating, boring, smoothing, and carving wood. The Indians, the Negroid races, and the Malayo-Polynesians were, each in its way, most excellent wood-workers. Living on the sea or in the interior, they achieved remarkable results.

"When the American Indians intended to fell a thick, strong tree," says Kalm, "they set fire to a quantity of wood at the roots of the tree. But that the fire might not reach higher than they would have it, they fastened some rags to a pole, dipped them into water, and kept continually washing the tree a little above the fire.

"Whenever they intended to hollow out a thick tree for a canoe they laid dry branches all along the stem of the tree, as far as it must be hollowed out. Then they put fire to

[1] Ling Roth, *J. Anthrop. Inst.*, London, 1892, vol. xxii. p. 51.

[2] Consul-General Wallace, quoting Rev. John Bulmer.

those dry branches, and as soon as they were burnt they were replaced by others. Whilst these branches were burning, the Indians were very busy with wet rags, and pouring water upon the tree to prevent the fire from spreading too far. The tree being burnt hollow as far as they think it sufficient, or as far as it could without damaging the canoe, they took their stone hatchets or sharp flints, or quartzes, or sharp shells, and scraped off the burnt part of the wood and smoothened the boats within. A canoe was commonly between twenty and thirty foot long."[1]

Dr. J. F. Snyder, who in 1850–1852, was living in California, saw the Indians of the north-western portion of the state fell a tree with stone axes. They began by hacking in through the bark, and a few of the annual layers with the edges of the axes, above and below a scarf two feet wide. With the butt end of the same axes they bruised these annual layers all around until they could work off a thin slab by means of elk-horn wedges. They then hacked in as far as possible at the top and bottom of the scarf, pounded with the butt of their axes, as before, and removed another slab. This process they continued until the tree was felled. The work was done by the combined and continuous labour of many men.

The joiner comes from a very ancient stock. In the Chapter on Tools mention is made of the inventions for clamping things temporarily, and also of the methods of effecting the permanent union of separate parts of an industrial product. Furthermore, the excavation of wood for all purposes was more fashionable at the first. This is so far true that all the north-west tribes of America do not make the sides of a box of separate pieces, but having taken a piece of board long enough for all the sides, they divide it by bevelled kerfs, cut crosswise nearly through. They then steam the board, bend it so as to have the kerfs inward, and unite the ends carefully with pegs. The work is very neatly done, so as to make the outside seem continuous.

[1] Kalm, *Travels, &c.*, London, 1771, vol. ii. p. 38.

The carpenter and the whole class of house and furniture makers must look a long way back for ancestors. One of the most instructive chapters upon the savage man in partnership with Nature in the evolution or elaboration of an invention is that upon house building among the Guiana Indians by im Thurn. "The houses are everywhere equally simple in structure, for the materials are everywhere much the same. Such differences as exist have evidently arisen in consequence of natural efforts to meet the special requirements of each kind of situation. Three chief types of houses are distinguishable. The Warrau built on piles over water; the Arawak and Caribs, sheltered from cold winds by the surrounding trees, built wall-less houses; on the open savannah the Macusi erected habitations with thick walls of clay as a protection against the cold winds from the mountains. On hunting expeditions the natives erect temporary shelters or benabs, which may be a few leaves of some palm, laid flat one upon another, and the stalks, which are bound together, are stuck in the ground at such an angle that the natural curve of the leaf affords some shelter. A more pretentious benab is made by sticking three poles upright in the ground in the angles of a triangle, joining the tops by means of three cross-sticks, and laying over these a bunch of palm leaves. Thus it is easy to trace the development of house building among these Indians.

"Their refined knowledge of plant craft is shown in the selection of the plants for thatch. The different leaves used do not signify different tribes at all, but each is the most easily attainable in its district. A trade was once carried on between the Indians and planters for the troolie palm (*Manicaria saccifera*) leaves. It is a significant sign of their profound botanical knowledge that the savages have learned to adulterate the laths to which the leaves are attached by substituting the booba for the manicole palm because the former is more easily procured."[1]

House building was a distinct trade in Samoa. On an

[1] Im Thurn, *Ind. Br. Guian.*, London, 1883, cap. ix.

average one among every three hundred men was a master carpenter. He had under him ten or twelve journeymen, who expected pay from him, and apprentices learning the trade. When a young man took a fancy to the trade he had only to go and attach himself to the staff of a master carpenter, and when he could point to a house that he had built, that set him up as a professed contractor. If a person wished a house built he offered a fine mat to a carpenter. If the latter accepted the mat he undertook the job. At an appointed time the carpenter came with his journeymen and apprentices, armed with axes and adzes having blades of shell and stone. The house owner provided board and lodging, and he, assisted by his neighbours, did the carrying and lifting. No price was fixed beforehand; it was left to the judgment, generosity, and means of the employer; but it was a lasting disgrace on any one to have it said that he treated the carpenter shabbily. The entire tribe or clan was his bank, upon whom he might make demands. If the carpenter from any cause decamped, no other carpenter would finish it. The employer must come to terms with the contractor.[1]

An example of aboriginal versatility is furnished in the efforts of the Tahitians to improve their dwellings to suit the exigencies of their new mode of life under the missionary. Ellis describes the lime-burning from coral in open ovens like those used in cooking in the Leeward Group. "It was no easy task for them to build houses of this kind. Every man had to go to the woods or the mountains and cut down trees for timber, shape them into posts, &c., remove them to the spot where his house was to be built, then erect the frame with doorway and windows. This being done, he must again repair to the woods for long branches of hibiscus for rafters. The leaves of the pandanus were next gathered and soaked and sewed on reeds, with which the roof was thatched. This formerly would have completed the dwelling, but he had now to collect a large pile of firewood, to dig a

[1] *Cf.* Turner, *Samoa*, London, 1884, p. 157.

FIG. 41.—WOODCARVING IN WEST INDIES, ONE-TENTH SIZE.
(*U. S. Nat. Museum.*)

lime pit, to dive into the sea for coral rock, to burn it, to mix it with sand so as to form mortar, wattle the walls and partitions of his house, and plaster them with limes. He then had to ascend the mountain again for trees which he must either split or saw into boards for flooring his apartments, manufacturing doors, windows, shutters, &c.

Their invention and perseverance overcame the lack of nails, and they constructed their doors by fastening together three upright boards by means of three narrow pieces across by strong wooden pegs. The substitutes for hinges were also worthy of the most primitive inventors. Ellis's description of the new style of architecture in progress is equal to Virgil's account of New Carthage.[1]

The cabinet-maker and the wood-carver in house-building were actively at work in the polished stone age at least. An excellent example of fully-developed, geometric ornament is to be seen in a paper by James Sibree on types of carved ornamentation in wood employed by the Betsileo Malagasy in their burial memorials and their houses.[2] In discussing this paper Mr. Balfour said that "an examination of a large number of examples may reveal an interesting series of transitions, showing the evolution of the conventionalised, purely meaningless though decorative forms." But these earlier forms were not forthcoming, so the series appeared as the end of a development.

In the great timber belt of South-eastern Alaska and British Columbia were developed handy lumbermen, and, indeed, the same is true of any other well-timbered area lying near the sea. The houses of these various stocks are communal. They are built of immense logs and puncheons. The trees, after being felled, were carefully split into planks

[1] Ellis, *Polynesian Researches*, London, 1859, p. 345–49. The Navajo Indians have been modifying their hogans to harmonise with the fashions set by different army headquarters built about their reservation.

[2] Sibree, *J. Anthrop. Inst.*, London, 1892, vol. xxi., pls. 16 and 17. These marvellous patterns should be compared with those of Polynesia and middle America.

and dressed down with adzes and chisels of stone. The carpenter and the wood-carver had full opportunity for the development of their talents. But it was upon their canoes that the natives spent most time and skill. Among the boat-building Californians and West Coast people, as soon as the tree was felled, the top was burned off at a proper distance to allow plenty of log for the dug-out canoe.

The outside of the craft was hewn to proper shape by means of stone axes and adzes, and all who have seen the work of these stone tools have been astonished at the regularity of the little polygonal scars looking like an engraving over the entire surface.

The hollowing out was done by burning. Fat pine knots were gathered in the greatest abundance, and little fires were kindled on the upper surface of the log. To feed them, to check their course by means of green bark and mud and water, to remove the fires when the ashes at the bottom checked their course, to broom away the *débris* of the flames, and, with flat, circular, or leaf-shaped flints, to dig out the charred portions down to unburnt wood, constituted a round of labour whose quick repetition would rough out the canoe.

Nowadays we should proceed with augers instead of little fires, and adzes of steel and mallets and chisels; but every one of us has seen the country blacksmith boring holes with a hot iron rod, and, furthermore, there were no augers in those days. The borer with fire had to come before the auger.

The log once hollowed, or during the last steps in hollowing it, the naval architect busied himself about his lines. He knew by a kind of cruel "selection" which the sea had been practising upon his ancestors that the fittest crafts survive. He did not reason it out in that way; he thought that his gods required him to build thus and so, or they would be angry with him and send him to the bottom. It amounts to the same thing; the voice of Nature is the voice of God, and the ship carpenter went to work to shape his craft. An old sea captain, J. W. Collins, of the United States Fish Commission, who is considered the best authority in the world on the building of fishing vessels, informed the writer one day, as we were looking at a splendid specimen of these West Coast cedar dug-outs, that he could hardly improve on her lines for the water in which she had to work.

To effect this object, that is, to get the hollow log into ship-shape, the boatwrights required these—plenty of water and hot stones. The log was filled with water, and all her chinks stopped with shredded bark and hot pitch. Red-hot stones were thrown into this queer cauldron, and the water kept at the boiling point until, with spreading and contracting, the gunwale had exactly the right curves at every point, and was securely lashed. The water and stones were removed, the vessel dried out, and the polishing and painting completed the operation. This craft was moved by means of paddles with crutch handles, giving the rower great power as he dug his way through the water. The number of rowers was limited only by their convenience in standing or kneeling. No rudder was used, and generally no steering paddle different in form from the rest. For sails, mats made of the shredded bark of the cedar, similar in form and texture with those laid on the floors of the long houses, were fastened to a cross-yard, and at their lower corners were held by sheets of cedar bark rope. These sails were used only in going before the wind, and the navigators were never so expert as those of the Polynesian Islands. The outfit of the craft will be noticed in a chapter on the capture of animals. It will be sufficient to say that the fish-hooks are carved from wood; and the club for killing halibut is often a work of art. The lines are excellent twine of native hemp and cedar bark and spruce root. The boxes for holding implements and clothing and the images of the gods on the bow and the stern of the boat do credit to the skill of the cabinet-maker and the sculptor.

"The probable cause of the absence of boats in Central California is the scarcity of suitable, favourably-located timber. Doubtless, if the banks of the Sacramento and the shores of San Francisco Bay had been lined with large straight pine or fir trees, their waters would have been filled with canoes." [1]

In the canoe or pirogue country of Columbus and his compatriots, the wood-workers were men. The axes, scrapers, and chisels of stone, which once formed their whole stock of wood-working implements, have given place to the steel axe, the cutlass, and the knife. And the Indian is capable of building a house, hewing a beautiful neat boat, stool, or other such article from a block of wood without the use of any other implement beyond his axe and cutlass. When a canoe is to be made a suitable tree is carefully sought in the forest, often as much as two miles from the nearest water. The tree is felled, and roughly hewed on the spot into the shape of the required canoe. It is then hollowed partly by chopping and partly by burning. A path through the bush to the waterside is then cleared and laid with cross pieces as runners, or like a corduroy, and the canoe is dragged to the waterside. The sides of the boat are forced apart in several ways. Sometimes the canoe is inverted over a fire till the action of the heat spreads the sides; sometimes it is filled with wet sand, the weight of which eventually forces the sides outward; and sometimes the canoe is sunk in running water, and when the wood is pliant, the sides are forced asunder by driving large wedge-shaped pieces of timber in between them. As soon as the sides have been spread, bars of hard wood, about an inch and a half in diameter, are fixed firmly across within the canoe from side to side, so as to prevent the sides from approaching each other. Two triangular pieces of plank-like wood are then cut and fitted into the gaps at bow and stern. The sides and ends are raised by the addition of a plank or extra "streak." The seams are caulked with shreds scraped from the inner bark of certain trees, and patched with resin (from *Icia hetaphylla*) or with karamanni, an adhesive

[1] Bancroft, *Native Races*, New York, 1874-6, vol. i. p. 385.

pitch or glue from the *Siphonia bacculifera*. The Caribs of St. Vincent make canoes in much the same manner.[1]

The making of a canoe, from the first act of selecting a tree in the wilderness to its final consecrating and launching when fully rigged, was in Hawaii, at all times and at every step, under the watchful eye of the *kahuna*, whose duty it was to see that no pains or expense was spared, no ceremony omitted, to propitiate the favour of the gods who had the power, if so desired, to bring good luck to the *waa* and all who might sail in it.

It should be borne in mind that the various migrations or *hekes* of the ancient Polynesians, and their progenitors, must have been accomplished in canoes, and that the *waa*, the *pahi*, &c., of historic and modern times are the lineal descendants of the seagoing craft in which the early ancestors of these same people made their voyages generations ago.[2]

The Polynesian canoe is described by Ellis. The keel was formed with a number of tough pieces of temanu wood (*Inophyllum callophyllum*) twelve or sixteen inches broad, and two inches thick, hollowed on the inside and rounded without, so as to form a convex angle along the bottom of the canoe; these were fastened together by lacings of tough elastic cord made from the fibres of the cocoanut husk. On the front end of the keel, a solid piece, cut out of the trunk of a tree, so contrived as to constitute the forepart of the canoe, was fixed with the same lashing; and on the upper part of it, a thick board or plank projecting horizontally in a line parallel with the surface of the water. This front piece, usually five or six feet long, and twelve or eighteen inches wide, was called the nose of the canoe, and without any joining comprised the stem, bows, and bowsprit of the vessel.

The sides of the canoe were composed of two lines of

[1] *Cf.* im Thurn, *Indians of British Guiana*, London, 1883, p. 292.

[2] Consult N. B. Emerson, "The Long Voyages of the Ancient Hawaiians," *Hawaiian Hist. Soc.*, May 18, 1893. Papers, No. 5.

short plank, an inch and a half or two inches thick. The lowest line was convex on the outside and nine or twelve inches broad : the upper one straight. The stern was considerably elevated, the keel was inclined upwards, the lower part of the stern was pointed, while the upper was flat, and nine or ten feet above the level of the sides.

The whole was fastened together with sennit, not continued along the seams, but by two, or, at most, three holes made in each board, within an inch of each other, and corresponding holes made in the opposite piece, and the lacing put through from one to the other. A space of nine inches or a foot was left, and then a similar set of holes made. The joints or seams were not grooved together, but the edge of one simply laid on that of the other, and fitted with remarkable exactness by the adze of the workman, guided only by his eye ; they never use line or rule. The edges of the plank were usually covered with a kind of pitch or gum from the breadfruit tree, and a thin layer of cocoanut husks spread between them, which swell when in contact with water, and fill any apertures that may exist.

The two canoes were fastened together by strong curved pieces of wood, placed horizontally across the upper edges or gunwales, to which they were fixed by strong lashings of thick coiar cordage.[1]

In Samoa any one could make a common fishing canoe to hold one or two men. But the keeled canoe was the work of professed carpenters. The keel was laid in one piece twenty-five to fifty feet long, and to that boards were added, by sewing each close to its fellow, until the sides were raised about three feet from the ground. These boards were a number of pieces and patches eighteen inches to five feet long, as the wood split up from the log might suit. Each board was hollowed like a trough, leaving a rim all round the edge which was to be inside. And through these rims

[1] Ellis, *Polynes. Researches,* London, 1859, vol. i. p. 155, with further account of the varieties of canoes and of the manner of propulsion.

where they joined they bored holes, and with sennit they sewed one board to another. The sewing, therefore, appeared only on the inside. All was well fastened together, and, with the help of gum from the breadfruit tree, made perfectly water-tight. Timbers, thwarts and gunwale were added to make all firm, and a deck built over the bow and the stern, under which things could be stored. As the width of the canoe was only eighteen to thirty inches an outrigger was necessary. This was made by fastening beams across the canoe, so that they might project some distance out from one side. To the end of each projecting beam were made fast small sticks descending toward the water, and to their lower ends was fastened a long thin piece of wood sharp at the end to cut through the water, and floating on the surface parallel to the boat.

The canoes were propelled by paddles, not by oars, the rowers facing the bow. The sails of matting were triangular, set with the base upward. Rows of white shells were used in decorating the craft, and carved images of human beings, animals, or mythic beings, formed the figure-head.[1]

The Dyaks hollow their canoes out of a single log by means of fire and adzes. They are guided only by the eye. When the shell is completed thwarts are inserted, and planks or gunwales are stitched to the sides, the seams being caulked with sago stems which are porous and swell when wet. Each of the gunwales is laced on by flaxen cords, and united to the opposite plank by the thwarts. The canoe is alike at both ends, which are pointed and curved upward. There is no keel, nor ribs, nor figure-head.[2] Man states that the average time spent on the Andamanese dugout canoe is that of about eight men for a fortnight.

[1] *Cf.* Turner, *Samoa*, London, 1884, chap. xiv.

[2] Ling Roth, *J. Anthrop. Inst.*, London, 1892, chap. xxii. p. 51.

CHAPTER VII.

THE TEXTILE INDUSTRY.

> " Who can find a virtuous woman? for her price is far above rubies.
> She seeketh wool and flax, and worketh willingly with her hands.
> She layeth her hand to the spindle, and her hands hold the distaff."
>
> *Proverbs of Solomon.*

THE Apaches call the Navajos spiders, in allusion to their beautiful weaving ; and on the breasts of skeletons in the mounds of Tennessee have been found shell gorgets upon which spiders are engraved.[1] It is not known that the person whose skeleton was there entombed had been really a weaver, but copper implements wrapped in coarse cloth were found hard by, and on much of the pottery exhumed from the Tennessee graves are marks of basketry. There is no objection to calling the spider gorget the trade badge of the dead one.

The textile art is older than the human species. For not only spiders and many caterpillars drew out extremely fine threads, but birds wove nests long before man's advent on earth. And, most significant of all, in tropical lands especially, trees and plants fabricated cloth, which men have worn from time immemorial, and on it they have also preserved their thoughts. There is no reason to doubt that the very first women were weavers of a crude kind, and that the textile art has been with us always in one form or another.

[1] For illustration of spider on shell gorget, see Holmes in *Third An. Rep. Bur. Ethnol.*, p. 466, fig. 141.

This department of industry is to be studied in its materials, in the tools and processes employed, in its products as to their form and use. Without committing ourselves to any theory of evolution in the art, it will be convenient to notice first the various types of bark cloth, basketry, and matting as not involving the spindle. Then the textile art proper may be considered under the topics of yarn, thread and braid, weaving, looping, netting, and embroidery. All of this will serve to show that the savage has not been idle in the development of fibres.

Mela says that the Germans were clothed in winter only with the *sagum* (a kind of poncho), or with the bark of trees.[1] It is difficult to understand what is meant by this phrase. The Germans lived too far north to be successful in making their shirts from the bast, or inner bark, of trees indigenous to Northern Europe. The Aryan race as a whole are not known to have clothed themselves thus. The bast of temperate zone plants, hemp, flax, cedar, has a fibre that shreds easily into filaments which may be spun.

Bark cloth may be seen in any museum from Central and South America, from Africa and from Polynesia. The art of pounding the bast, or inner bark of a tree, whose fibrous strands do not lie parallel, as in textile plants, but are interlaced inextricably, is of very wide extent racially. Three out of the five great types of mankind practice it—the American, the Negroid, and the Malayo-Polynesian. The cloth, so called, manufactured in this way is quite durable, and much more expeditiously made than any produced by weaving.

In Polynesia the bast of the paper mulberry and of the breadfruit is chiefly used. The process involves several discoveries and inventions worthy of notice. The outer rind was scraped off with a shell ; the inner bark was then slightly beaten and allowed to ferment, or was macerated in water. Upon a long, smooth log the bast was then beaten with a heavy mallet of *Casuarina* wood. This mallet was

[1] Mela, *De Situ Orbis*, vol. iii. p. 3.

square in cross section, and each side grooved—one with very coarse, the opposite with fine ridges; a third side was sometimes plain, a fourth cut into checker patterns. Each of these sides was useful in spreading out the texture, removing the pulpy matter, and giving the appearance of textile work. The cloth was dried and bleached in the sun. For colouring a variety of vegetable dyes were used. Nature supplied the pattern. The natives selected some of the most delicate and beautiful ferns or the hibiscus flowers. When the dye was prepared the leaf or flower is laid carefully on the dye. As soon as the surface was covered with the colouring matter the stained leaf or flower was fixed on the cloth and pressed regularly down. Many of the patterns were printed on in regular designs, worked out on a surface of palm leaf with little bits of stem sewed on in geometric figures. The stronger kinds of cloth were covered with a brown or black varnish, which made the texture tougher, and also waterproof.[1]

For colouring their bark or tapa cloth, the natives of the Society Islands used a variety of vegetable dyes. Among them the bark of the *Casuarina* and *Aleurites*, giving a dark red or chocolate colour. Brilliant red was prepared by mixing the milky juice of the berry of *Ficus prolixa* with the leaves of a species of *Cordia*. When prepared the dye was absorbed on the fibres of a kind of rush and dried for use. When covered with a varnish of gum it retained its brightness until the garment was worn out. Yellow was prepared from the inner bark of the root of *Morinda citrifolia*. An infusion of the bark in water was made, and the cloth soaked therein and then dried in the sun.[2]

The tapa beating of the Hawaiians is minutely described in a catalogue of the Bishop Museum in Honolulu. The material is the bark of the paper mulberry (*Broussonetia papyrifera*), but half a dozen other plants are occasionally

[1] Ellis, *Polynes. Res.*, vol. i. pp. 178–185, giving a detailed and excellent account of tapa beating.

[2] Ellis, *op. cit.*, vol. i. p. 182.

substituted. The apparatus consisted of a log or anvil, a series of mallets, calabashes of water and of mucilage, dye stuffs, and tools or stamps for decorating. The anvil was a log of hard wood, about six feet long, hewn to a flat surface three inches wide at top, and hollowed longitudinally under neath. This was supported on two stones. Of clubs there was a variety ; round for the first beating, and flat for the finishing. Ornamented beaters had longitudinal grooves cut at varying distances apart on the four sides, and in others these parallels were cut by other grooves at right angles, or undulating patterns were engraved. From scions of the paper mulberry the bark was stripped in lengths of about six feet, and two inches wide. These strips were dried until the sap was wholly evaporated, and they were then stored for future use, either with the outer bark still on or after this had been removed by scraping on a smooth board with a plate of shell or of bone.

When to be used the strips were soaked in water and beaten with the round club on a smooth stone until the fibres were felted together. The strips were then soaked again and beaten on the log, strip was welded to strip until sheets of a surface 125 square feet was obtained. The peculiar "water marks" to be seen in tapa were given by the patterns carved on the decorated beaters. The colouring of the tapa was effected in various ways. Previously pulped material of various colours was beaten in, or the fabric was dyed after it was beaten. When the pigments were to be applied locally they were ground in oil in a stone mortar and applied by cords, by pens, by brushes, and by dies. Waterproof kapa or tapa was saturated with the oil of the cocoanut.[1]

The Hawaiians excelled all other peoples in the world in their bark cloths. This remark applies to the fineness of the product, the diversity of lines beaten into the texture, and to the variety of ornamentation added to the surface. Many of the varieties have special names, and a

[1] *Cf.* Brigham, *Cat. Bishop Museum*, Honolulu, 1892, p. 23.

collection of them all would fill an enormous album, giving one page to each.

The bark cloth of Africa and of Tropical America are much simpler in construction than that of Polynesia. In neither of the continents is there any overlaying or uniting of pieces. Each garment is made, so to speak, of the whole piece. "The Warraus of Guiana make the *lap* and the *queyu*, the breech clout and the 'fig leaf' apron, of the inner bark of the *Lecythis ollaria*, which is beaten until it is comparatively soft, and of the texture of thick rough cloth."[1] All over the Andes the tribes use this bark cloth as a body on which they sew feathers, teeth, beetles' wings, bones, and other decorative objects.

In making the African varieties a piece of bark about six feet long, and as wide as possible, is detached from the trunk of a tree. The outside rind is pared off with a lance-head used with two hands, like a cooper's drawing knife. The bark is then laid upon a beam or wood, on which it is hammered with a mallet grooved in fine cuts, so that the repeated blows stamp the bark with lines somewhat resembling corduroy. This cloth turns brown by exposure, and is dyed or ornamented in black with water from iron springs. Uganda is celebrated for this curious production."[2]

By the general term "basketry" is meant all kinds of woven vessels in which the materials are not spun. But there is also a large class of flat textiles, made up precisely after the same fashions as basketry, commonly designated "matting." Basketry and matting together constitute a most important division of savage invention. They are the one art that is more beautiful among the uncivilised. Enlightened nations express their æsthetic conceptions in lace and cloths and embroideries, the savage woman gives vent to her sense of beauty in basketry.[3]

[1] Im Thurn, *Indians of British Guiana*, London, 1883, p. 194.

[2] Sir Samuel W. Baker, *Ismaïlia*, New York, 1875, pp. 328, 350.

[3] For a minute description of all the styles of savage basketry see the author's paper in *Smithsonian Report*, 1883, part ii., pp. 291–306, pl. l.–lxiv.

To the unobservant, a basket is a basket, and that is all there is about it. But to the technographer the materials, methods, and products of this art form an excellent guide to peoples or tribes. Almost every type of basketry is confined to a single tribe, or to a very restricted area. From a textile point of view baskets are divided into two great classes—the woven and the sewed or stitched—and it will be necessary to look with a little care at each to comprehend how much original human thought has been bestowed upon this industry, both in very ancient times and in our own.

The most simple form of woven basketry can be seen in the work of the Algonkian and Iroquoian Indians of the Eastern United States, made of thin strips or splints of ash, beech, oak, or hickory, all of uniform thickness and width, and forming a rude warp and woof like the threads in common Manchester cotton stuff.

FIG. 42.—Plain Weaving in Cedar Bark, coast of British Columbia.

Improvement on this very plain style, seen to perfection in the cedar-bark stuffs of North-west America, may be made by varying some of the strips or splints in width and colouring a portion of them. This would give the beautiful Polynesian palm and pandanus ware, a great deal found in America, besides much of the mat-work the world over.

The next step, the next patent we should say, consisted in carrying the weft splints over and then under two or more warp splints at a time. Beautiful examples of this are to be found in Guiana and Northern South America, and it is exquisitely done in African matting. The Salishan tribes of Washington make thin, narrow splints of birchwood, from which to weave their wallets for holding fish ; while the Cherokees, Choctaws, and other Southern Indians of the United States employed the split cane dyed in two colours. But they all understood the overlapping, producing a diagonal or diaper pattern.

Still keeping within the limits of plain weaving, there is a quite different effect, called wicker-work, which one may see, in some material, in any market-house in the world. It seems to be the universal heritage from savagery.

This work consists in using a rigid warp and a flexible weft as in corded silk weaving. The Moki Indians of the Pueblo of Oraibi gather the twigs of the "grease bush" and strip off the rough bark. This yields filaments only a few inches long, but by colouring these and weaving them in and out over a warp of twigs the Moki produce a basket-tray which is highly prized by collectors ; but the manipulation is precisely similar to that in our coal and oyster baskets and the old-fashioned farm crates, or the wicker-work around a demijohn. The Zuñi Indians make their peach-baskets in a similar fashion, and there is little doubt that they and the Moki were instructed by the same teachers.

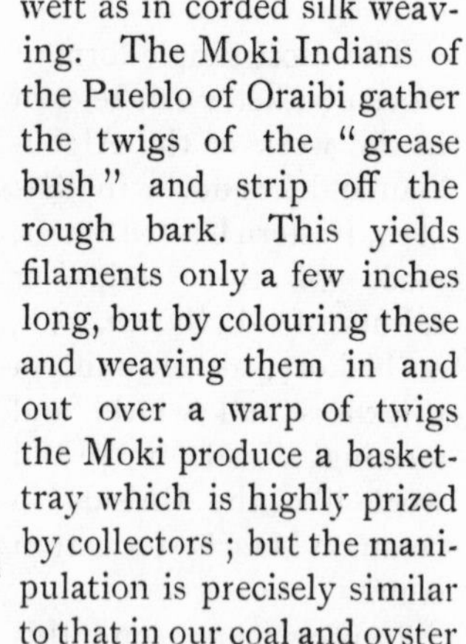

FIG. 43.—Diagonal Weaving, common in many American tribes.

It is a fact worthy of attention here that this one Moki Pueblo of Oraibi in North-eastern Arizona is the only spot west of the Rocky Mountains where this wicker-work is practised. The people belong to the Shoshonean stock, who, outside this Pueblo, practise another style. In fact, the Oraibi Indians make two kinds of baskets essentially different from those of their blood kindred, the Utes, and they do not weave the Ute basket at all.

There is a style of woven basketry that a patent examiner might declare to be derived from the wicker just described. Indeed, it is a kind of double wicker. But its stitch or mesh is found on the oldest pottery, in very ancient graves, and in the Kilima-Njaro country, as well as all over America. The most simple and rudimentary specimen of

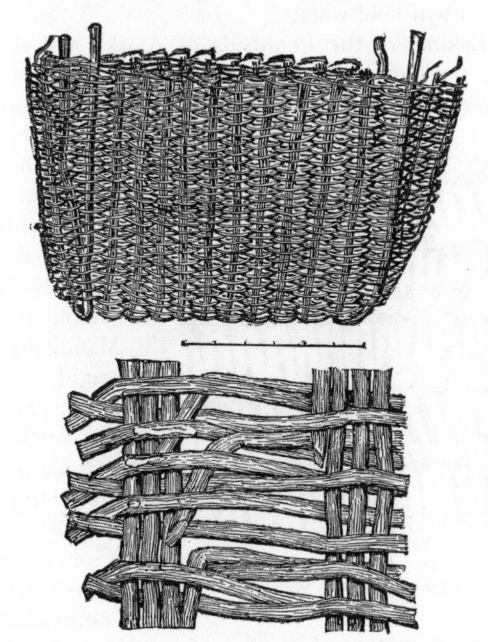

FIG. 44.—Wicker Basket of the coarsest type. Zuñi Indians, with detail. (*U. S. Nat. Museum.*)

this type is the wattling fence, in which two pieces of brush are woven among a row of stakes and twisted into a two-ply rope at the same time. In basketry this is called twined-weaving. Two weft fillets or twigs are carried along at the same time between the warp elements, only they alternate in passing each other above and below so as to make a twine. If all the warp sticks were pulled out, these two weft strands would be twined together, as in a two-ply thread, continuously from beginning to end.

If the elements are whole, or split osier or other twigs, the work will be open and strong, as in the Pacific Coast large basket. If the elements are fine root, or grass, or bast, or spun thread, or yarn, the work will be fine and close, like cloth, and will hold water.

The Eskimo of the Peninsula of Alaska and the Aleuts produce most dainty ware in this stitch from the stem of the *Elymus*, while their coarser work resembles so closely specimens brought home by Dr. W. E. Abbott from Kilima-Njaro that only an expert would detect the difference by the material employed. From the Mandingo country also rigid baskets are in the same style.

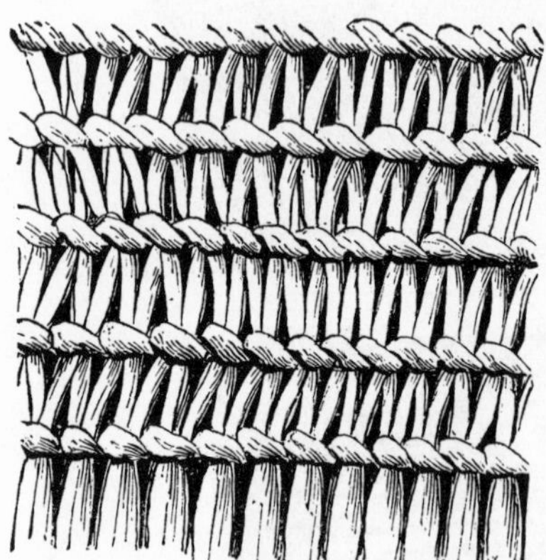

FIG. 45.—Aleutian Basketry, showing twined weaving on split warp filaments.

The North American Indians, living wherever the *Apocynum cannabinum* grew, made soft pliable sacks or wallets in this stitch, and impressions of it are found on pottery fragments from the Atlantic states, and robes of turkey down and of rabbit skin in the same pattern come from the cliff dwellings. But the most exquisite productions in this stitch are to be seen on the West Coast of the United States, from Mount St. Elias to the Bay of San Francisco, where it suddenly gives place to a variety entirely different in structure.

The T'lingit ware about Sitka may be taken as the best representative of twined basketry. It is made of the roots of the spruce, carefully cleaned and split. Those who have seen the operation say that the women use no other tool than a mussel-shell and no other gauge than their thumb-nail to secure uniformity in the splints. This is almost incredible, so regular do the surfaces of the best pieces of basketry appear. The warp is set up with a view to the size and form of the wallet to be made. The weft splints are then twined between the warp splints and pushed as close home as possible, producing a watertight structure. Bands and lines of ornament are produced by means of dyed splints. When two strands of different colours are worked together the alternate appearance and disappearance of each one gives a spotted line, and the management of the succeeding lines with respect to this one gives the artiste an unlimited resource of decoration.

But this is not the end of her tether. An inspection of a piece of this ware will show that every stitch in the weaving is double, one part being outside, the other inside, the warp splint. It is possible, therefore, by means of coloured grasses and bark to embroider the outside of the vessel without affecting the inside, and this indeed is done in such manner as to produce most wonderful effects.

The tribes of Vancouver Island and of Cape Flattery vary from this style somewhat, in that they have three sets of elements; but two are rigid and only one is flexed in weaving. The process is exactly that of wire bird-cages. An upright series of fillets form the warp, a second piece is carried around inside and coiled against the warp, and the third is wrapped around the other two spirally at every point where the coiled piece intersects one of the warp pieces. This gives to the outside the same appearance as is to be seen on the back of a watch. By using different coloured grasses, geometric patterns of great variety and beauty are produced.

In all the range of basketry there is none other so pretty as that of the northern tribes of South America, made nowadays by men. The beauty of the work lies in its extreme chasteness of design and colour. The body is done in splints of the reed-like stem of maranta (*Itchnosiphon*). These are very uniform in size and are woven in diaper and diagonal fashion, the brownish or amber-colour of the wood being variegated with simple but very decorative geometric figures in black. There are multitudes of tribes in other parts of the world who make finer baskets and put more work upon them, but for chaste beauty those of this region excel.

The Andamanese basket is made of the rattan or cane of the country. The stalks are cut into lengths of about four feet, the cuticle is peeled off and divided into narrow strips of uniform width, and the remainder is split up into convenient pieces. The style of weaving is similar to the American. The maker gives a "kick" to the bottom of his basket by scooping a hole in the ground, pressing the framework into it with his heel until he has proceeded far enough, the frame sticks are then reversed and the work goes on to completion. The rims are finished off with a piece of *Uvaria micrantha*, and the handles made of the bark of *Melochia velutina*.[1]

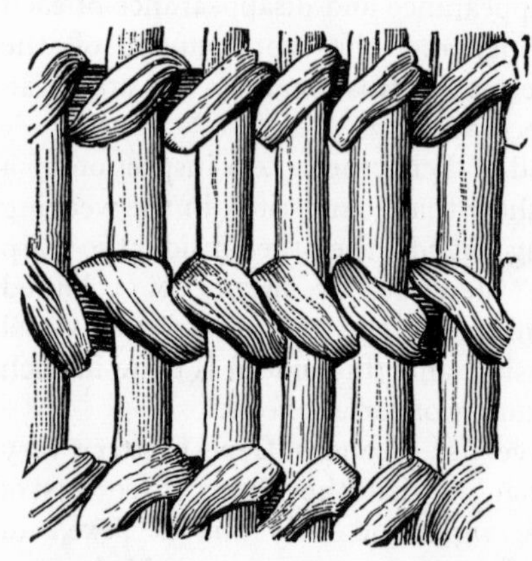

FIG. 46.—Detail of "Birdcage" Stitch in Basketry, Vancouver Island. (*U. S. Nat. Museum.*)

The second class of basketry work is the coiled or sewed ware. Its affinities are not with weaving at all, but with sewing. The savages, in making garments of skin, whip two edges together, as carpet-sewers do, by means of sinew thread. That is, the sewing progresses in a continuous spiral of thread.

[1] Man, *ut supra*, p. 162.

After this fashion the peoples of the Far East, of every stock in Africa, and in all parts of America make coiled basketry, the material each time modifying the method of working and the appearance. The best idea of this class of work can be gathered from the simplest examples, the first patents, to use the modern phrase. These may be seen in Further India or in North-western Canada among the Athapascans. The elements are a stiff root or rod for the fundamental coil, and a soft splint or strip of the same material for the sewing. In making her basket, the woman starts in the centre of the bottom, coiling the rod and wrapping it as she proceeds with the split root or rattan, so as to bind it to the preceding turn, drawing her splint between the spirals. When the rod comes to an end, she neatly splices the end to that of a new one and proceeds as before, carefully concealing the joint. When the splint is exhausted, the end is tucked in behind the spiral and another one started in the same manner, but so carefully joined as to defy detection. The Siamese do not decorate ware of this kind, and much of the spruce-root coiled ware from Canada and Alaska is severely plain. But further south in British Columbia and Washington, bits of birch-bark, straw, or quill are doubled over the splint on the outside of the basket and the two ends concealed under succeeding stitches so as to give an imbricated effect in many colours. Nothing in basketry could be more beautiful, and when it is remembered that every stitch is covered by one of these loops, some conception may be formed of the immense number in a single piece.

Now, to vary the foregoing, some tribes insert a strip of tough material between the coiled rods and reeve the sewing splint between this and the preceding coil. This makes a water-tight joint ; so ware of this sort is commonly used for boiling food by means of hot stones. The Eskimo use a small wisp of grass for a rod. The Oregon Indians use osier and *rhus*, and the California tribes, who make the most beautiful ware in the world, employ *Vilfa*, *scirbus*, *salix*, and pine-roots. All their basket botany has not been made out.

The wonderful uniformity of the coil and the sewing splint in the California basketry is not due to the possession of delicate machinery for dressing the material. Delicate machinery was devised to make things cheap. Knack and a strong thumb-nail achieve the result. A bone awl is the needle, a true eye, a genuine pride in her work, and a skilful hand do the rest. Patience indeed has her perfect work, for there is one of these beautiful baskets in the United States National Museum which employed a Hupa woman every spare moment during three years to finish it. In the Moki Pueblos, all over Middle America, and throughout Africa, finely stripped yucca or palm-leaf is used in sewing, but the method is the same. Decoration is effected by dyeing portions of the strips used in sewing, by employing the front and the back of certain leaves alternately and by choosing straws and leaves and other sections of plants of different colours. Maidenhair fern, martynia pods, rushes, grass stems, fibre soaked in muddy water, and all sorts of expedients are resorted to in order to produce the greatest embellishment.

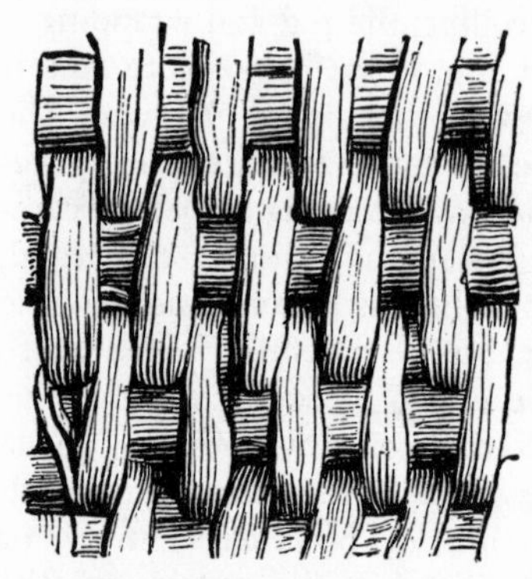

FIG. 47.—Detail of Coiled Basketry, South-western United States. (*U. S. Nat. Museum.*)

A variety of stitch is carried out in the spiral sewing by taking a half-turn around the splint or fillet at each round between the coils. This is seen not only on pretty Japanese lunch baskets, but also on the fish-baskets made of rushes by the Fuegians. They are the only American savages who employ this style of basket-work.

"The Panamint squaws, in Death's Valley, California,"

says Coville, "make their basketry at the cost of a great deal of time, care, and skill. The materials are the year-old shoots of tough willow (*Salix lasiandra*) and aromatic sumac (*Rhus trilobata*), the horn of the mature pods of the unicorn plant (*Martynia proboscidea*), and the long red roots of the tree yucca (*Yucca brevifolia*). These give three colours, the red, the black, and the white. Sumac and willow are thus prepared. The bark is removed from the fresh shoots by biting it loose at the end and tearing it off. The woody portion is scraped to remove inequalities, and is then allowed to dry. These slender pieces are for the warp or foundation. The weft splints are prepared from the same plants. A squaw selects a fresh shoot, breaks off the too slender upper portion, and bites one end so that it starts to split into three nearly equal parts. Holding one of these parts in her teeth, and one in either hand, she pulls them apart, guiding the splits with her fingers so dexterously that the whole shoot is divided into three equal even portions. Taking one of these, by a similar process she splits off the pith and the adjacent less flexible tissue from the inner face, and the bark from the outer, leaving a pliant, strong, flat strip of young willow or sumac wood. The weaving of the basket is begun at the bottom with two layers of warp twigs fastened by their middles at right angles. The free ends are bent upward, and in and out between them the splints are woven. [Mr. Coville fails to notice that the twined style of weaving is followed.] The free ends are bent upward and concealed in the weaving. As the basket widens new warp twigs are inserted. Ornamentation is produced by retaining the back, or by staining them, and by varying the manipulation in the weaving. A squaw commonly occupies an entire month constructing one basket." [1]

The Panamints make their pot baskets and plates as follows: they are built up with willow and sumac strands as above described, but narrower and of finer quality.

[1] Coville, *Am. Anthrop.*, Washington, 1892, vol. v. p. 358.

Similar strands of martynia pods, and the long-jointed, slender stems of a native grass (*Epicampes rigens*) are also needed. The grass is particularly adapted to the use from its firm texture and the fact that the portion above the uppermost joint is very often eighteen inches long.

Starting from a central point, a bundle of two or three grass stems and one very slender withe are sewn by a willow splint to the part already finished. At the proper point the bundle is drawn more tightly, so that the remainder of the spiral forms the sides of the basket. The punctures for the sewing are now made with an iron awl, but the aboriginal tool is a stout horny cactus spine from the devil's pincushion (*Echinocactus polycephalus*), set in a head of hard pitch. The grass stems, when the stitches are drawn tightly, make a perfect packing, and the basket when finished is watertight. Patterns are produced by substituting strands of martynia pod for willow in the sewing.[1]

Mr. Coville is here describing the method of making coiled basketry. If the sewing takes place always on the outer edge of the finished part of the coil, the work will be flat like a mat. But these textile artistes understood narrowing, by sewing always a little above the outer edge making the tour a little shorter. The consequence is a bowl, a jar, or a pot, just as the maker wills. I have elsewhere described minutely the cleverness with which these savage women secure a water-tight vessel, light and strong.

Mat-making, hat-plaiting, and all such work as does not require a loom, is in the nature of basketry. The Chilkat Indians of Alaska weave a ceremonial blanket of cedar bark, and wool from the mountain goat. These are covered with totemic devices in yellow, black, and white. But there is no shuttle employed. The warp threads are set up in a frame, and the weft is wrought in by twined work, after the manner of a tapestry worker. It is, in fact, the T'lingit twined basketry in pliant material. All over the

[1] Coville, *op cit.*, p. 359.

North American continent cloth and matting were thus produced.

Hand-weaving, without the aid of loom or frame of any kind, was perfected among the Polynesians. Besides the beaten or tapa cloth they made robes and sleeping mats, with or without fringe or pile from the bark of the hibiscus and *Phormium tenax*. The bark was peeled from young shoots in strips four or five feet long. These were scraped with a shell and dried. From these ribands split the mats were woven, the work commencing at one corner, and a fabric nine feet long and four feet wide being wrought by the fingers of the workman alone. These were of a beautiful white colour, and were worn only by men.[1]

The so-called Panama hats and similar work from Africa, though looking like the work of a loom, are produced by hand. Either in Mexico or in Africa the natives may be seen seated on the ground with the split filaments at their sides working away almost unconsciously, and scarcely looking at what they are doing. The method of drawing the working filaments alternately above and below what may be called the warp is ingenious. Each filament is doubled near the point of working, and the nimble fingers place one warp strand above and one below as the loop is drawn along so dexterously that the eye can with difficulty follow the operation.

The true textile art begins, however, with spinning or the making of yarn. This involves the separation of the fibrous tissue from starchy and other foreign matter, and the twisting of the fibre so as to make a strong yarn. Or it involves the removal of hair or wool from animals, and subjecting them to the same operation.

In its beneficial results this art is surpassed by none other invented by savages. When one considers the millions of flying spindles now whirling in all the factories of the world, he does not wonder that the Fates or controllers of human destiny were worshipped in the form of three

[1] Ellis, *Polynes. Researches*, London, vol. i. p. 186.

very plain women, one making yarn, the second spinning it out, the third with the fatal shears. It is easy to believe that the first yarn was twisted between the palms of the hands or on the thigh by means of the palm. The cobbler, in untwisting his thread, keeps alive the latter process. But the spindle is a very old device. The simplest form in use to-day is a stick or rod of wood. The one who used it sat on the ground with legs extended. The yarn was fastened by one end to the middle of the stick. The spinner held the bunch of fibre in the left hand, and rolled the stick along on the thigh quickly with the right hand, catching and carrying it back to the groin, where it stopped twirling. The spun yarn was wound on the stick as soon as it was sufficiently twisted, and this made a sort of fly-wheel.

It was a very easy step in advance to put some weighty object upon this stick, inventing thereby the spindle-whorl. And if the spinner wished to get up and walk around, it would be necessary to have a spindle-stick with a hook or notch on the upper end. Stick, whorl, notch—that is all there is in spinning. All further inventions were for the purpose of doing the work faster and finer. Indeed, in Finland the spinning-wheel is nothing more than a large spindle laid horizontally, and worked in two upright sticks that serve for bearings.

Every region of the earth has its own string. The Arctic peoples prepare thread and twine of sinew, some of them as fine as our best cotton, only very much stronger. The Japanese make excellent string of the mulberry paper, and the Chinese, as well as many peoples south of them, use bamboo splints, while the silkworm goddess is the patroness of the Far East. All over the Pacific Islands the coir, or prepared fibre from the outer husk of the cocoanut, is the staple from which string is made, not by twisting, but chiefly by braiding. This braided coir serves every conceivable purpose. Houses, boats, and implements are tied, not nailed or rivetted together. Its preparation occupies the leisure of men and women, and great rolls of black and brown sennit, for that is the native name, may be seen in museums. In Mexico and South America the pita fibre and cotton furnish the principal staples, but all over temperate North America the *Apocynum cannabinum*, or Indian hemp, was made into yarn and twine, and woven into cloth. The hair of ruminants and of the dog easily lent itself to the spindle, and among some tribes skins with the fur on were cut into very thin strips, and these were twisted and woven into blankets. Bast from trees is frequently twisted into a kind of single ply twine, and used even for bow strings.

Sinew dressing is a textile art. The long and tough bundles of sinew are removed from the legs of the larger mammals, very carefully cleaned of any flesh or fat and dried. At convenient seasons these bundles are shredded just as men and women pick oakum. This shredded sinew is used without further preparation for seizing or wrapping thousands of things together. For instance, when a savage would firmly attach the feathers to an arrowshaft, he takes the shaft under his left arm with the nock end in his left hand. He then puts a shred of sinew in his mouth, while he lays on his feathers carefully. As soon as the sinew shred is thoroughly softened, the wrapping is neatly done by holding the shred tight with the right hand and revolving the shaft with the left. At both ends the shred is tucked under and rubbed down so as to render the fastening invisible.

This shredded sinew answers another purpose, namely, for strengthening the backs of bows. A quantity of this material is well moistened and mixed with animal or fish glue, and little by little built up on the back of a wooden bow previously prepared. This is so neatly done as to resemble the bark of the wood, and must be very carefully managed to avoid overstraining the wood backward in drying, and to give at the same time the elasticity demanded. Some of these bows will do terrible execution.

The sinew, when very finely shredded and combed out, makes an excellent yarn or thread or cord. The eastern Eskimo very seldom twist the sinew except for bow strings, but the western Eskimo are extremely clever in its management. They have invented a kind of swift which enables the spinner to run out a much longer filament than can be done by hand. The spindle is seldom used in sinew twisting. For bowstrings and cord where strength is needed, the North American Indian had no substitute for this.

The shredded cedar and willow bark and some grasses or the Pacific coast of America make excellent twine for nets, fishing lines, or domestic use. The hackling is done with a very dull bone knife in the shape of a kitchen chopper, and in some cases the filaments are quite short. The twining is done altogether with the fingers, and very skilfully, after the manner of twisting a whip-cracker. The woman holds the twined part in her left hand between thumb and forefinger, and presses her middle finger against the ball of the thumb to hold a strand, while with her right hand she gives the other strands a few turns. She deftly turns the strand, passes it to the middle finger of the left hand to hold, at the same time seizes the other strand, gives that a turn or two, twining the two strands each time. It is said that Sicilian women make twine for chair bottoms in the same way from rushes.

The basket-makers of Guiana are men, but the spinning and weaving, with slight exceptions, are done by women. The string is of three kinds of fibre—cotton, tibisiri (*Mauritia flexuosa* fibre), and crowia (*Bromelia* and *Anannassa* fibre). Cotton is grown and spun by almost all Indians, but especially by the Arecunas. The fibre having been picked and freed from the seeds is pulled out into a long, uneven loose band, and this is wound around the right wrist. One end is attached to the end of a common spindle and the thread is carefully drawn out and twisted and wound about the spindle shaft. In making twine two or three of these spindles are used, as in common twine. The gathering of tibisiri is a unique process. The young leaf spike furnishes the fibre. Each leaf or spike is taken singly, a sharp dexterous rub at the top separates the outer skin, and the whole is then torn off. It is further prepared by boiling, drying in the sun, and twisting into strings by means of the palm of the hand and the thigh, after the manner of the cobbler. The fibre from a dozen spikes is sufficient to make a large hammock.

The crowia fibre is thus obtained. A noose or slip-knot is passed over the end of a leaf tightly, the other end of the string being attached to a tree. The Indian then takes the point of the leaf in his hand and forces the fibrous portion through the noose by a sudden and strong pull. The green skin and soft matter are removed by the loop. The fibre is then washed and dried in the sun. The crowia is also twisted on the thigh by means of the palm of the hand. Mr. im Thurn explains the existence of thigh-twisting and spindle-twisting in the same area by the statement that the latter is confined mostly to Carib tribes, and that the thigh-twisting is probably the aboriginal method, and that subsequently the stocks had borrowed customs from one another.[1]

These Indians construct their hammocks by netting the tibisiri fibre after the manner of an old-fashioned silk purse. The square wooden frame on which they are made lies on the ground, and the whole is netted of one continuous string. The Caribs weave their cotton hammocks in a frame. After setting up the warp, they weave at intervals or braid bands across with three shuttles, taking up the warp strands alternately in the plait. The work on these is done by women, from the planting to the finishing off in the loom, but the "scale lines are put on by the men."

The Andamanese cordmaker uses the bark fibres of trees and shrubs for material. For harpoon lines and turtle nets he resorts to the *Melochia velutina*, removes the bark, and with a Cyrena shell scrapes the cellular integument until the fibre which it encloses is laid bare. These are then placed

[1] *Cf.* im Thurn, *Indians of British Guiana*, London, 1883, pp. 283–290.

in the sun to dry. When ready for use the ropemaker ties several of the filaments to his toe and winds another strand round them spirally, adding fresh lengths when needed. When about thirty yards have been made, a large portion of yarn is wound around the kutegbo, or reel made of two cross sticks. The operator then seats himself, stretches his legs apart, places a stick or cane between his two big toes, and over it passes his reel. His yarn or fibre is placed at his right side and drawn behind his neck and over his left shoulder as he proceeds. The man converts himself thus into a kind of ropewalk.

The women of this race also make a less durable twine for fishing-nets and sleeping mats. And bowstrings are manufactured by twisting fibre on the thigh. Even in net-making the finger is used as a mesh stick, though the netting-needle of bamboo is in vogue. No sewing, as we understand the word, is to be seen. In repairing their dug-out canoes they bore holes above and below the crack and reeve strips of cane through these, filling the interstices with wax of the black honeycomb.

Dr. Faurot says that among the idle men about the village on the island of Kamarane, south of Arabia, may be seen some walking about and spinning. The spindle consists of a shaft and a whorl on top, the latter pierced near the outer border and having an eyelet extending above the centre. The spinner holds the thread high up with the left hand, and with the right palm sets the spindle whirling by striking the palm against the edge of the whorl.[1] Livingstone observed the same implement in Africa, and in the first volume of his explorations refers to a figure of Wilkinson's. In the Egyptian group women are doing the spinning, one twirling as just described, a second is rubbing the shaft against her thigh, as a shoemaker does now, and a third is using both palms, having suspended the thread from the fork of a tree.

[1] Dr. L. Faurot, *Rev. d'Ethnog.*, Paris, 1887, vol. vi. p. 435. Compare with Tibetan spindles, Rockhill, *Smithson. Rep.*, 1892.

ZUNI WOMAN WEAVING BLANKET. (*Photo in U.S. Nat. Museum.*)

Livingstone says that the markets or sleeping-places of Angola are well supplied with provisions by great numbers of women, every one of whom is seen spinning cotton with a spindle and distaff. A woman is scarcely seen going to the fields, though she may have a pot on her head, a child on her back, and the hoe over her shoulder, but she is employed in this way.[1]

Among the Mendi negroes on the Niger, the men do the heavy work and clear the bush, they also weave, sew, and make their own clothing. The women till the ground, fetch water, go fishing, prepare and cook food. They spin cotton thread, dye it, and make mats.[2]

The Polynesian race, as well as the negroid peoples of Oceanica, make a braided cord from the husk of the cocoanut. In Samoa, the sennit is braided chiefly by the men. They sit at their ease in their houses and work away very rapidly. At political meetings also, where many hours are spent in formal palaver and speechifying, the old men take their work with them.[3]

Loom-weaving is a savage invention. In the Mexican Codices a mother is pictured giving instruction to her daughter in the art of weaving.[4] The warp is fastened to sticks at either end and the alternate threads are lifted and depressed by means of a very simple harness. The Africans also had looms, as well as the Polynesians, involving in a primitive way the parts and the processes of our more pretentious machines. One of the latter, however, will do the work of one hundred savage women, and a well-equipped factory would weave more cloth in a day than ten thousand African experts.

The simplest form of weaving is a plain checker in which the same kind of thread is both warp and woof, and both

[1] Livingstone, *Travels, &c., in South Africa*, New York, 1858, p. 433, with drawing copied from Wilkinson's *Ancient Egyptians*.

[2] Garrett, *Proc. Roy. Geog. Soc.*, London, 1892, p. 436.

[3] Turner, *Samoa*, London, 1884, p. 170.

[4] Bancroft, *Native Races*, New York, vol. ii. p. 484, with authorities.

are drawn equally tight so as to appear on either side. As before mentioned, this is a very common style of basket-making, but it is not so easy to manage in a rude loom, as we shall see. In many parts of the world, savages set up frames, very much like the old-fashioned quilting frames, only of very rude sticks laid parallel on the ground or fastened to some stable objects just as far apart as the fabric is to be long. In a continuous long spiral they wind the warp yarn backward and forward around these two sticks until the warp is as wide as the blanket or other fabric is to be. The threads are thus adjusted at equal distances on the sticks above and below. A long rod is then laid against the warp, and by means of a continuous yarn this harness is made fast to the warp threads farthest from it, the back threads, if the loom is standing. This can be done by simply winding the yarn round the stick and passing it between the front warp fillets and around the back ones, as a hurdle or heald, until every back thread is attached to the harness stick. The yarn of the weft is wound on a long stick by wrapping it around one end once or twice, carrying it to the other end, wrapping it there and so on, backward and forward until enough is wound. The weaving consists in drawing the harness stick toward the weaver, which pulls the back set of warp threads forward between the other or front set. The primitive shuttle is then passed between the two sets of warps, the end of the yarn having been fastened to the outside warp, and enough yarn unwound to go across. With the two hands this first weft fillet is drawn taut, any inequalities adjusted with a pointed bone or stick, and then it is driven home by a wooden sword, lightly if the texture is to be plain, with some force if it is to have a corded appearance. The sword is then withdrawn, the harness stick slackened, the back set of warp threads forced into their places by the sword and another weft thread carried across. This constitutes the action of the most primitive loom. There is no machinery of much importance in savagery, so we must not look for flying shuttles of the most primitive sort. But the harness does become more complicated. However, so long as there are no true heddles the weaving must necessarily be of the plainest kind. Different colours are easily introduced into this work by having sets of bobbins or reels for each colour and drawing them through the requisite number of stitches each time. The apparatus cannot be set for such performance, but the weaver must carry her pattern in her mind and count properly at each turn. Stripes are easily made, but geometric patterns require great skill and close attention. A curious harness is found among the North American Pueblos and in Finland. It consists of a number of small wooden rods, or heddles, made into a rack by lashing their ends to two parallel rods, after the manner of a ladder, only the rounds are so close together as just to allow the warp thread to pass freely from one cross-bar to the other. The small rods or heddles are all perforated in the middle to form the eyes or "mails." When the warp is set up, threads are passed through the "mails" and between the rods. This enables the weaver to push, or "shred" one half of the warp threads past the other half quickly. It also allows the weaver to "darn" the weft thread through the warp threads that are uppermost and create geometric figures and diaper effects *ad libitum*. The Chinese have a large block of wood with saw cuts inclined so as to throw the warp up and down in weaving the Canton matting.

In the African grass and palm fibre looms a harness is made by a single set of heddles acting precisely as do the perforated rods in the Zuñi belt-weaving.[1] In the manufacture of the garters worn in their ceremonial dances the Pueblo Indians turn their bodies into a very convenient stretching frame. The woman sits on the ground with legs extended, and holds one of the warp bars with her two great toes while the other rests against her stomach, and is made fast to a belt passing around her back. By moving her toes

[1] Matthews, "Navajo Weavers," *Third An. Rep. Bur. Ethnol.*, Washington, 1884, pp. 371–391, pl. xxxiv.–xxxviii., figs. 42–59.

outward and straightening her legs she gets all the tension she desires, and can relax it instantly.

The first attempt at weaving cloth in a long piece, afterwards to be cut up, finds its counterpart in the cotton looms of Liberia and other portions of West Africa. The warp is measured off by driving stakes in the ground and walking around them with the thread as often as there are to be filaments in the narrow cloth. Sometimes it is necessary to go around the house or the whole group of structures. The warp is held taut by a large stone, and the narrow strips are afterwards sewn together. It is not here affirmed that this is altogether a native art, but native processes have survived in it sufficiently to make the study instructive.

A style of weaving controlled largely by the abundance of cat-tail and other great rushlike stems remains to be described. A number of the stalks laid parallel and very close together were joined by sewing at short distances a cord of native hemp straight through the whole series. The slender wing-bones of birds served as needles, and a double crease following the lines of the uniting threads gave an ornamental effect to the surface. This style is described in Smith's *History of Virginia*, and examples of the work with all the apparatus were sent to the United States National Museum by Mr. Willoughby from Washington State.

Plain sewing among the lowest peoples is an affair of the skin dresser. They do not, as has been said, make cloth in long pieces to be cut up and sewed into garments and other useful things. This being the fact, the best tailors ought to be sought in the Arctic regions. And this is true as any one knows who has examined the garments of caribou skin, of seal-skin, of the pelts of the little fur-bearing animals, of the intestines of the larger mammals, wrought by the Siberians and the Eskimo.

Parkas or blouses, trowsers or boots, are cut out with stone or metal knives. The edges of the parts are whipped together with sinews so as to be water-tight. Bits of different coloured fur are inserted for ornamentation, and, frequently, to save every scrap, the sempstress will have a hundred pieces of skin in a single garment. Her needle is a tough bit of bone working like an awl, and her sinew is drawn through with a true needle made of bird bone. Her thimble is a bit of tough seal hide drawn over the end of the forefinger, though in modern times they imitate in ivory the white woman's thimble. Lighter goods, such as the intestines of seals and the more delicate skins are run together by a basting stitch of wonderful uniformity, and bits of feather are caught between the parts of the seam for ornament. As far south in America as the country of the loom weaver and the bark-cloth beater, sewing women abounded. Especially in modern times were they skilful and active in the buffalo country, where they constructed by hundreds the huge teepees or tents as well as the clothing of their tribes. Indeed, the whole work of skin-dressing and manufacturing devolved on them.

Netting among savages is difficult to study because there is much dispute as to whether it has been introduced among them ; but any one who has examined the knots of Polynesians, of Eskimo, of the ancient Peruvians, has no difficulty in believing savage textile artisans capable of making any kind of nets. The costly feather cloaks of Hawaii are founded upon nets, the quill of the feathers being caught systematically into the knot of each mesh. In a collection of Eskimo objects will be seen netting needles, shuttles, bobbins, spacers for meshes of all sizes and materials, wound with twine and babiche, or fine raw-hide string.

Net-making for salmon in Polynesia was an affair of state. "The salmon net is seldom possessed by any but the principal chiefs ; it is sometimes forty fathoms long and twelve or more feet deep. As is customary on all occasions of public work, the proprietor of the net required the other chiefs to assist in the preparation. Before he began two large pigs were killed and baked. When taken from the oven they were cut up and the governor's messenger sent with a piece to every chief. If it was accepted, the chief agreed to perform the part assigned him. The cord was about a quarter of an inch in diameter, and made with the tough bark of the maté (*Ficus prolixa*), which, next to the romaha, or flax, is considered more durable than any other vegetable substance. The cord was twisted with the hand across the knee, in two or three strands or threads. The meshes were about four inches square, made with a needle not unlike those employed by European workmen. As the parties brought in their portions the chief and his men joined them together. The floats were dried pieces of hibiscus, and the bottom was hung with smooth stones enveloped in pieces of the matted fibre of the cocoanut husk tied together at the ends, and attached to the lower border of the net."[1]

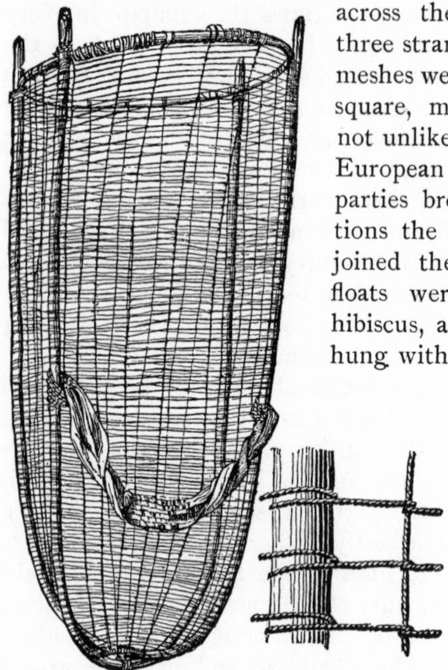

FIG. 49.—Mohave Burden Basket, weft simply wrapped about the warp threads. Compare Andaman patterns.

Loopwork is a fabric made by the interlocking of loops in a continuous string, like crocheting. Hammocks are often thus constructed. There are no knots as in netting nor double motion as in knitting, but the loops are drawn through as in spiral basketry, and the row now forming is kept from ravelling by having the next row of loops drawn through it. The best and purest

[1] Ellis, *Polynes. Res.*, London, 1859, vol. i. p. 141.

forms of this work are to be seen in the wallets and open net sacks of the African tribes. The same stitch may now be seen in Central America, and the query is whether the negroes taught the Indians the stitch. When the work is done with a bone needle and a rod for a spacer, the end

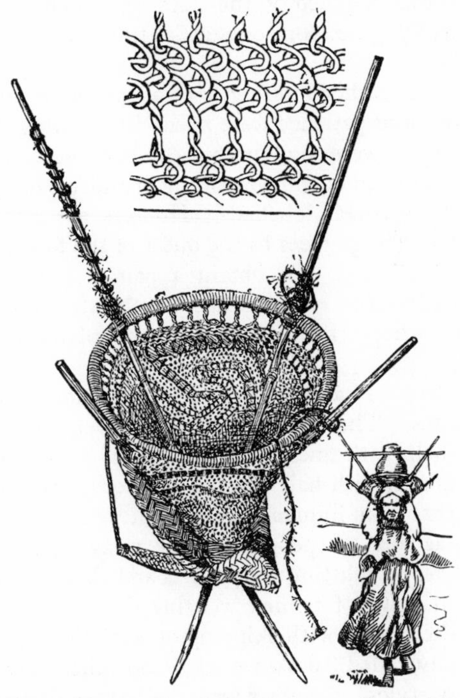

FIG. 50.—Pima Burden Basket, Arizona. Detail showing beginnings of lace making, or loopwork. (*U. S. Nat. Museum.*)

may be drawn through the loops and form a link between looping and netting.

In addition to the weaving of feathers among their cotton fabrics, the ancient Mexicans practised to perfection an art which may be called feather-mosaic. Even in our day attractive examples of this work may be bought in Mexico, but these bear no comparison with those made by the ancient. To prepare for the feather-painting, the *amanlica*, or artist, arranged his feathers in small earthen dishes, stretched a piece of cloth on a board before him, and provided himself with a pot of glue and a pair of tweezers. His design was sketched on the cloth, and then the feathers were carefully glued on one at a time with exemplary patience.[1]

In Hawaii, feather-hunting was a special vocation, and much labour and patience were spent in catching the birds. Nets and snares were sometimes used, but, more frequently birdlime, composed of the gum of the breadfruit or the viscid milk of the arboreal lobeliads. Hunters are said to have transplanted strange trees to the midst of the forest to excite the birds' curiosity. To obtain a pair of tail feathers of the Koae (*Phaeton rubricauda*), the hunters climbed the steep palis where the birds nested and plucked the long feathers.[2]

Embroidery was also a savage art long before the coming of the whites. The surfaces of textiles were covered with beads of shell, with finely stripped and dyed quills of birds and porcupines, with hair of moose and other mammals, not rejecting that from human heads. A little above the lowest savagery, as soon as people became weavers, by omitting weft threads, by splitting warp fillets and changing the parts included in rows of twined weaving, by a figure of eight weaving alternating with skipping of warp threads, by what is technically called "drawn work," and other devices, they established styles of embroidery that are imitated by the most cultured.

Superconscructive features, so important in the decoration of fabrics are the result of devices by which a construction already capable of fulfilling the duties imposed by function has added to it parts intended to enhance beauty, and which

[1] Bancroft, *Native Races*, referring to many authors, chap. ii. p. 488, *seq.* Also Mrs. Zelia Nuttall, in Peabody Museum Papers.

[2] Brigham, *Cat. Bishop Museum*, Honolulu, 1892, p. 10.

may or may not be of advantage to the fabric.[1] They constitute one of the most widely used and effective resources of the textile decorator, and are added by *sewing or stitching*, *inserting*, *drawing*, *cutting*, *applying*, *appending*, *&c.* These methods of over-laying and added decoration are seen in their perfection among the northern skin-workers. The weaving people of Peru and Mexico had come to this stage of their art, but plain weavers had not. But the clever little Eskimo woman could herring-bone with shredded quill, let in a gore cut from the ankle of the caribon, and cut out parts or "pink" the edges of a garment according to her mind.

In all primitive weaving definite reticulated patterns are produced by variations in the spacings and other relations of the warp and woof. The production of reticulated work is the especial function of netting, knitting, crocheting, and certain varieties of needlework, and a great diversity of relieved results are produced, no figure being too complex, and no form too pronounced to be undertaken by ambitious workmen.[2]

The decoration of basketry and textiles is, after all, a kind of chess playing. Each stitch is restricted to a definite area, and if the maker has been skilful, the area may be indefinitely small. The decoration of basketry is the development of geometry, producing straight lines on wallets and curved lines on true baskets and jars. These lead on to the creation of triangles and rectangles and polygons of every sort, to herring-bone, chevrons, and frets or meanders, in short, to everything that can be made out of dots and small figures and lines. Basket-making also introduces and keeps before the mind the elements of arithmetic. It would be very difficult to find another savage occupation which exacted so extended a count and such a ready use of figures. The basket-maker must hold in her memory and count in a twinkle any number of stitches, certainly up to twenty.

[1] Holmes, *Sixth An. Rep. Bur. Ethnol.*, pp. 211–232. With many illustrations. [2] *Ibid.*, p. 210.

The Zuñi belt-weaver, introducing the same design over and over after an interval of ten or a dozen passes of her shuttle is a tolerably fair mistress of rapid counting.

This primitive counting and geometry laid the foundation for decorations innumerable on other material. The potter has never ceased to copy them though the transfer has been accomplished with the greatest difficulty, and has led to modifications made possible by the softness of the material.

On the other hand, the ambitious basket-weaver, working in extremely fine stitches, and instinctively guessing that a curved figure is only a polygon with infinite number of sides, soars away from right-lined geometry, and attempts animal forms. These animal forms, curtailed and abbreviated, become at first a quaint pictography, which is lost by and by in other geometric forms of a higher order. These are borrowed and multiplied from land to land, and form the stock-in-trade of the designers.

It is of the utmost importance that the stitch in basketry and the mesh in weaving be correctly understood in their relation to art in textile and also in pottery, which, as was seen in Chapter V., is a child of basketry. The one thing sought after by the skilful weaver in savagery is uniformity of dimensions in the stitch. The most cultivated persons have marvelled in looking at a Yuki Indian's or a Congo negro's textile work to see the uniformity of the plaiting or the weaving. Children and young women struggle and struggle on until they acquire the knack, and become old in the pursuit until they attain it. This once learned with any degree of nicety, the young artist is ready for the second lesson, that is, to give variety to the surface by means of shading or by colour. The Panamint Ute woman working with splints of rhus or willow, and with the split pods of Martynia, which are jet black, has now the means of branching out into form. But do not forget that the exigencies of her material and of her method preclude the possibility of her ever achieving aught but geometric figures. It is one, two, three in black, and then as many more in white, and the thing is done. The next time she comes around, the blacks and whites will face opposite colours in the preceding row. That is all there is in it to begin with.

The colouring of textiles, both basketry and weaving, is an ancient art. In the first place, Nature assisted the weaver by supplying brown, black, red, green, yellow filaments in a multitude of shades. The rest is art or invention. Some vegetable substances assume new colours when buried in marshy places. Others are changed in contact with mineral or vegetal or animal substances. The California Indians immerse splints in muddy places, and secure a permanent chocolate brown; but everybody knows that vegetal dyes need the addition of a mordant to make the colouring matter adhere. This part of the art of colouring, however, was worked out in savagery.

The Navajo Indians make their native dyes as follows:

Black.—*Rhus aromatica*, yellow ochre, gum of the piñon (*Pinus edulis*). The sumac leaves and stems are boiled five or six hours. The ochre is finely powdered, and roasted to a light brown colour. It is then mixed with an equal quantity of piñon gum, set on the fire, and stirred until the mass is reduced to a fine black powder. When it has cooled somewhat it is thrown into the decoction of sumac, when it instantly forms a blue-black fluid, the tannic acid of the sumac combining with the sesquioxide of iron in the roasted ochre, the whole enriched by the carbon of the calcined gum.

Yellow.—The flowering tops of *Bigelovia graveolens* are boiled until a decoction of deep yellow is produced. Some almogen (an impure native alum) is heated over the fire and added to the decoction, and the wool is put into the mixture to boil. This produces a tint nearly a lemon yellow.

A second process consists in crushing the fresh root of a plant, as yet undetermined, upon a metate, and in using the almogen as a mordant. The cold paste is rubbed between the hands into the wool.

Red.—A reddish dye is made of the bark of *Alnus incana*, var. virescens (Watson), and the bark of the root of *Cercocarpus panifolius*, the mordant being fine juniper ashes.[1]

The Lacandons of Guatemala used as dyes, indigo for blue, cochineal for red, and indigo mixed with lemon-juice for black. The Nicaraguans obtained a highly prized purple by pressing the valve of a shell-fish found on the seashore. They take the material to the seaside, and, after obtaining a quantity of fresh colouring matter, dip each thread singly into it, and lay it aside to dry. From the aloe and pita they obtain a very fine thread. Reeds and bark give material for coarser stuff, such as ropes and nets.[2]

The ancient Mexicans, in preparing dyes and paints, used mineral, vegetal, and animal substances. Of plants, they used the wood, the bark, the leaves, the flowers, the fruits, and many of their dyes have, since the conquest, been introduced throughout the world. Chief among these was the cochineal, *nochiztli*. The flower of the *matlalxihuitl* supplied blue shades ; indigo was the sediment of water in which branches of the *xiuhquilipitzahuac* had been soaked ; seeds of the *achiotl* boiled in water yielded the red, the French *roucou ;* ochre, or *tecozahuitl*, furnished yellow, as did also the plant *xochipalli*, the latter being changed to orange by the use of nitre ; other shades were produced by the use of alum ; the stones *chimaltzatl* and *tizatlalli* being calcined, produced something like Spanish white ; black was obtained from a stinking mineral, *tlaliac*, or from the soot of a pine *ocotl*. In mixing paints they used chian-oil, or sometimes the glutinous juice of the *tzauhtli*. The numerous dye woods of the *tierra caliente*, now the chief export from that region, were all employed by the native dyers.[3]

The oldest books speak of cloth and nets and embroideries

[1] Matthews, *Third An. Rep. Bur. Ethnol.*, p. 376.
[2] Bancroft, *Native Races*, New York, 1874, vol. i. p. 699.
[3] *Ibid.*, vol. ii. p. 487.

and dye stuffs. Indeed, there are some types of hand-work in the textile art that no machinery can be made to imitate. This body of industries, like others of which we have been speaking, seems to have been invented and developed long ago, and when the curtain rises on the drama of written history, the spindle, the distaff, the loom, the needle are there on the stage in place. This chapter relates especially to woman's work throughout. It ought not to depreciate the inventors of the textile art in the eyes of cultivated women when they learn that the delicate stitches and patterns which they employ were invented so long ago by their own sex.[1]

[1] Havelock Ellis, *Man and Woman*, London, 1894, Walter Scott; O. T. Mason, *Woman's Share in Primitive Culture*, London, 1894, Macmillan.

CHAPTER VIII.

WAR ON THE ANIMAL KINGDOM.

"Listen to the words of warning
From the lips of the Great Spirit!
I have given you lands to hunt in;
I have given you streams to fish in;
I have given you bear and bison;
I have given you roe and reindeer;
I have given you brant and beaver;
Filled the marshes full of wild fowl,
Filled the rivers full of fishes."

LONGFELLOW, *Hiawatha*, i.

IN his contact with the animal kingdom, the primitive man developed both militancy and industrialism. He occupies two attitudes in the view of the student, that of a slayer, and that of a captor and tamer. Omitting now the inquiry whether the very first men were carnivorous or vegetarian, we may apply our thoughts to the general subject of man in his relation to the animal world. It is important to ask how our species came to be masters of the brute kingdom, and what intellectual advantages were gained in the struggle.

The first of our species were poorly provided with apparatus for contending with their fellow-creatures, or even for defending themselves therefrom. The lower forms of terrestrial and marine life were accessible to them, and the young of many higher mammals; but the conflict must have been slowly and feebly waged at first.

The creatures potentially useful to early man in ways

innumerable, belonged to every family of the zoological kingdom. There was no want of his that could arise which there was not some being to serve.

The account of the ways along which this animal world has gradually succumbed to our species would involve the whole history of civilisation. All this conflict and enterprise was in front of the first men. No other animal started ever on such a mission. And yet men could not fly, like the rapacious birds ; nor burrow, like the bear, the fox, or the mole ; nor swim, like the fish. They had no hairy covering, their teeth and nails were the weakest ; most animals were more fleet than they. The bear, the lion, the tiger, the wolf, the serpent, the gorilla, could easily overpower them.[1]

Their problem was to invent missiles that could fly faster than the objects of their pursuit ; to create apparatus for digging and burrowing, and for compelling underground animals to quit their dens ; to pursue the aquatic creatures in their own element ; to lay tribute on all hairy and fur-bearing species ; to devise engines that would strike harder than the paw of the lion, pierce deeper than the tiger's fang, wind their victims in more deadly folds than the embrace of the serpent, and burn more effectually than the stings of all venomous beings combined.

Just as the modern inventor is ever seeking for sharper eyes in his optical instruments, more delicate muscular sense in his refined metric apparatus, the genius of the first men was engaged in adding speed to their feet, momentum to their fists, the strength of withes and ropes and thongs to their grip.

By and by they turned the artillery of Nature on herself. The dog raised a flag of truce and came in to join the hosts of man against the rest. The mountain sheep and the wild goat descended from their rocky fortresses, gave up the contest, and surrendered skins and fleece and flesh and milk to clothe and feed the inventor of the fatal arrow.

[1] *Cf.* J. Hampton Porter, *Wild Beasts*, New York, 1894, Scribner.

Tired of deadly weapons and decoys and snares and pitfalls set by the most cunning of enemies too long ago for any historian, the llama, the camel, the horse, the ass, the elephant, the cow entered into a solemn and everlasting treaty to lend their agile feet, their patient backs and necks and shoulders, their milk, their flesh, their hides, their hair, their very bones, to minister to men's wants. How well this treaty has been observed on both sides let all domestic creatures bear witness.

Those that refused to enter in any way into these stipulations are doomed sooner or later to extinction, and many species have already disappeared or withdrawn to the waste places of the earth in despair.

Savagery, barbarism, civilisation, the three general periods into which sociologists divide the evolution of culture, may well be marked off in the progress of men in relation to animals. It is possible to follow any one animal up through the three periods, or to mark the increasing number or genera and species that have been thought necessary to human happiness at each stage of its upward career, or, finally, to note how many parts of the animal frame may be brought into the industrial currents, and the multitudinous functions which a single part of the animal may come to serve.

From the lowest savagery to the highest civilisation of men and animals, the progress of both from nature to artificialism, in culture and domestication and breeding, is now studied seriously as one of the most promising divisions of anthropology. It will be seen that both the quality and the rapidity of refinement have always been conditioned by the animals in the foreground.

It is a false notion that savage or primitive men knew little or nothing of zoology. Inasmuch as their brains were nearly equal to ours, as their pulses beat as fast and their senses were normal, as they passed their daily lives in pursuing or escaping from the animals, their knowledge concerning them was extensive. The author has lately gone carefully over the list of the higher animals known to North American savages, and the result is astonishing. The Indians were not naturalists in the modern sense, but they had uses for all the species they knew.[1]

To have a proper conception of the time when the contest began between men and beasts, it is necessary to imagine a state of things when there were no sportsmen nor professional hunters, no peltry and plumage collectors, no lighthouses nor locomotive headlights nor telegraph wires, no great field and forest fires, no smoky and noisy cities. The natural food and places of refuge for the animal creation were disturbed only in the smallest degree by man, who simply helped himself in an unobtrusive manner.[2]

In addition to its destructiveness, the gun wrought incalculable changes in the psychology of the animal kingdom. Sir John Lubbock quotes Mr. Galton on the subject of conscious danger among animals in a savage state as a type of the anxious life which savage man lived. If "every antelope in South Africa has literally to run for its life once in every day or two, and starts under the influence of a false alarm many times a day"[3] simply through fear of its natural enemy, one may imagine, at least, the quadrupling of this dread which adds the terror of the ear to those of the eye.

In this very connection, continuing the study of the seeming impassable gulf between the wolf and the faithful dog, one may wonder whether wolves themselves were as savage once as they are now, and whether on the destruction of their abundant natural supply their suspicions have not increased their ferocity. Lately this subject has been reviewed by a writer in the *Popular Science Monthly*,[4] and taken up in *Science* by Theodore B. Comstock.[5]

[1] *Report U. S. Nat. Museum*, 1888–89, p. 555.

[2] Consult Gibbs, *Science*, New York, Sept. 30, 1892, p. 183.

[3] Galton, *Trans. Ethnol. Soc.*, vol. iii. p. 133, quoted in Lubbock, *Preh. Times*, New York, 1872, p. 595.

[4] Notes, Sept., 1892, p. 719, New York.

[5] *Science*, New York, Sept. 16, 1892, p. 155.

Mr. Hudson, in his *Naturalist in La Plata*, says that "the puma never attacks men except in self-defence. In the pampas, it is said, a child may sleep on the plain unprotected in equal security." Mr. Comstock says "the puma, or American panther, and the jaguar, its South American representative, are not regarded by experienced hunters as animals to be feared, excepting under circumstances which leave no avenue of escape open to the beast." He also says that venomous reptiles and insects—such as the rattlesnake, tarantulas, scorpions, centipedes—have reputations beyond their desert for bloodthirstiness. The boa-constrictor, the alligator, bears, skunks, and other dreadful creatures all come in for a good word.

In addition to ferocity awakened by being cornered, there is no doubt that the pairing season works the same change in animals. In America those who have tamed carnivorous and ruminant pets know what watchfulness has to be exercised over their pupils at such times. The female in charge of her young also learns that man may be a coward, and gathers reassurance from her own fright. The point I am making is that the psychological endowments of wild creatures were profoundly modified by man even before he began to domesticate them. Killing them, taking away their natural supply, corralling them on reservation increased their savagery. The gun, more than all other causes combined, puts birds to flight, causes the mammalia to hide away in terror, and even reptiles and fishes and insects have taken on new and artificial behaviour. Finally, the better qualities of animal nature are reassured in civilisation or zootechny, and the artificially cruel manners of beast and bird, first intensified by man, become afterwards dormant, and are bred out of them in succeeding generations.

The popular superstition concerning the venomous nature of many insects may be recalled. "In Arizona," says Comstock, "the bite of a certain small species of skunk is very much dreaded, owing to the belief that hydrophobia is a probable result." The writer can recall when he was a child the constant dread of his life in which he lived from all sorts of creatures, several species of snake among them, which his brother used to carry about and handle with impunity. Dragon-flies, the cicada, the common lizard of Virginia, the garter snake—all were really as harmless as a fly. Overcome by these superstitions and tales, young and old pursued these creatures remorselessly, until hunger, surprise, fear, danger, and all the category of ills made them revengeful.

On the contrary, the prejudice against taking life of any kind and for any purpose has been the cause of many thousands of people dying from snake-bites in India and other Buddhistic countries. Indeed, the prejudice against taking life of any kind has curiously modified the industrial aspect of all regions where Buddhism has once had full sway.

The great migrations of men by which they have finally distributed themselves over the earth as we see them have been governed largely by the distribution of animals. If any one will consult the Fish Commissioner's map of the United States for the places where food fishes most resort for spawning, he will at the same time be on the track of the most prolific old Indian camp sites. Men have migrated at the beck and call of animals ; they have also been driven from vast regions of the earth by pestiferous insects.

This most intimate association of man with animated nature is well exemplified in a remark of Boas concerning the movements of the Eskimo in Baffin land. " All depends on the distribution of food at the different seasons. The migrations or the accessibility of the game compel the natives to move their habitations from time to time, and hence the distribution of the villages depends to a great extent upon that of the animals which supply them with food. In Arctic America the abundance of seals found in all parts of the sea enables man to withstand the inclemency of the climate and the sterility of the soil. The skins of seals furnish the material for summer garments and for the tent ; their flesh is almost the only food, and their blubber the indispensable fuel during the long, dark winter. Scarcely less important is the deer, of whose heavy skin the winter garments are made, and these enable the Eskimo to brave its storms and cold."[1]

To overcome the animals whose bodies they desired, primitive men set out to kill them by brute force. For this purpose they first invented weapons for despatching or seizing their prey ; after that they gave their thoughts to devices by means of which the animal would be its own destroyer or captor.

The apparatus employed in this pursuit may be classified in accordance with the manner of using it, or in accordance with the result upon the victim. The latter is preferable in this connection as the chief concept, the former being used in subdivision. In this case the following arrangement results :—

IMPLEMENTS OF THE CHASE AND OF FISHING.

1. Implements for striking.
2. Implements for cutting.
3. Implements for piercing.
4. Implements for seizure.
5. Entangling apparatus.
6. Baited apparatus.
7. Co-operative hunting and accessories.

As to their operation, each device may be either held in the hand, thrown from the hand, or left to do its own work. These classes are arranged somewhat in an evolutionary series, the last of the series indicating the greatest ingenuity and the procurement of the largest result with least effort.

In a bulletin published by the Director of the United States National Museum, Dr. G. Brown Goode, to illustrate the animal resources and fisheries of the United States, a list of the animals beneficial and injurious to man is given. The wonderful part of this enumeration is that nearly all

[1] Boas, "Central Eskimo," *Sixth An. Rep. Bur. Ethnol.*, p. 419.

the species there enumerated were known by name to the aborigines, who had invented some ingenious way to capture or slay them. There is also in the same publication, prepared by a corps of able assistants, a classified catalogue of the apparatus employed in the destruction of these animals.[1]

The beginning of this series of inventions is a club. Mr. Swan says that every fisherman in Alaska carries a club, and, on hauling a fish to the surface, knocks it on the head to prevent it from jumping about in the canoe.[2]

But the old-time fishermen, before there were any canoes, knocked both beasts and fishes on the head and broke their bones with sticks. From that rude starting-point the hunting club, for striking and for throwing, was differentiated. In the chapter devoted to warlike apparatus, it becomes a whole class of important implements for bruising flesh and breaking bones of men.

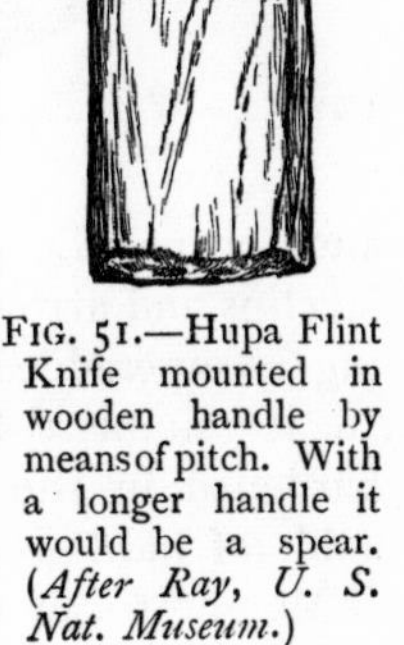

FIG. 51.—Hupa Flint Knife mounted in wooden handle by means of pitch. With a longer handle it would be a spear. (*After Ray, U. S. Nat. Museum.*)

Hand implements for cutting include hunters' knives and fishermen's knives. The history of the hunter's knife, as distinguished from the dagger, commences with the flint flake. The leaf-shaped blade, when hafted at the point or butt, may be either a stabbing or a skinning knife. The most primitive hafting of this sort is a long strip of fur wrapped carefully around one end.[3] Other examples occur, in which one point is driven into the end of a piece of wood or antler and further secured by gum. And in still other examples a "saw cut" is made in the end of a handle, the truncated blade let in and secured by lashing or glue.

[1] Goode, *Bulletin* 14, *U. S. Nat. Museum*, Washington, 1879.

[2] *Bulletin U. S. Nat. Museum*, 27, p. 833.

[3] See p. 35.

If the leaf-shaped blade be let into a handle laterally, it gives the universal and indispensable scaling, scraping, cutting knife used by women, especially about both game and fish, and called by the Eskimo, "ulu." As soon as trade brought savage men in contact with civilisation, iron blades took the place of those made of stone. These knives must have been universally used, because they have come down to us in the kitchen mincing knife and the saddler's round knife.

Implements for piercing game may be divided, according to the length of the handle, or grip, into hunting knives

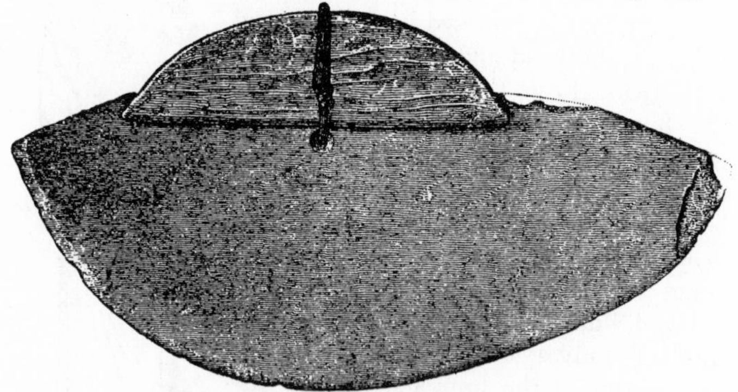

FIG. 52.—Woman's Knife, showing inserted and lashed handle, Alaska. (*U. S. Nat. Museum.*)

and spears. Despatched from the hand, they are classed as javelins and arrows. In some examples the idea of retrieving is superadded, but these will be considered separately. In its simplest form the piercing implement is a staff hardened in the fire and ground sharp, or, simpler still, a bit of hard wood ground sharp.

Many African tribes affix the horn of the antelope to a pole, and the American savages had no end of chipped blades, with sizes varying for the deer, bear, buffalo, and whales. Their bayonets were also tipped with ivory, antler, and bone. The simple lance has no retrieving function. Its purpose is to penetrate the soft, vital organs of the game at such distance as to place the hunter beyond the reach of teeth and horns and murderous claws. It lengthened the arm.

A moment's reflection will show that all savages had a practical knowledge of anatomy. They knew where to strike with the club to paralyse the brain, to slash with the cutlass for the shallow arteries, to pierce with the spear to reach the fountain of life.

Among the Central Eskimo and Athapascans, the method of hunting the deer is to attack it in the ponds when swimming from one side to the other. In many places the natives lie in ambush with their kayaks at the narrow parts of lakes, where the animals are in the habit of swimming across. In other places they are driven into the water by the Eskimo, and attacked by the drivers or by hunters stationed on the lake. Favourite places for such a chase are narrow peninsulas. The Eskimo deploy into a skirmish line, and slowly drive the herd to the point of the peninsula, whence the deer, the retreat being cut off, take to the water.

If the shore be too straight to permit this method of hunting, they drive the deer to a hill stretching to the lake. A line of cairns is erected on the top, intended to deceive the deer.. They take to the water as they see no retreat. If there are no hills, a line of cairns is erected on some part of the plain.

As soon as the deer are in the water, the natives pursue them in kayaks and kill them with the spear. Sometimes the wounded deer will turn upon the boat, in which case the hunter must escape with the utmost speed, else he will be capsized, or the skin of the boat will be torn to pieces by the animal's antlers.

In some of the narrow valleys with steep faces on both sides, the deer are driven towards the hunters, who lie in ambush.[1] If the deer cannot be driven into the water, the Eskimo either stalk them or shoot them from a stand. In a plain where the hunter cannot hide himself, it is easier

[1] Boas, *Sixth An. Rep. Bur. Ethnol.*, 1888, p. 501.

to approach the herd if two men hunt together. They advance, the second man hiding behind the first one by stooping a little. The bows or guns are carried on the shoulder, so as to resemble the antlers of a deer. The men imitate their grunting and approach slowly, now stopping and stooping, now advancing. If the deer look about suspiciously, they sit down, the second man lying almost flat on the ground, and both at some distance off greatly resemble the animals themselves.[1]

"The common deer are far more dangerous to approach in canoes, as they kick up their hind legs with such violence as to endanger any birch-rind canoe that comes within their reach; for which reason all the Indians who kill deer upon the water are provided with a long stick that will reach far beyond the head of the canoe."[2]

This assertion of Hearne's is in a line with the previous remarks that the ingenious savage knew just how long he should make his arm to give the deadly thrust and keep himself out of harm's way. Practically in these expeditions one man manages the canoe, while the other, standing in front, handles the lance.

"When the Central Eskimo hunt the musk-ox," says Boas, "the dogs are let loose as soon as a track is found. The musk-oxen form a circle of defence, in which they are kept at bay until the hunter approaches. While the dogs continue attacking and dodging, the musk-oxen try to hit them with their horns, and do not heed the Eskimo, who assail them at close quarters with a lance to which a thong is frequently attached. When an ox is wounded it makes an impetuous attack on the hunter, who dodges to one side. The dogs being at hand again immediately keep it at bay, thus enabling the hunter to let fly another arrow or throw his lance again. Thus the struggle continues until the

[1] Boas, *Sixth An. Rep. Bur. Ethnol.*, 1888, p. 508; quoting also Ross, *Second Voyage*, London, 1835, Webster, p. 252; and Parry, *Second Voyage*, p. 512.

[2] Hearne, *Journey*, &c., London, 1795, Strahan, p. 257.

greater part of the herd is killed. In rare instances an ox dashes out of the circle and escapes from the pack." [1]

It is worth while to notice here the dog coming in as a helper of mankind in running down and killing his own fellow-creatures. The *cheetah*, or hunting leopard (*Felis jubata*), of the Deccan, will also occur to readers in this connection. Hunter says, "The speed with which it bounds from the cart after the antelope exceeds the swiftness of any other wild animal." [2]

The Central Eskimo pursue the polar bear in light sledges, and when they are near the game the traces of the most reliable dogs in the team are cut, when they dash forward and bring the bear to bay. As the hunter gets sufficiently near, the last dogs are let loose, and the bear is killed with a spear or with bow and arrow. The best season for hunting bear is in March and April, when they come up to the fjords in pursuit of young seals.[3]

For killing elephants the Batonga tribes, on the Zambesi, erect stages on high trees overhanging the paths by which the elephants come, and then use a large spear with a handle nearly as thick as a man's wrist and five feet long. When the animal comes beneath they throw the spear, and if it enters between the ribs above, the motion of the handle, aided by knocking against the trees, makes frightful gashes within, and soon causes death. They kill them also with the dreadful spear inserted in a block of wood suspended above, and released by a cord stretched across the path.

To the simple lance the ingenious savage added many devices. Chief among them was the hand-rest or stop on the side of the shaft, to prevent the cold or gloved hand slipping on the shaft when the plunge is made.

[1] Boas, *Sixth An. Rep. Bur. Ethnol.*, 1888, p. 509.

[2] *Imp. Gazetteer of India*, London, 1886, Trübner, vol. vi. p. 653.

[3] Boas, *Sixth An. Rep. Bur. Ethnol.*, 1888, p. 509. For the ingenious apparatus of the Central Eskimo in hunting deer, musk-ox, and bears, see Boas, *Sixth An. Rep. Bur. Ethnol.*, 1888, pp. 508–10, figs. 438–51. Piercing weapons of the Pt. Barrow Eskimo are exhaustively treated by Murdoch, *Ninth An. Rep. Bur. Ethnol.*, 1892, pp. 240–44, figs. 238–45.

An elaborate device is to be seen on many Eskimo whale lances as well as on the harpoons, which astonishes every one who sees it for the first time. The iron blade of the lance is inserted in an ivory tang or blade-piece, which is rounded at the inner end to fit into a shallow socket in the ivory socket-piece at the end of the shaft. When the tang is in place a raw-hide line is passed through both tang and shaft once or twice, so that when dry it holds the head straight out. But when the weapon is driven home into a large sea mammal, and much thrashing about in the water precedes death, the tang is slipped out of its shallow socket and the breaking of the shaft prevented. In later times, the boar spear, the lance, and even the bayonet are the descendants of these primitive devices for thrusting through the game. But every one of them is simplicity itself compared with the intricate apparatus devised by the Arctic Highlanders.

Each one of the classes of weapons before-mentioned may become missiles. The Moki Indians have a rabbit stick, which they throw at the legs of game running from them. The Australian has his boomerang and club, and the African his knobbed stick.

Some of these have edges for cutting, but the African throwing-axes, under various names, are marvels of casting and slashing weapons. The Plains Indian and the trapper hurl their hunting-knives and tomahawks with great dexterity.

Among ancient missiles for destroying animals none can compare with the arrow. The bow and the arrow have been the pride of the warrior also, and the inventions of the bowyer and the fletcher might better be described in the Chapter on War.

The bow is the same in both activities, excepting that the Eskimo, who are most ingenious bowyers, never go to war. But the arrow on American soil has been more highly developed in the arts of hunting and fishing. The possible parts of a most complex arrow are shaft, fore-shaft, barb-piece, head, feather, and nock. In addition to these parts occur the seizing, blood-streaks, shaftment-streaks, and owner-marks. The simplest arrow is a mere rod or scion of wood, with blunt head for knocking down birds. Each tribe of savages, and each kind of hunting and fishing, and each region has its peculiar arrow. The most complicated form in America was the sea-otter arrow of the Alaskan Indians, which might be thus described: shaft of cedar, thirty inches long and three-eighths of an inch thick; fore-shaft of bone, six inches long; feathers three, daintily laid on and trimmed; "cock-feather," white; nock large, bulbous, and deeply notched; head, a dainty little barb of bone or native copper, fitting loosely into the outer end of the fore-shaft, pierced on the shank for the fastening of a braided martingale of sinew-cord; martingale tied into the head at one extremity and at the other divided for

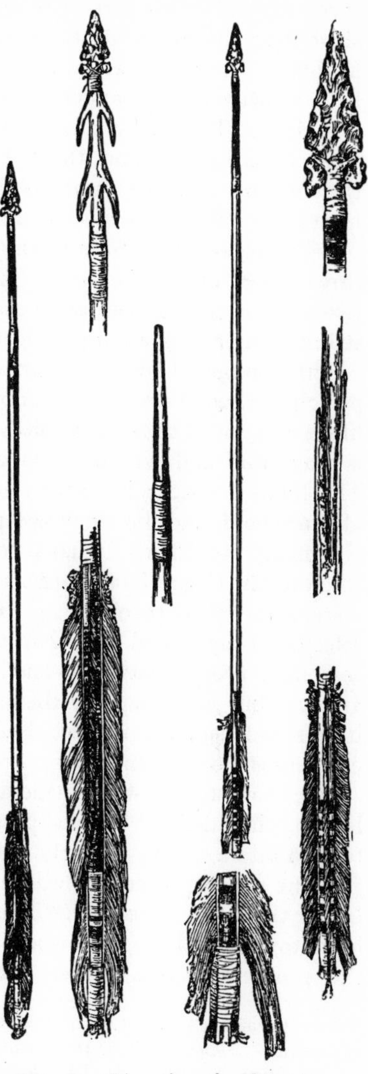

FIG. 53.—Dissection of a Hupa Arrow. (*U. S. Nat. Museum.*)

three feet into two parts, the end of one part tied to the shaft near the fore-shaft, the other end made fast near the feather. When this arrow is shot at the otter, the little barb is driven quite under its skin and is pulled from the fore-shaft. The sinew martingale unwinds, the bone fore-shaft sinks in the water, the tell-tale feathers bob about in the air, the shaft acts both as drag and buoy, aiding the hunter to follow and retrieve his game.[1]

"The bows of the Deer Horn (Athapascan) are formed of three pieces of fir, the centre piece alone bent, the other two pieces lying in the same straight line with the bow-string; the pieces are tied together with sinew."[2]

This compound bow is even in our day framed on the same principle, though in some examples the elasticity is in the limbs and not in the central piece. The horn of the caribou is often used, and even whalebone in place of the fir-wood. The limbs are securely lashed to the central piece by means of sinew-cord, and the whole weapon is always clumsy.

Among the Plains Indians bows were made from the wood of the Osage Orange (*Bois d'arc*), and long journeys were often taken to obtain it. Only the best stocks were selected, straight, and as free from knots as possible. The seasoning process was slow and thorough, a little scraping and cutting and shaping, then a rubbing with fat, and it was laid aside for weeks. Each warrior had several in different stages of completion.

The bow-strings were made of sinew, cut out in full length, shredded as fine as possible and then spun and twisted into a string, perfectly round, and uniform in size.[3]

"The making of an arrow," says Dodge, "requires more labour than that of a bow. The Plains Indians used any hard, tough, straight-grain wood. It was scraped down to the proper size and shape. Under the most favourable

[1] See the author's minute description of North American bows, arrows, and quivers. *Smithson. Rep.*, 1893, many figures.

[2] Franklin, *Narrative, &c.*, London, 1824, vol. ii. p. 180.

[3] Dodge, *The Plains of the Great West*, New York, 1887, p. 348.

circumstances the most skilful Indian could not hope to complete more than a single arrow in a hard day's work. In a short fight or an exciting dash after game, he will expend as many arrows as will keep him busy for a month to replace."[1]

Something in the nature of a land harpoon, arrow, or retrieving apparatus for burrowing creatures is found in both hemispheres. The Indians of Arizona and Southern California make barbs on the side of the arrow-shaft near the head, so that when a prairie dog is shot and runs into his hole he may be retrieved. The Utes make a hook which they thrust into the hole to fish out small mammals, and the Australians do the same. Though made to capture fruit and not animals, the Andamanese hook on a pole fifteen feet long, to pull down jack-fruit, is in the same line of invention. It lengthens man's arm, and enables him to retrieve things out of his reach.

Colonel Dodge declares that the Plains Indians had not the slightest knowledge of trapping. They seem to be the only aboriginal people in the world who have not some pit-fall, spring, or native trap. They had no knowledge of angling. The Indian had no "necessity," and his invention was therefore never born. I attribute this lack to the plentiful supply of large game.

The same author describes graphically the pursuit of this large game. "With his head covered by a cap of grass or weeds, the Indian will lie for hours on his belly, noiseless as a snake, watching the game ; now perfectly motionless, now crawling a few feet ; no constraint of position, no fiercest heat of the sun, seeming to affect him in the least. He will lie for a whole day at a water-hole, waiting for the game to come to drink, in such position that the wind will not reveal him.

The Plains Indian hunts but little in winter. Every year "the great fall hunt" is made for the purpose of killing and curing the supply of meat for winter's use. Great prepara-

[1] Dodge, *Plains of the Great West*, New York, 1887, p. 349.

tions are made in advance. Runners are sent out to seek the most eligible position for the camp. It must be near water, there must be timber for tent-poles and drying scaffolds, and level sward for the stretching and drying hides, and the location must be, above all, in the centre of a region abounding in game. The spot being decided on, the whole band move to it, and everything put in order for work.

The "dog soldiers" are masters now. All things being ready the best hunters are sent out before dawn. The herd is selected for slaughter whose position is such that the "surround" will least disturb the others. A narrow valley with lateral ravines is favourable. If the herd is unfavourably situated the hunter waits for it to go to water, or by discreet appearance at intervals drive it to the best spot. During all this time the whole active masculine portion of the band is congregated out of sight of the buffalo, silent and trembling with excitement.

The herd being in proper place, the leaders tell off the men and send them, under temporary captains, to designated positions. Carefully concealed, these parties pour down the valley to leeward, and spread gradually on each flank of the wind until the herd is surrounded, except on the windward side. Seeing that every man is in his place and all ready, the head hunter rapidly swings in a party to close the gap, gives the signal, and with a yell that would almost wake the dead, the whole line dashes and closes on the game.

The buffaloes make desperate rushes, until, utterly bewildered, they almost stand still and await their fate. In a few moments the slaughter is complete.

When bows and arrows were used each warrior, knowing his own, had no difficulty in positively identifying the buffalo killed by him. These were his property, except that he was assessed a certain portion. If arrows of different men were found in the same dead buffaloes, the ownership was decided by their position.

Since the use of firearms the identification of the owner

has been impossible, and new laws of division have been invented.

The slaughter completed, the "soldiers" return to camp, while the women skin, cut up, and carry to camp almost every portion of the dead animals. As soon as the women's work is done other "surrounds" are made until enough meat and skins are obtained. The work of the woman is most laborious during the fall hunt. If the buffaloes are moving the success of the hunt may depend upon the rapidity with which she performs her work on a batch of dead buffaloes. These animals spoil very quickly if not disembowelled. The men do not wish to kill in any one day more than the squaws can skin and cut up on that same day.

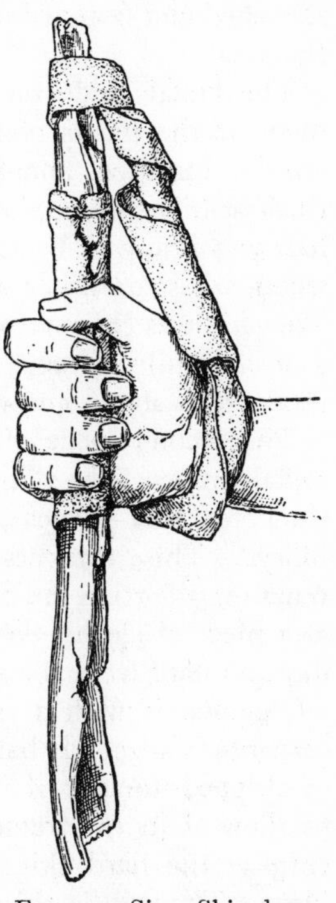

FIG. 54.—Sioux Skin-dressing Tool, made from an old waggon skein. (*U. S. Nat. Museum.*)

No sooner are the buffaloes dead than the squaws are at work. The skin is removed with marvellous celerity. The meat is cut from the bones, tied up in the skin, and taken to the camp. The entrails, emptied and eaten raw, form the principal food during the hunt. Marrow-bones and ribs roasted on the coals serve for delicious suppers. All these are prepared by women and brought to camp.

The meat is thoroughly dried on the pole scaffolds until it is as hard as a rock. It is then pounded into meal by means of stone mauls, and packed in cases made of raw-hide (parflèche cases). Melted tallow is poured over the whole, which is kept warm until the mass is thoroughly saturated. When the meat, now called pemmican, is cold, the parflêches are closed and tied up. The contents so prepared will keep in good condition for several years.

The skins, as soon as they are emptied of their freight of meat, are spread, flesh-side upward, on a level piece of ground, small slits cut in the edges of each, and it is stretched and fastened down by wooden pegs driven through the slits.

The buffalo hide received three different kinds of treatment at the hands of these aboriginal skin-dressers. No tannin was used, and no leather was really made. The thickest hides were selected for shields, parflêches, &c. The hair was removed by soaking the skins in water in which wood ashes, or other alkaline substance has been mixed. The skin was then cut into the required shape and put on a form while green. When it became dry raw-hide, it retained its shape, and was almost as hard as iron.

The second mode of treatment is the production of the buffalo robe. The skin in its natural condition is much too thick for this purpose, being unwieldly and lacking pliability. This thickness must be reduced one-half, the remaining portion must be uniform throughout, and as soft as a piece of cloth. When the stretched skin has become dry and hard from the action of the sun, the woman goes to work upon it with a small iron instrument, shaped like a carpenter's adze. It has a handle of elk horn, and the blade of chipped stone or of iron is lashed on with raw-hide, so as to allow of its easy removal for sharpening. With this she chips at the hard skin, cutting off a thin shaving at each blow. The skill in this process is shown in cutting the skin and not cutting through it, and in obtaining a perfectly smooth and even inner surface and uniform thickness. To render this skin soft and pliable, every little while the chipping is stopped and the surface smeared with fat and brains of buffalo, which are thoroughly rubbed in with a smooth stone.

The third process on the buffalo hide is for making

lodges. The hair is taken off, the skin reduced in thickness, and the whole made pliable as above.

In addition to the buffalo hide work, the same tribes dress the hides of deer beautifully, producing soft and indispensable material for clothing.[1]

The differentiation of aboriginal work is well shown in the foregoing description. The men appear as organised, intelligent, obedient to a leader, observant, self-possessed, quick-sighted, brave, strong, enduring. The women assume the industrial *rôles* of butchers, meat-packers, cooks, purveyors, carriers, hide-dressers in three forms, tent-makers, clothiers, trunk-makers, shoemakers, modistes, common carriers, and house-builders.

The apparatus of the Andamanese for pursuit are the S-shaped bow, with its arrows of five varieties, the pig spear, recently introduced, and the fish or turtle spears. Their arrows have no feather, and are held steady in flight by a fore-shaft of hard wood. They use no arrow straightener but their teeth and fingers to keep the shafts in line. Mr. Man draws attention to one "improvement" which would now entitle them to a patent. In one style the blade of the arrow, the barbs, and the seam into which the tang is inserted, are in the same plane, and the seam is used as a "sight." In the seamless arrow the barb which is most in line with the blade is placed uppermost in shooting.

In the Malay Archipelago the *Sumpitan*, and in South America the *Sarbacan* (*Zarabatana*), and the *Pucuna* stand in the place of our air-guns and rifles. They both discharge projectiles from a tube by means of the sudden expansion of a gas of some kind. The ingenuity of the savage is put to its most efficient exercise in this apparatus.

Nature supplies the Indian with material for blow-pipes all over America where canes of any kind grow. In the

[1] Colonel Richard Irving Dodge, *The Plains of the Great West*, New York, 1877, pp. 353–359. This quotation is necessarily abridged, the whole account should be read. Parkman, in *The Oregon Trail*, gives splendid accounts of old-time buffalo hunting.

Southern States of the Union the Muskhogean stock were familiar with them, and the Attacapas and Chetimachas lashed several reeds together, thus anticipating the revolver. "The quiver for the darts is a neat affair generally, in shape of a dice-box. Attached to the quiver is a lid made of the tough hide of the tapir. Inside the quiver is a bundle of darts, the lower jaw of a perai fish (*Serasalmo niger*), for preparing the missiles and some crowia fibre for wadding. The darts, each about eight inches long, are made of splinters of the woody midrib of the cockerite palm (*Maximiliana regia*), as sharp as needles, which are dipped in urari poison. These darts are fastened together, palisade fashion, by means of two parallel plaits of string, and wound around a spindle, on the top of which a few sticks are tied together in the form of a wheel. This is to protect the hand from any chance of contact with the poison-smeared points of the darts. When about to use the weapon the Indian withdraws one of the darts, wraps around the butt end enough wadding to fill the end of the tube cleverly, and then, pointing at his game, with a quick puff of his breath he drives the dart from the tube. In the lands where noisy guns have not frightened the life out of birds and beasts, it is easier to steal close upon the game, so as almost to bring the point of the tube in contact with its body." [1]

The blow-tube is a tropical invention, confined to areas where the cane abounds. Though it was used by Indians of the Southern States of the Union, Colonel Lane Fox is correct in saying that the two areas of the full development of the apparatus are South America and Southern Asia. Four varieties are mentioned by him, the *Zarabatana* and the *Pucuna*, in South America; the *Sumpitan*, of Borneo, and the *Tomeąng*. Each of these, as will be seen, has reference to the work to be done and the materials at hand. The *Zarabatana* is formed of two separate pieces of

[1] *Cf.* im Thurn, *Ind. of Brit. Guiana*, London, 1883, p. 300. Compare Wood's *Uncivilised Races*, Hartford, 1870, vol. ii. pp. 465; 583–587. Good figures in Wood, though rather dark.

wood, in each of which is cut a semicircular groove by means of the incisor teeth of rodents, so that when they are placed in contact with each other they form a long wooden rod, pierced with a circular bore. The two halves are bound together by means of long stripes of *Jacitara* wood. The arrow is made of the leaf rib of the *Concourite* palm, wound round at the near end with wild cotton, in order to make it fit the bore. It is pointed by scraping it between the sharp teeth of the *Piraï* fish. The arrow is poisoned, and before shooting it the Indian cuts the shaft almost in two near the point. The Yameos are said to shoot thirty or forty yards with them.[1]

The *Pucuna* is constructed of two portions, the inner reed, called *Ourah* consists of the first joint of the *Arundinaria Schomburgkii*, which grows on the sandstone ridge of the Upper Orinoco. This is inserted in an outer tube, called *Samourah*, which consists of the stem of the palm *Ircartia setigera*, the interior pulp of which is previously removed, and the spaces between the inner and outer tube tamped with a black wax made by a wild bee and mixed with a pitchy substance. There are varieties of both these classes, for the weapon is a widely diffused one in Tropical America.[2]

The blow-pipe of Borneo, called *Sumpitan*, is of one piece, constructed of various kinds of wood, bored with great care, like a gun-barrel. The arrows are made of the thorn of the sago palm and have a conical piece of pith, or soft wood, either solid or hollow, attached to the barbed end. The arrows are poisoned and are carried in a bamboo quiver.

The *Tomeang* is the blow-tube of the Mautras—aborigines of the Malay peninsula. It is made, after the fashion of the *Púcuna*, of two bamboo tubes, one inside the other. The outer one is usually ornamented.[3]

In Copan even the children go armed with a *sarbacan*,

[1] Lane Fox, *Catalogue*, *&c.*, pls. i. and ii., p. 148.

[2] *Ibid.*, pp. 148–149.

[3] *Ibid.*, p. 151.

or blow-tube, an instrument which they use very dexterously and which they have inherited from their earliest ancestors.[1]

Fish and fowl inhabit elements inaccessible to man. He cannot seize them if he would. As for insects and many smaller land mammals, he may pick them up with his hands. The first seizers were hands. When men desired to take their game from the air or from the waters, they had to elongate their arms with poles and to put fingers of bone or wood at the end with palms of network. Savages were good swimmers and knew how to dive into the depths to bring up their treasures, but they also knew how to make rakes for oysters and clams and even for pearls.

But for the fish he had to devise the dip-net, and uses it still. Moreover, with longer hands he secreted himself where the migratory birds congregated in vast quantities and dipped them from the air much as an entomologist secures his insects. In the good old days when wild fish were plentiful it was only necessary to wade in the water with almost any kind of basket or network to secure a dinner. The Quilleute Indians of Cape Flattery make a scoop-net like a great barn shovel. On the appearance of the fish they rush into the surf and press the outer edge of the net down firmly on sand or shingle, the swash of the breaker forces the smelts into the net ; then as the water recedes the fishermen turn round quickly and hold the net so that the undertow will force more smelts into it. In this way at least a bushel are taken at a single scoop. This is rather better than picking up things with the hand. The

[1] Morelet, *Travels in Central America*, New York, 1871, p. 334, quoting, *Taladran sutilmente las zabratenas con puas muy largas.* Herrera, Dec. iv., l. x., c. 14. Also, "Among the presents which he [Montezuma] sent to Cortez were a dozen of these implements, painted with considerable skill, &c." Lorenzana, l. ii. p. 100. See also Bancroft, *Native Races of the Pacific States*, for the use of the blow-tube after the manner of the modern peashooter, vol. i. p. 627 ; for the style of poisoned arrows in vogue on the Isthmus, vol. i. p. 762 ; on clay pellets as blow-tube missiles in Guatemala, vol. ii. p. 720.

FIG. 55.—FISHING FOR SMELTS IN THE SURF, WASHINGTON STATE.
(*From Smithson. Report.*)

dip-net is used by savages quite as often as by civilised man in taking fish from the surface of the water after they have been drawn up or captured by the hook. Whatever length of arm or kind of artificial fingers were demanded they were forthcoming.

The Dyak women join in the sport of wholesale fishing and scoop up the small fry with their nets. The scoop-net is used chiefly by the women, who are fond of wading up the shallows, net in hand and basket slung from the shoulder, scooping up the prawns and periwinkles, &c., that come in their way.[1] The most recent modifications of this simple process are terribly destructive of marine life. The great purse seines worked by steam for gathering fish to be used as fertiliser have played havoc with one food supply, though they do put back on the land in this way a deal of waste from cities.

Among implements of seizure the hook stands pre-eminent, and nothing would be more interesting than a series of fish and animal hooks arranged by a patent examiner and an ethnologist. The evolution of the hook, the adaptation of structure to function, the control of environment over material and form would furnish an interesting topic for a natural history study.

The simplest form of a hook is really a barb, though barbs on hooks come later. In the same way the Makah Indians of Cape Flattery attach a hook to the end of a long pole which is held down in the water until a salmon is felt against it, when with a quick pull the fish is hooked and hauled on board. The modern fisherman cannot dispense with the gaff-hook, which is only the savage method improved.

The negroes of the Southern States, and perhaps their kindred elsewhere, split the end of a stick, run it into a hole wherein a coon or an opossum is hiding, and having twisted the fork into the long hair of the animal withdraw it from its hiding-place with ease. It used to be said in plantation

[1] Ling Roth, *J. Anthrop, Inst.*, London, 1892, vol. xxii. p. 50.

days that the same implement was useful in stealing cotton from the warehouse through crannies and knot-holes.

For the purpose of securing a great many candle-fish or oulachon (*Thaleichthys pacificus*) at once, the Salish tribes about Port Townsend employ a rake or coarse comb with many teeth at the end of a long handle. The fisherman draws the rake steadily through a school of the fish and impales several at once. The catch is shaken from the teeth quickly into the boat and the process repeated. The squid gigs of our civilised fishermen are in the same line of invention.

The barbed spear for fishing is used by the Passamaquoddy Indian. In this case we have two motions, a thrust and a pull. The Fuegians employ a similar apparatus with a detachable barbed point set in a socket and fastened by a lanyard to the head of a long shaft. In many Oceanic examples the barbed point of a long spear points to the double function of thrusting and retrieving. There is no reason why this device may not have been useful on land, but it blossomed out more vigorously on the sea, and led to the invention of many-pointed spears in great variety. The Norton Sound Eskimo employ a salmon spear having three points attached to the head of the shaft. The central point is a plain piece of ivory, but the lateral points are barbed. As we follow down the West Coast of America, each region presents some curious modification of striking and retrieving, adapted to the depth of water, the game and the material available. Some allowance must be made for intelligence in the tribes, but it is difficult to conjecture how they could have done better, though the Indians of this coast represent a great variety of linguistic stocks. The Makah Indians of Puget Sound attach as many as five barbed points to the head of a long slender pole for killing ducks. "At certain times during stormy weather the wild fowl congregate in vast numbers in Neah Bay. The Indians go out in their canoes with a bright light from torches of pitchwood placed in the stern. The canoe is paddled stern first among the flocks of wild fowl. The birds, bewildered by the light, are killed in great numbers. The prongs of the spear get entangled among the feathers and hold fast. A bird is hauled in the canoe, its neck wrung, and others in succession are quickly speared. Sometimes as many as a hundred canoes will be out at the same time and the light from the torches is an interesting sight.[1]

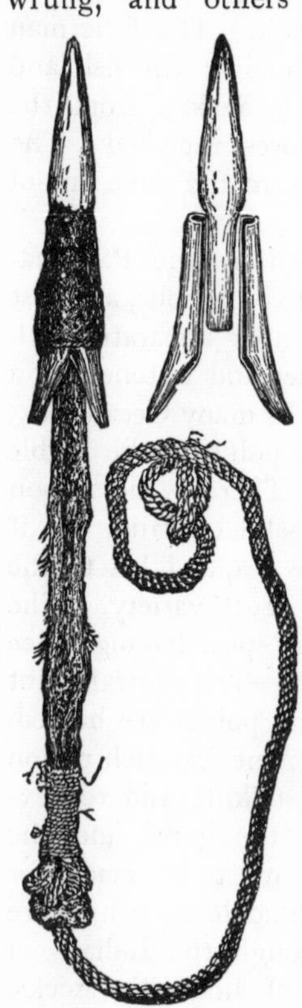

FIG. 56.—Toggle Harpoon of Salish Indians, complete and dissected.

There is a class of hunting and fishing implements which forms the connecting link between those for striking and retrieving and the class of missiles. Indeed some of the harpoons are held in the hand when the thrust is made, others are hurled in some fashion.

As to their heads, this class of weapons is divided into toggle harpoons and barbed harpoons. The latter are very simple, consisting of a slender shaft, a fore-shaft of bone or ivory, a barb fitting loosely into a socket in front of the fore-shaft, a line running from the barb and attached to the shaft by a "bridle" at two points. These are either plunged into the seal or other game, or hurled from a "throwing-stick." The barb fastens under the skin and slips from the fore-shaft. The line unrolls from the shaft and the latter stands up in the water to act as a drag and a buoy. A similar device is used by fishers after swordfish, only a keg serves for float and a flag for the buoy. The whalers substitute also the explosive power of gunpowder for the kinetics of the arm.

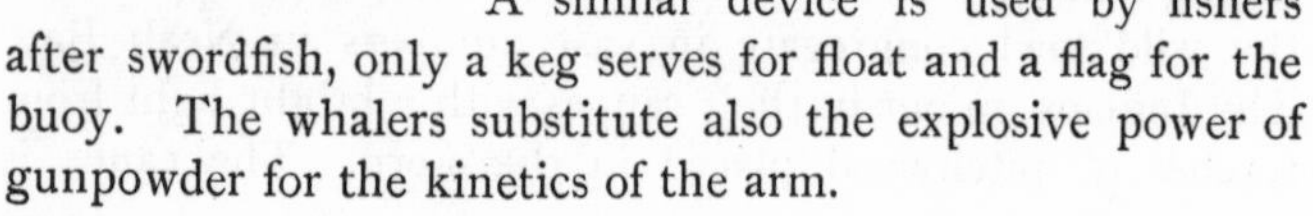

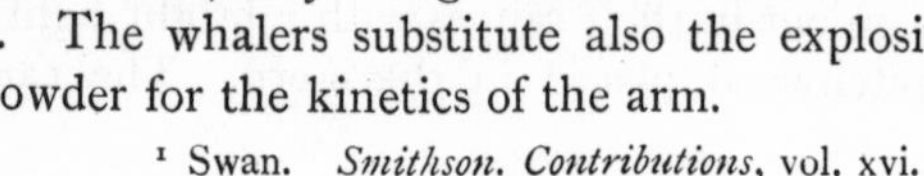

[1] Swan. *Smithson. Contributions*, vol. xvi.

The toggle-head harpoon is a most complicated affair. Its parts may be shaft, fore-shaft, loose shaft, toggle-head, ice-pick, assembling line, toggle-head becket, leader, hand-rest and float. The shaft is a stout staff of wood, one end of which is attached to the ice-pick or spud, and to the other the fore-shaft of ivory or bone. Upon this shaft will also be found the hand-rest and the line-hook. The loose shaft is a rod of ivory, dull pointed at both ends, the one fitting into the socket of the fore-shaft, the other into that of the toggle-head. It is also securely held in place by a line passing through a perforation near its base, and one through the fore-shaft. The object of this most ingenious piece is to assist in mounting the toggle-head for action, and to prevent the breaking of the weapon by a sudden lateral strain, in which case the loose shaft slips from its shallow socket. The toggle-head is the part that enters the body of the animal, is turned at right angles by the line passing through a hole in its middle, and acts as a toggle in preventing escape. This line is

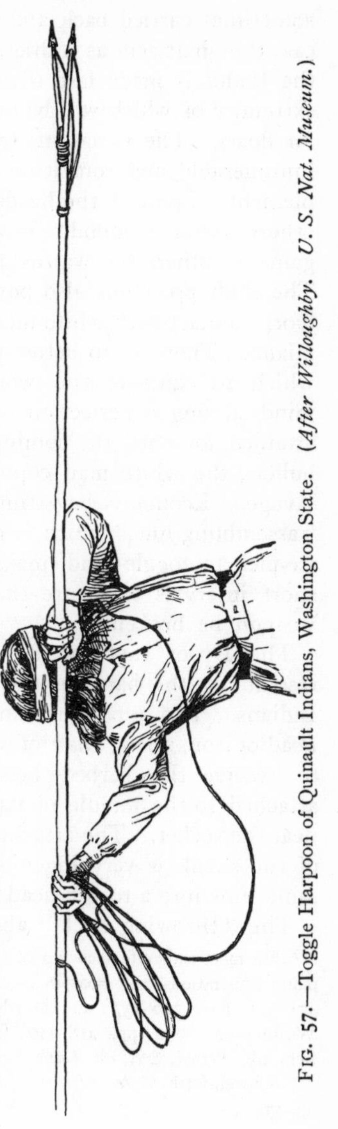

FIG. 57.—Toggle Harpoon of Quinault Indians, Washington State. (*After Willoughby. U. S. Nat. Museum.*)

sometimes carried back and attached to the shaft, in which case the shaft acts as a drag, and a buoy at other times, or the leader is made fast to a long walrus-hide line at the extremity of which will be one or more seal-skins blown up for floats. The variations from these general features are innumerable, and constitute the natural history of the implement. Some of the heads are of tiny proportions, while others weigh a pound. In one locality the whale is the game, in others the walrus, the seal or even the sea otter. The shaft proclaims also poverty or abundance of wood or ivory, contact with white men, and even trade with Indians inland. There is no better piece of savage mechanism on which to compare the workings of different aboriginal minds aiming at perfection. And this seemed to have been attained, for until the coming of gunpowder and explosive bullets, the white man copied implicitly the work of the savage. Economy begetting caution is set forth in the "assembling line," which is a raw-hide string, passing from ice-pick to toggle-head, making knots and half hitches at short intervals to insure the salvation of the parts if the weapon be broken.[1]

The Dyoor capture the crocodile with an apparatus resembling the harpoon with bone barb of the American Indians. The apparatus consists of three parts, a barbed head of iron, a long shaft of wood, with a socket at one end to receive the barbed head loosely, and a stout line attached to the middle of its shank and to the long shaft near the socket. The attachment of this line to the middle of the shank is very ingenious, since it converts it at the same time into a toggle-head.[2]

The "throwing stick" above mentioned, as a hurler of

[1] For most elaborate accounts of the harpoon and of seal-hunting, with many illustrations, see Boas in *Sixth An. Rep. Bur. Ethnol.*, 1888, pp. 471–501, figs. 390–437; and Murdoch, *Ninth An. Rep. Bur. Ethnol.*, Washington, 1892, pp. 218–240, figs. 206–240, and compare with von Schrenk, *Reisen, &c., im Amur-Lande*, St. Petersburg, vol. iii.

[2] Schweinfurth, *Artes Africanae*, London, 1875, Sampson Low, pl. ii., fig. 17.

the harpoon, is found in both Australia and among the Eskimo everywhere, as well as in many parts of Middle and South America. The essential parts of this implement are the groove for the dart, the hook for the end of the dart, and the provisions for the fingers. Almost all the Eskimo specimens are right-handed, while those further south fit either hand.

With this dart-sling the hunter adds an extra joint to his arm, and this hurls his javelin with more deadly effect. Inseparable from this weapon in North America is the bird trident, which consists of a slender shaft, to the end or to the middle of which three barbs are lashed. In the latter case an extra barbed point is placed on the front end. The purpose of this weapon is to cripple ducks and other birds on the wing.

The throwing stick is used to hurl the barbed harpoon, but the Greenland Eskimo attach it to the side of the great harpoon, which new device indicates that the throwing stick travelled from West to East.

The sling is of the same nature as the throwing stick, only the velocity is intensified by the whirling of the sling around the head. Here and there it is in use in pursuit of game, but it is not found to be a universal favourite in savagery.[1]

For the seizure of swift animals, the lasso, in its various forms, must not be omitted. It is a noose at the end of a long line. This form of capture occurs also in the series of traps, where the victim more or less consciously places its head into a noose. But the lasso is virtually a long tentacle, which the hunter thrusts out with great dexterity, and seizes the poor creature by the horns or the legs. It is known all over the Americas. Even the Eskimo boys are said to be expert in the use of it. But it is virtually an

[1] See Mason, *Report U. S. Nat. Mus.*, 1884, and *Proc. U. S. Nat. Mus.*, Washington, 1893, p. 219. Both articles illustrated and authorities given. See also Lane Fox, *Catalogue*, p. 38; Murdoch, *Ninth An. Rep. Bur. Ethnol.*, Washington, 1892, pp. 210–217, figs. 195–205.

instrument of domestication, and was most probably introduced into the Western World from abroad.

Quite similar to the lasso is the bolas, an apparatus for

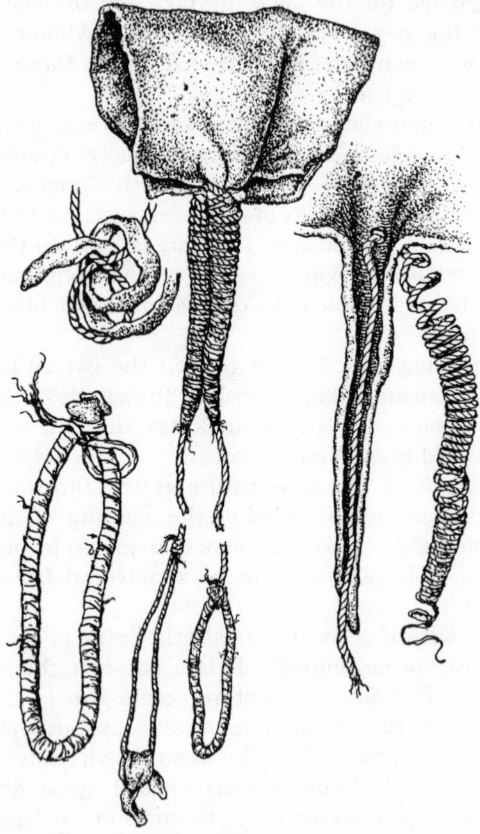

FIG. 58.—Sling of Hupa Indians dissected. (*U. S. Nat. Museum.*)

tripping or entangling game. The Eskimo tie little balls of ivory or bone at the ends of strong sinew cords at least a yard long. Half a dozen of these are joined together at the ends away from the balls. The hunter whirls one over his head and hurls it among a flock of geese or ducks, so that the balls will spread out in their flight. One of them is sure to entangle itself around the wing or limb of a bird and bring it down.[1]

In South America the balls of the bolas weigh a pound each, and the apparatus is used on ostriches, the guanacos, and even upon horses and cattle.

"When the Patagonian goes out hunting he carries no weapon except a bolas and a knife. . . . Should he see a herd of guanacos, he makes silently towards them, imitating the cry of the young one in distress. . . . When a small herd is seen, they can generally be enticed within range by a hunter on foot who plays various antics, such as lying on his back and kicking his legs in the air, waving a bunch of feathers. The inquisitive creatures seem unable to resist the promptings of their curiosity, and though they are really afraid of the strange object, come closer and closer until the hunter is able to hurl the terrible bolas at them."[2]

One of the curious devices of the seal hunter is his little probe with which he ascertains the presence of the seal at the breathing hole. It is a rod of antler or ivory, not larger than the knitting needles for coarse worsted. A thread of sinew is attached to one end, and the other end is allowed to drop down through the hole. The seal approaching pushes up the rod with its nose, very much as a perch plays with the fisherman's bait, and informs the hunter just when and where to strike.

The arrow has some forms in each country devoted to the exclusive purpose of killing and retrieving fish and game, and never used to kill men.

The Point Barrow natives have an arrow called "sleep-a-night-and-die." It is a long, slender, barbed head of bone or ivory, very loosely fitted into a socket in the end of the shaft. Shot into a reindeer or bear, this arrow goes searching for some vital part at each step of the creature, and at

[1] Murdoch, *Ninth An. Rep. Bur. Ethnol.*, p. 245, figs. 247, 248.

[2] Wood, *Unciv. Races*, Hartford, 1870, vol. ii. p. 532.

last rankles in the wound until the game succumbs. Fishing arrows with barbs are universal. The South American turtle arrows combine some of the features of the Eskimo barbed harpoon, only in the South American turtle arrow the socket is in the barbed head. This barbed and socketed head is found on fish spears along the West Coast of the United States.

The Aleutian islanders have an arrow which operates in precisely the same manner as the barbed harpoon with a bridle. Nothing could exceed the delicate complexity of this apparatus with its bulbous nock, its neatly trimmed feathers, its slender fore-shaft, into which fits a dainty barb of bone or native copper attached to the shaft in two places by a line and bridle of beautifully braided sinew cord. Of course the Papuan arrows are more ornate, but these are the most highly finished projectiles in the world.

Among the Siouan as well as among other tribes dwelling on the plains of the Great West, men went out singly or in small numbers to hunt. But there existed, besides this method, a more general practice of co-operative hunting worthy of mention.

In the first place, these expeditions had their special seasons, when the corn and pumpkins had been planted, the beans gathered, and the like. They terminated, says Dorsey, when the wind blew upon the sunflowers, which was about the first of September. It was then the whole people camped in the tribal circle on the open prairies. The fall or winter hunt was organised first in accordance with the weather, or, some would say, the weather gave form to the autumn hunt. Again, the state of the game had somewhat to say along this line, for it was then that the hides were covered with thick hair, or at this season the food was in good condition.[1]

In the next chapter will be studied the conflict of mind between men and animals.

[1] Consult Dorsey, *Third An. Rep. Bur. Ethnol.*, for an extended account of the Omaha hunting customs, pp. 283–302.

CHAPTER IX.

CAPTURE AND DOMESTICATION OF ANIMALS.

"The nature of wild animals doesn't change like the nature of men; we have grown wiser, while they have remained the same."—HALSEY THRASHER.

IN the foregoing chapter on the killing and seizing of animals, the inventive faculty was engaged in a duel between men and brütes. The present chapter will be concerned with the outwitting and the enslaving of animals, a branch of our subject which may be called *zootechny*. The successful hunter of fish and birds and beasts must needs know a great deal about their homes, their habits, their times, and seasons. He is indeed a naturalist. But he acquires his knowledge for the sake of using force upon them. In this his intellect is stimulated, inventions are developed, great rewards are secured, and tribal genius is established.

But when the hunters or the fishermen had in mind to outwit their prey and to induce them to ensnare themselves or commit self-slaughter, the thoughts involved were of a higher order, the social structures brought about thereby was more highly organised, the apparatus used had more parts and came nearer to the idea of machinery. Indeed, the outcome of this series of activities has been the connection of animals with machines of many kinds. In the trapping of animals the intellect of the man plays a game of skill with the instincts and thoughts of the brute.

A convenient classification of devices for inducing animals to offer themselves to human comfort is into *tackle* and *traps*. The former is a general term for apparatus used in the water, the latter in land pursuit.

For ensnaring water animals the fisherman has two plans which he follows, namely, (1) to set his meshing or his labyrinthine nets where the marine animals will unconsciously swim into them in the ordinary routine of their lives; (2) to bait his hooks with such things or in such fashion as to tempt the aquatic epicure to swallow hook and all.

The savage man's skill in fishing is undoubted, and has always been the admiration and the envy of the civilised. The gill-net, the fish-trap, the weir, the pound, the tide-trap are well known to the aborigines of all the continents. The fish-hooks of savages are generally without barbs. The great abundance of these animals before their wholesale destruction in recent times enabled the fisherman to have his choice without much trouble. On the coast of California shell-hooks have barbs, and such is the case with some Polynesian examples. But in the Madisonville Cemetery, Ohio, Putnam found only barbless hooks. The usual patterns are the toggle and the angular form made in two pieces.

In his work on aboriginal fishing, Dr. Charles Rau gives great attention to fish-hooks. From our point of view, the baited hook can only be here considered. This is, of course, a device for assassination. The appetites and habits, the weaknesses and idiosyncrasies of the animal are thoroughly studied, and the fisherman comes to his victim as an angel of light. The bait is eagerly seized, the lurking spine does its treacherous work, and the fish is high and dry. The full discussion of the subject involves hooks without barbs, hooks with barbs, lines, sinkers, floats, reel, line holders, live pens, and primitive anglers' outfits of every sort.[1]

The Dyaks have many ways of fishing:

[1] *Cf.* Rau, "Aboriginal Fishing," *Smithson Contribution to Knowledge.*

1. Angling, which they commence at an early age.
2. Diving into rocky pools and pulling fish from the holes.
3. With scoop-nets, chiefly by women, in shallow pools.
4. With the casting-net.
5. With barbed spear and with pronged spear.
6. With traps, large and small.
7. By torch-light, with spear and with hand-net.
8. With poison.[1]

Some of these methods are for killing the fish outright, or for arresting them, and belong in the last chapter ; but the Dyaks are very ingenious, and some of the processes described by the author exhibit a careful study of natural history and habits of mind.

The Samoan islanders in making their fish-hooks cut a strip off the pearl-shell two to three inches long, and rub it smooth on a stone so as to resemble a small fish. On the under-side they fasten a fluke made of tortoise-shell. Alongside the hook, concealing its point, in imitation of the fins of a little fish, they fasten two small white feathers.[2] This delicious bit of deception is not unknown to modern fishermen, who find it extremely difficult to devise a more enticing lure than the pearl-backing of a Polynesian fish-hook. Ellis tells us that in no part of the world, perhaps, are the inhabitants better fishermen. He then describes a variety of hooks, each worthy of study, in that the maker's motive is to have the fish catch itself so far as this can be done, by pandering to the weaknesses and foibles of the creature. They troll for the bonito as we do for blue fish, but they had also an invention still more ingenious. "A pair of light, swift canoes are selected. Between their bows a broad, deep basket, constructed of fern-stalks interwoven with the tough fibre of the *icie*, is fastened to contain the fish and not impede the rowers. To the fore-part of the canoes a long curved pole is fastened, branching in opposite

[1] Ling Roth, *J. Anthrop. Inst.*, London, 1892, vol. xxii. p. 50.

[2] Turner, *Samoa*, London, 1884, p. 168.

directions at the outer end. From each of the projecting branches lines with pearl-shell hooks are suspended, and so adjusted as to be kept near the surface of the water. To that part of the pole which is divided into two branches, strong ropes are attached, extending to the stern of the canoe, where they are held by persons watching the seizure of the hook. The *tira*, or mast, projects a considerable distance beyond the stern of the canoe, and bunches of feathers are fastened to its extremities. This is done to resemble the aquatic birds that follow the course of the small fish, and often pounce down and divide the prey which the large ones pursue. As the bonitos follow the course of the birds as much as that of the fishes, when the fishermen perceive the birds they proceed to the place, and usually find the fish. When the fish perceives the pearl hook it dashes after it and is caught. The men in the stern of the canoe immediately haul up the tira and drag in the fish, suspended as it were from a kind of shears. When the fish is removed the shears are lowered and the rowers hasten after the shoal.[1]

The Tahitian outwits the cuttle-fish in the following manner. "These creatures resort to the holes in the coral rock and protrude their arms for the bait of the unsophisticated fisherman, while they themselves remain firm within the retreat. The instrument employed is a rod of polished wood, half an inch thick and a foot long. Near one end of this a number of beautiful pieces of shell are fastened, one over another like the scales of a fish, until it is the size of a turkey's egg. This is lowered by a strong line from a canoe. The fisherman jerks the line until the cuttle-fish darts out one arm and winds it about the shell and fastens among the opening between the plates. The fisherman continues playing with the line until the animal has fastened every one of its rays to the shell, when it is drawn up into the canoe and secured." [2]

[1] Ellis, *Polynes. Researches*, London, 1859, vol. i. pp. 147–149, fig. on p. 148.

[2] *Ibid.*, p. 144.

In Sarawak an "alir" is a stick of tough wood, say ten inches long, sharp at both ends and grooved around the middle. To the alir is attached a short bit of line braided from soft, tough bark, and to the end of this a long rattan rope, and to the extreme end of the rope a cocoanut-shell float. A bait covers the alir, and the line is suspended from a limb so that the former may hang just above the water. The crocodile swallows the bait, the alir toggles in his stomach, the cocoanut shows the hunter where to find his game, and by means of the rattan-line he drags it to land.[1]

The Polynesians used to catch the *au*, or needle-fish, in the following manner. They built a number of rafts, each about fifteen feet long and six feet wide, from the light branches of the hibiscus wood. At one edge of each raft a screen was raised four or five feet high by fixing poles laid horizontally one above another to upright sticks. Men on the rafts went out so as to enclose a large space of water, having the screen on the outside of each raft. They gradually approached until the rafts touched one another, forming a connected circle in some shallow part of the lake. One or two persons then went out in a canoe toward the enclosed space, with long white sticks, which they struck on the water with a great noise, driving the fish towards the rafts. On approaching these the fish darted out of the water, and in attempting to spring over the raft struck against the screen and fell on the surface of the raft, where they were gathered into baskets or into canoes outside.[2]

In the old plantation days the negroes about the Chesapeake Bay used to go out in dugouts with torches. The light attracted the weak fish, as the fishermen called them, and they were caught in great numbers in trying to leap over the boat; and I have heard it said that the torch at times had to be extinguished to keep the fish from sinking the craft.

[1] Hornaday, *Two Years in a Jungle*, New York, 1885, p. 305, fig. on p. 307. [2] *Cf.* Ellis, *Polynes. Researches*, London, 1859, vol. i. p. 140.

"Angling for fish under the ice in winter," says Hearne, "requires no other process than cutting round holes in the ice from one to two feet in diameter, and letting down a baited hook, which is always kept in motion, not only to prevent the water from freezing so soon as it would do if suffered to remain quite still, but because it is found at the same time to be a great means of alluring the fish to the hole; for it is always observed that the fish in those parts will take a bait which is in motion much sooner than one that is at rest."[1]

No sooner had our savage learned to knit than he began to entangle the beasts of the fields, the fowls of the air, and the fish of the sea. That is, he set his nets in the woods for the beasts, in the fields for the birds, and in the waters for the fish. His designs on these creatures may have been either to encircle them or to entangle them. In the one case the victim was in a trap, in the other case it was in a mesh.

The operations included in net-making are those of the spinner, the net-maker, the rope-maker, the wood-worker, and the stone-worker, though the savage did not put so much labour on his sinkers as he did upon the clever devices by which he fastened a common stone to his net.

Hearne says the spinning of the twine and forming the net is a textile art, and all other parts of the manufacture have their mechanical side. The drag-net, the purse-net, and other encircling devices, are ancient. Even the gill-net is prehistoric. "To set a net under the ice it is first necessary to ascertain its exact length by stretching it out upon the ice near the part proposed for setting it. This being done, a number of round holes are cut in the ice, at ten or twelve feet distance from each other, and as many in number as will be sufficient to stretch the net at its full length. A line is then passed under the ice by means of a long light pole, which is first introduced at one of the end holes, and by means of two forked sticks this pole is easily

[1] Hearne, *Journey*, &c., London, 1795, Strahan, p. 15.

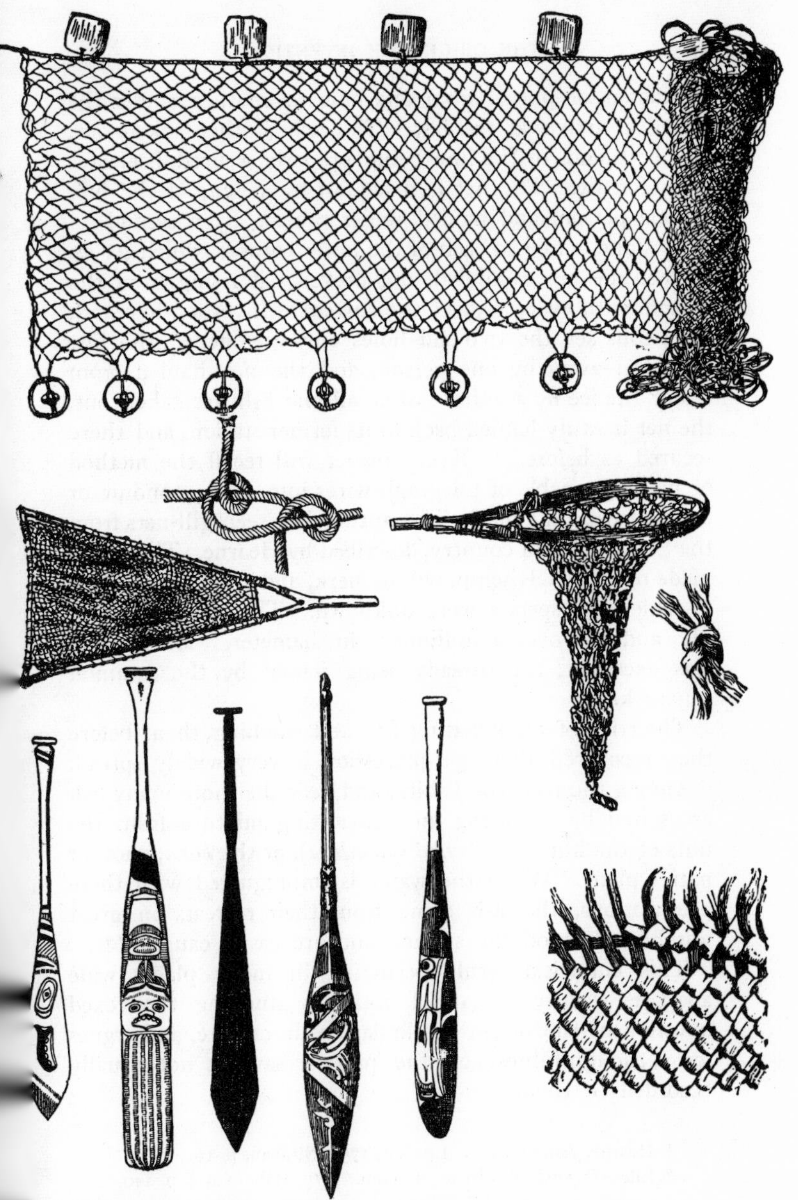

9.—STYLES OF PADDLES ON NORTH-WEST COAST. NET AND DETAILS OF WAKA-HAN BASKETRY. (*Rep. U. S. Nat. Museum*, 1888, pl. xxxii. *After Niblack.*)

conducted, or passed from one hole to another, under the ice, till it arrives at the last. The pole is then taken out, and, both ends of the line being properly secured, is always ready for use. The net is made fast to one end of the line by one person, and hauled under the ice by a second ; a large stone is tied to each of the lower corners, which serves to keep the net expanded, and prevents it rising from the bottom with every waft of the current. In order to search a net thus set, the two end holes only are opened, the line is veered away by one person, and the net hauled from under the ice by another ; after all the fish are taken out, the net is easily hauled back to its former station, and there secured as before."[1] Every reader will recall the method of hauling a cable of telegraph wires into a city conduit or the halyards on a flag. The author has seen gill-nets from the Saskatchewan country, described by Hearne. They were made of the wild hemp, willow bark, and from grass fibres of which the species were unknown. The twine was two ply, and not over a millimetre in diameter. The netting was excellent, the threads being joined by the common square knot.

The trick of asphyxiating fish and catching them before they recovered their self-possession is very widely spread. "Among the reefs [of Tahiti] and near the shore many fish are seized by preparing an intoxicating mixture from the nuts of the huteo (*Betonicá splendida*), or the *hora*, another native plant. When the water is impregnated with these preparations, the fish come from their retreats in great numbers, float on the surface, and are easily caught."[2]

This empirical result occurring in many places wide apart is worthy of careful notice in studying the vexed question of the origin of similarities in culture, and argues for a thoughtfulness on the part of savages not usually accorded to them.

[1] Hearne, *Journey, &c.*, London, 1795, Strahan, p. 16.

[2] Ellis, *Polynes. Researches*, London, 1859, Bohn, vol. i. p. 140.

The favourite mode of fishing among the Dyaks is with the *tubai* root (*Menispermum*), the juice of which is baled into a stream to poison its waters, and to cause the fish to rise stupefied to the surface. Basket-work screens are first erected to prevent the escape of the fish into pure water. Each person brings a bundle of *tubai* root. A reach is selected where suitable stones abound. It must be two or three hours' pull from the entrance, or the sport will be over too soon. The canoes line either bank, and at a given signal the entire party commence to hammer out the root and soak it in the water in the bottom of the boats. "A few minutes later, when all hands are ready, the poisoned liquid is baled into the stream, and the canoes, after a short pause, begin to drift slowly down the current, and as the fish rise to the surface they are speared with fish forks or captured with hand nets. The women join in the sport and scoop up the small fry with their nets." [1]

As a survival of a very ancient practice the following account from Thomson relates to a member of the Semitic stock, all of whose tribes have long since risen above savagery: "An old Arab sat on a low cliff and threw poisoned crumbs as far as he could reach [in the sea of Galilee], which the fish seized, and turning over dead were washed ashore and collected for market. The natives around Lake Hûleh sometimes cast into the water a fruit which so stupefies the fish that they are easily caught with the hand." [2]

It is true that they capture one fish at a time, but they are practising a declining art, and this always degrades the apparatus and the processes of more vigorous ancestors. A trap as distinguished from a weapon is a device whose function is to induce an animal to imprison itself or to commit self-destruction. The creature does not attack the trap—there is no war between it and the trap. The appa-

[1] Ling Roth, *J. Anthrop. Inst.*, London, 1892, vol. xxii. p. 51.

[2] Thomson, *The Land and the Book*, New York, 1880, vol. ii. p. 394, quoting Tristram, *Land of Israel*, pp. 429, 430.

ratus is the result of study, of induction, of adaptation of means to ends. The maker does not endeavour to enforce his ideas on his victim, he practically says, I shall consult your tastes and your wishes wholly. Indeed, if in the war of man on nature the destruction of animals by weapons may be likened to tactics, the capture of animals by any sort of trap may be called strategy. Every feeling and every thought of the animal is made to subserve the end.

Dr. G. B. Goode, in his report on animal resources at the Centennial Exposition in Philadelphia, divides traps into pen-traps, grasping or clutching traps, falling traps, missile traps, and adhesive preparations. The simplest of pen-traps is one into which the animal — fish, flesh, or fowl—enters, but it cannot go out. Pitfalls, salmon baskets, eel-traps, pockets of all kind, are of this class. The problem is to foreknow where the game will go and to take them unawares. No bait is used, but decoys are frequently employed to attract and absorb the attention of the unwary. Labyrinthine traps and weirs, lobster pots, set nets, fykes, and pounds, have all for motive to get the victim to incarcerate itself. It may be that the savage mind went no further, but men in civilised communities, as in Virginia, still set box and pen traps for rabbits, partridges, opossums, and coons, and boys drive whole flocks of quails into wing nets set in the fields by riding cautiously to the leeward on horseback.

The clutching or grasping trap is a device for seizing the victim either by the foot or by the neck. In the former case it simply sets its foot down into an artificial hand, made of cord, or other material. In the latter case, driven by hunger or a refined taste, it puts its head in a position to be gibbetted, and becomes its own executioner. To this class belong footpath snares, springes, snoods, spring-traps, and even our common mouse-traps. The custom of arresting by the foot and of garrotting are very ancient, therefore, and both are in full operation in our own day.

The fall trap is an apparatus which causes the victim to imprison or kill itself by crushing, by slashing, or by piercing. Dead falls, whether sprung by the hunter or by his victim, are well known to all savages. In some, the mere passing through a close place releases the fall when the animal brushes against a trigger. Whether the "figure four" or some other method be adopted, a stick is brought very delicately into a position where a nibble or the slightest touch will cause it to slip from its moorings. This releases or removes a support, and the whole weight comes down. An improvement on the use of a mere weight is seen in that of appliances in which a spear, a harpoon, or some deadly thing is attached. These last-named inventions are used chiefly in slaying those huge beasts against which early man's puny arms were unavailable, beasts that could never be tamed nor harnessed, but stood in the path of the higher life.[1]

The cross-bow trap was little known in savagery. It belongs to barbarism rather, and culminates in the tiger gun of Asia. In this line of invention is a trap used by the aborigines of Alaska and North-eastern Asia. They make a loop, or endless rope or becket, of sinew, and stretch it over the ends of two stakes driven firmly in the ground about a foot apart. A club of tough wood is twisted into the rope until not another turn can be taken, just as in tightening a woodsaw. In the other end of the wood is a sharp blade of stone or metal. When the wood is drawn back and fastened with a notched stick, a bait is set on the other side of the trap. The fox or other animal takes the bait, the knife is released, flies over and brains it. The Dyak pete or spring bow consists of a single bamboo lance attached to an elastic stem. It is laid in a horizontal direction above the ground about the height of the animal it is intended to transfix. A sapling bent for the purpose forms the spring by being held back. A string crosses the path, the least touch of which

[1] J. H. Porter, *Wild Beasts*, New York, 1894, Scribner.

loosens the spring, which forces the bamboo in a straight line across the path, and consequently through the animal.[1]

A Chinese rat-trap in the United States National Museum is furnished with a spring whose end moves horizontally on being released, precisely after the fashion of the Dyak trap. Two pieces of bamboo are hinged at the top like a pair of dividers. The rat puts his head between these, touches the bait, the strip of bamboo straightens, and the thief finds himself in the stocks.

Adhesive preparations like bird-lime were used in securing birds, chiefly for their beautiful plumage. Such sport would have little attraction for bustling savages. But in Hawaii, where the showy feathers were wrought into necklaces, capes, cloaks, helmets, and shields, the hunters were eager to catch the birds. In the islands the creatures were scarce. To slay them was to kill the goose that laid the golden egg. Therefore the gum of the breadfruit tree and the viscid milk of the arboreal lobeliads was smeared on the branches that the birds frequented.[2] It is astonishing how little resistance will prevent a bird from leaving the perch. The initial effort of flying is very great, and just a little resistance added to the weight of the bird will keep it at rest. The Zuñi Indians use a little horsehair snood on the top of a sunflower stalk in place of bird-lime to catch hawks, as will be seen further on.

The Crees in the Saskatchewan country captured large numbers of buffalo in pounds. In some enclosed dell at the end of a long valley they built a circular fence one hundred feet or more in diameter. Trunks of trees were laced together by means of withes, and braced by outside supports. Leading to an opening in this fence were two diverging rows of felled trees and bushes, extending some miles into the prairie. The herd were directed towards these wings by men in hiding, who appeared for a moment and waved their

[1] Ling Roth, *J. Anthrop. Inst.*, London, 1892, vol. xxii. p. 46.

[2] *Cf.* Brigham, *Cat. Bishop Museum*, Honolulu, 1892, p. 10.

robes in the right direction. In passing between the wings the buffalo were kept in their courses by men and women appearing at openings therein. At the entrance to the pound there was a log about a foot thick, and on the inner side an excavation sufficiently deep to prevent the buffalo from leaping back. The animals galloped about the pound, crushed one another, were shot by the hunters at the fence, gored one another until the whole number, say two or three hundred, lay in one common slaughter.[1] In this example there is a mixture of assault and strategy, and it marks a more complex type of inventions.

"When the Indians design to impound deer," says Hearne, "they look out for one of the paths in which a number of them have trod, and which is observed to be still frequented by them. When these paths cross a lake, a wide river, or a barren plain, they are found to be much the best for the purpose; and if the path run through a cluster of woods, capable of affording materials for building the pound, it adds considerably to the commodiousness of the situation. The pound is built by making a strong fence with brushy trees, without observing any degree of regularity, and the work is continued to any extent, according to the pleasure of the builders. I have seen some that were not less than a mile round, and am informed that there are others still more extensive. The door, or entrance of the pound, is not larger than a common gate, and the inside is so crowded with small counter-hedges as very much to resemble a maze, in every opening of which they set a snare, made with thongs of parchment deer-skins well twisted together, which are amazingly strong. One end of the snare is usually made fast to a growing pole; but if no one of a sufficient size can be found near the place where the snare is set, a loose pole is substituted in its room, which is always of such size and length that a deer cannot drag it far before it gets entangled among the other woods, which are all left

[1] Hind, *Canadian Red River, &c.*, London, 1890, Longmans, vol. i. pp. 357-359.

standing except what is found necessary for making the fence, hedges, &c. The pound being thus prepared, a row of small brushwood is stuck up in the snow on each side the door or entrance ; and these hedgerows are continued along the open part of the lake, river, or plain, where neither stick nor stump besides is to be seen, which makes them the more distinctly observed. These poles or brush-wood are generally placed at the distance of fifteen or twenty yards from each other, and ranged in such a manner as to form two sides of a long acute angle, growing gradually wider in proportion to the distance they extend from the entrance to the pound, which sometimes is not less than two or three miles; while the deer's path is exactly along the middle, between the two rows of brushwood."[1]

The wolf trap of the Central Eskimo is similar to the one used to catch deer. The hole dug in the snow is about eight or nine feet deep and is covered with a slab of snow, on the centre of which a bait is laid. A wall is built around it which compels the wolf to leap across it before he can reach the bait. By so doing he breaks through the roof, and as the bottom of the pit is too narrow to afford him jumping room, he is caught and killed there.[2]

Livingstone gives many ways of overcoming the African elephant by strategy. They dig deep wedge-shaped pitfalls, carefully cover them over and plaster them so as to have the appearance of the rest of the path. Many females and young are destroyed by this last means. These methods are often rendered futile by one elephant helping another out of a pitfall, or by the sagacious beast snuffing danger and quitting the country. Even when successful it can only be with one animal, for the others at once forsake the district if one of their number falls a victim.[3]

[1] Hearne, *Journey, &c.*, London, 1795, Strahan, p. 78.

[2] Boas, *Sixth An. Rep. Bur. Ethnol.*, 1888, quoting Rae, *Narrative, &c.*, London, 1850, Boone, p. 135.

[3] Livingstone, *J. Roy. Geog. Soc.*, London, 1855, vol. xxv. p. 220; also *Trav. in S. Africa*, New York, 1858, Harpers, p. 82.

A second method described by many travellers for waylaying the elephant is by means of a log of wood, having a poisoned spear-head inserted. It is suspended above the elephant's path by means of a cord, which is secured to a small wooden catch on the ground. When the catch is touched by the foot of the elephant in passing along, the beam falls on his back and the barbed spear-head remains. In this case the trust of the hunter lies in his poison, as well as the hope that the barbed head will rankle in the wound or move on to reach some vital part.

"In the Dyak pitfall the bottom of the excavation is staked with bamboo or ironwood spikes, in order to impale the victim who falls therein."[1] In Africa it is common to leave a cone of earth in a pitfall, or to make the pit double, so that the victim will rest on the belly and have its legs dangling on either side away from the bottom. If the animal is not killed by the fall, the hunter has no difficulty in despatching his helpless victim.

Somewhat similar in structure and function to the gill-net is the whole class of snares, snoods, springes, which do for the land animal what the former does for the fishes, namely, catch them about the necks and choke them to death.

To snare partridges, the Indians of North-western Canada, says Hearne, make a few little hedges across a creek or a few short hedges projecting at right angles from the side of an island of willows, which those birds are found to frequent. Openings must be left in each hedge, and in each of them a snare must be set. The partridges hopping along the willows to feed are soon caught. But the following more extended account is excellent, inasmuch as it brings out the entire details of the intellectual duel between men and birds :—

"To snare swans, geese, or ducks, in the water, it requires no other process than to make a number of hedges, or fences, project into the water, at right angles, from a bank of a river, lake, or pond ; for it is observed that those birds generally

[1] Ling Roth, *J. Anthrop. Inst.*, London, 1892, vol. xxii. p. 47.

swim near the margin, for the benefit of feeding on the grass, &c. Those fences are continued for some distance from the shore, and separated two or three yards from each other, so that openings are left sufficiently large to let the birds swim through. In each of those openings a snare is hung and fastened to a stake, which the bird, when entangled, cannot drag from the bottom; and to prevent the snare from being wafted out of its proper place by the wind, it is secured to the stakes which form the opening with tender grass which is easily broken. This method, though it has the appearance of being simple, is nevertheless attended with much trouble, particularly when we consider the smallness of their canoes, and the great inconveniency they labour under in performing works of this kind in the water. Many of the stakes used on these occasions are of a considerable length and size, and the small branches which form the principal part of the hedges are not arranged without much caution, for fear of oversetting the canoes, particularly where the water is deep, as it is in some of the lakes; and in many of the rivers the current is very swift, which renders this business equally troublesome. When the lakes and rivers are shallow, the natives are frequently at the pains to make fences from shore to shore. To snare those birds in their nests requires a considerable degree of art and, as the natives say, a great deal of cleanliness; for they have observed that when snares have been set by those whose hands were not clean, the birds would not go into the nest. Even the goose, though a simple bird, is notoriously known to forsake her eggs if they are breathed on by the Indians. The smaller species of birds which make their nests in the ground are by no means so delicate; of course less care is necessary to snare them. It has been observed that all birds which build in the ground go into their nest at one particular side, and out of it on the opposite. The Indians, thoroughly convinced of this, always set the snares on the side on which the bird enters the nest; and if care be taken in setting them, seldom fail of seizing their object.

For small birds such as larks, and many others of equal size, the Indians only use two or three hairs out of their head; but for larger birds, particularly swans, geese, and ducks, they make snares of deer sinews twisted like pack-thread, and occasionally of a small thong cut from a parchment deer-skin."[1]

The Similkameen Indians of British Columbia snared deer by bending down two saplings, one on either side of a deer run, and suspending a noose between their tops. The deer were then driven down the run by men and dogs. Bounding heedlessly along, the frightened animal was involved in the noose, and its struggles releasing the saplings, the deer was hung by the neck.[2]

Among the same Indians, remarks Mrs. Allison, before the days of shot guns, birds were snared in slip nooses set in trees which they frequented. The Fool bird, a species of grouse quite deserving its name, was caught with a loop tied to the end of a long pole. This loop was thrown over the bird's head, and the victim was "yanked off the tree or bush on which it sat."[3]

The greater familiarity of wild creatures with mankind before the terrible fright produced by the noise of firearms is suggested by this statement, and must be always borne in mind in studying the relations of true savagery to the animal kingdom.

The Zuñi Indians have a funny way of catching a hawk by lassoing its foot. Places for lighting are scarce in the fields where the hawks hunt the field mice. So the Zuñi drive stalks of sunflower into the ground, on the top of which they fix horsehair nooses. The hawk comes down and kindly puts its foot into the noose. As soon as it makes an effort to fly the noose brings it to perch; it can't let go, it can't fly, so after an effort or two it waits calmly for the Zuñi to come along and knock it on the head.

[1] Hearne, *Journey*, &c., London, 1795, Strahan, p. 275.

[2] Mrs. S. S. Allison, *J. Anthrop. Inst.*, London, 1892, vol. xxi. p. 306.

[3] *Ibid.*, p. 307.

The eagle feathers are the most dear of all objects to an Indian's heart, so in Eastern California he digs himself a little pit and gets into it, having first put over it a covering of brush, and on this a dead rabbit or some toothsome bait. The eagle lights down, and before it can gather up its powers to fly away, is dragged through the brush and its heart crushed between the hunter's knees.

These two methods of taking the eagle involve the same principle, namely, that of an unseen hand grasping the foot of the bird at the moment of absolute rest between lighting and soaring again. This should be compared with the use of bird-lime before mentioned.

The ancient inhabitants of Copan caught the quetzal bird in snares, and after having plucked out their beautiful tails set them at liberty again. To kill them was a crime punishable by law.[1]

This custom, as well as that of the Hawaiians, is a step towards domestication. It belongs to the same series as the gathering of rubber or maple sugar from the natural trees, and is one grade upward above the destroying the gifts of nature irrevocably in the first use of them.

A curious example of capture by noose is one in which Wallace describes a duel between a Bouru man and an immense serpent which had invaded the house of the naturalist in Amboyna. The brute was about twelve feet long, and capable of swallowing a dog or a child. The man made a strong noose of rattan, and with a long pole in the other hand poked at the snake, who began to uncoil itself. The man then slipped the noose over the snake's head, and dragged it from the roof; then catching the snake by the tail he rushed from the house, and running quickly, dashed its head with a swing against a tree.[2]

The Dyaks ensnare deer in a kind of compound noose or land trawl. It is a long cane cable with a continuous series

[1] Morelet, *Trav. in Cent. Am.*, New York, 1871, p. 335, quoting Herrera, Dec. iii., l.x., c. ii.

[2] Wallace, *Malay Archipel.*, New York, 1869, p. 303, plate op. p. 304.

of cane loops or nooses depending from it. This trawl is stretched across a narrow neck of land, and upheld so as to intercept the deer making a stampede into the bush. The hunting party then divides, some to frighten the animals toward the *jarieng*, as it is called, and others to despatch them as soon as they are caught.[1]

The reader should mark this example of borrowing between arts as a stage of invention. There is not, perhaps, a savage or civilised people on the earth that does not use the trawl in fishing. Well does the author remember old uncle William's "trot-line" in Virginia, in the days of slavery. Whether he brought it from Africa, no one can tell. But he used to stretch a clothes line across the creek at sundown, with a great many hooks dangling from it into the water to which he had tied chicken gizzards and other tit-bits from the kitchen. In the morning, before "sun-up," uncle William could be seen coming up the hill with a basket of cat-fish for missus. But the Dyaks, watching the fishermen, have carried the art over to their land activities and set "trot lines" for deer, substituting nooses for hooks.

The universal spring is also in use among these people, who have excellent material not only in the bamboo, for they also manufacture fine and strong twine from the inner bark of several kinds of trees.[2] After man-power, spring-power comes second in the list of natural forces to serve our race.

The Congos have an ingenious contrivance for catching wild buffalo. A stout stick, one metre long, is strung as a bow with a double cord; a stick inserted between the parts of this double cord is turned several times to increase the tension, and the farther end then catches on the bow, as in the old-fashioned device for tightening a wood-saw. This is planted in the track of the animal, which thrusts its foot between bow and cord, the recoil throwing the bow

[1] Ling Roth, *J. Anthrop. Inst.*, London, 1892, vol. xxii. p. 47.
[2] *Ibid.*, vol. xxii. p. 47.

upward.[1] It is a little difficult to see what advantage such a device has over the common noose.

All savages are expert in the use of decoys and lures. They know exactly the dainty things each animal likes best, and have gone so far as to cut from pearl shell and stone and bone forms that deceive the very elect. They can imitate the peculiar whistling sound of the deer, the linnet and other birds. Even odours known to be pleasing to certain animals are faithfully reproduced. For every swimming bird forms are created that cannot be distinguished from the original. Each duck has its separate decoy, and frequently the savage donned the hide of the animal pursued. The modern hunter, with his wooden plovers and tin ducks, is only borrowing from the ancients.

In one of their myths called the Mountain Chant, the Navajo speak of dead-fall traps set at night near the burrows of small animals, such as rats and prairie dogs. For each trap they buried a flat stone with its upper side on a level with the surface of the ground, and on this they sprinkled a little earth, so that the rat would suspect nothing; over this they placed another flat stone leaning at an angle and supported by a slender stick, to which were attached berries of the aromatic sumach as a bait. The author does not say, but it is evident that the creature, in gnawing the bait, either disturbed the prop or gnawed it in two, and thus was crushed.[2]

The old Navajo in the Chant tells his two sons how to make a decoy for deer. Cut the skin around the neck; then carefully take the skin from the head, so as to remove the horns, ears, and all other parts, without tearing any part. Leave such an amount of flesh with the nose and lips that they will not shrivel and lose their shape when dry. They prepared the skin according to the old man's directions. To keep the skin of the neck open they put into it a wooden hoop. They sewed up the mouth, left the

[1] Schweinfurth, *Artes Africanae*, London, 1871, plate v. fig. 12.

[2] Matthews, *Fifth An. Rep. Bur. Ethnol.*, pp. 388, 390.

eye-holes open, stuffed the skin with hay, and hung it in a tree to dry, where it would get smoky and dusty. They cut places in the neck through which the hunter might see. The skin of a doe which the younger brother had killed they painted red to make it look like the skin of an antelope. They prepared two short sticks to enable the hunter to move with ease and hold the head (decoy) at the proper height when he crept in disguise on the deer. During four days the elder brother practised imitating the walk of the game.[1]

Sound decoys are found among all hunting peoples. The North American Indians imitated the partridge, the turkey, the deer, the crow exactly. Among the frontier hunters are men who so cleverly mock all the wild creatures as to deceive even the oldest.

Morelet tells of one of his companions who could reproduce to perfection the plaintive voice of the couroucou (quetzal bird), "a talent possessed in a greater or less degree by all the hunters of Copan."[2]

The Navajo seem to have connected the sweat bath with hunting in a practical way. Every good hunter knows that success depends quite as much on removing the things that offend as in presenting the things that allure. In the Navajo Mountain Chant myth, the old hunter and his two sons not only take four good sweats, but they lined the floor of the sudatory with all the plants on which the deer like most to browse. After the sweat they cleansed even their hair with soap rood, and had good success both with their traps and with their bows.[3]

The old beaver trappers had many wonderful stories to tell of their experiences, but the secret revolves itself into anticipation rather than treachery. The Indians knew that

[1] Matthews, *Fifth An. Rep. Bur. Ethnol.*, p. 391. For Eskimo sound decoys, see Murdoch, *Ninth An. Rep. Bur. Ethnol.*, Washington, 1892, p. 253.

[2] *Trav. in Cent. Am.*, New York, 1871, p. 338.

[3] Matthews, *Fifth An. Rep. Bur. Ethnol.*, p. 389.

the beaver makes for deep water when he is caught, so he fastened a heavy stone to the trap, which held the creature under the water until it was drowned. The hunter also knew that the beaver would amputate its leg when it found itself caught, so he must provide for that. Their objection to the smell of anything human is also strong, so the most aromatic substances were mixed with castor to sink the stronger into the weaker scent. The savages frequently made a co-operative onslaught upon a village, anticipated this plunging into the stream by rows of stakes driven close together, and killed the beaver trying to make its escape.

To catch a fox it is necessary first to win its confidence, and this the savage knows. So he prepares a trap that is perfectly harmless, and lets Reynard walk about over the ashes or fresh earth or chaff, picking up dainty bits until all sense of treachery is removed. Now is the time to conceal the trap. But all vestige of human hand or foot must be removed, and the apparatus must be cleaned and smoked most effectually. There can be no doubt that the fox has become more wary even in modern times ; but, while the wolf has surrendered to domestication, the fox continues his original warfare on human rights until his name is a synonym for untamable cunning.

Even the sports and pastimes of animals were taken advantage of to secure their capture. For example, the land otter travel over the snow chiefly by sliding. "They seek a steep place by the water side, crawl to the top of it, and then face about and go head first into the water. Then up they climb and at it again, having great sport. One of these slides is the best place to catch otter."[1] The trap is set immediately at the bottom, and the happy creature rushes headlong into self-slaughter.

In addition to the transfer of devices from one branch of any craft to another, and the acknowledgment of such a transfer to be a true invention, there are still further

[1] Thrasher, *Hunter and Trapper*, New York, 1868, Judd, p. 28.

adaptations. Many of the devices used in other crafts have undergone special modifications with a view to hunting and fishing.

The snow shoe, the harness of animals, boats of many kinds, designed for travel and transportation, were among primitive tribes used more by hunters and fishers than by all other persons combined, and were constructed to meet their wants. In our own day there is a very large class of boats with no other function than for fishing. The kaiak, the birch-bark canoes, the great dug-outs and other sea craft, piroques, bark boats, balsas, catamarans, proas, were primarily for reaping the harvests of the sea. Everything about the boat has been similarly modified, except perhaps the anchor. The bow becomes a spear thrower ; the spear, a harpoon ; the arrow, a retriever. The same is true in land hunting. House, clothing, apparatus must all conform to the absorbing occupations. Even the customs and laws were changed.

The hunting and capturing and domesticating of wild animals, specially those that go in droves, led in most primitive tribes to the laws of game and the declaration of the rights of owners. It seems to have been universally agreed upon that the man who first made his mark upon an animal was its owner. Mr. Cushing says that whoever afterwards killed the creature was obliged to bring it to its owner, and received his share of game from the man whose mark it bore. The same testimony is borne by Bourke, Dodge, and others who are familiar with the usages of savagery.

Among the Waboni, on the river Tana, East Africa, the etiquette of hunting is strictly observed : thus, if a dead elephant is found by a hunter other than the one who wounded it in the first instance, the owner is identified.

It is certain that law was developed in two ways about the question of animal capture, namely, in the direction of boundaries showing where each tribe might hunt, and about the game itself, declaring to whom it belonged. At this same point some curious solutions of anatomy were made. When two or three men struck the same animal, there were laws which decided which part of the body entitled to ownership, and these were not based on our knowledge, but upon the savages' notions of vulnerability and totemic precedence. Kennan, in *Tent Life*, calls attention to the anatomical lore of the Koraks.

The employment of cunning in place of brute force in the taking of animals alive soon led to the rudiments of domestication. There will always be persons who prefer wild strawberries to cultivated ones, a "high" snipe to spring chicken, and wild boar to Westphalia ham. Though the argument might be yielded to, it would do its advocates little good. Wild strawberries are almost extinct, snipe are already higher in price than they are in flavour, and, as for the "boar-sticker," that has been hung up with bric-a-brac for a century. There are no longer "as good fish in the sea as were ever caught out of it," unless we include those that are artificially propagated.

Darwin, in speaking of the part man has taken in the domestication of animals, adds to the natural selection upon which he dwells: 1. *Unconscious Selection*, following from men naturally preserving the most valued and destroying the least valued individuals, without any thought of altering the breed. 2. *Methodical Selection*, or that which guides a man who systematically endeavours to modify a breed according to some predetermined standard. Unconscious selection graduates into methodical and only extreme cases can be distinctly separated.[1] The same observation might be made regarding invention throughout every line of its life history. He further admits that he had fallen into the current belief as to the stupidity of savagery. "It appeared to me at one time probable that although ancient and semi-civilised people might have attended to the improvement of their more useful animals in essential points, yet that they would have disregarded

[1] *Variation of Animals and Plants*, New York, 1876, vol. ii. p. 177.

unimportant characters. But human nature is the same throughout the world : fashion everywhere reigns supreme, and man is apt to value whatever he may chance to possess."[1] A long list of examples is given by the distinguished naturalist, which must be omitted for want of space.

That Darwin's unconscious selection has been going on in savagery from the earliest period there has been abundant evidence. As in the case of plants, so every animal had to pass an examination. It is not for one moment to be supposed that by an act of inspiration the first men guessed that the dog, the cat, the horse, the cow, the sheep, the goat, the ass, the camel, the elephant, the hog, the llama would be useful to them. Nor is it supposable that these creatures came and offered their services in his naked and helpless condition.

That is a very interesting panorama hinted at in Genesis, where all the wild animals came trooping by to see what Adam would name them. Only I think the procession must be strung out over many lands and must occupy many, many centuries in the passing. Even now the species have not all reported to get their binomial Greek and Latin names, and the lists have frequently changed. Nevertheless, before a page of history was written, representatives of most of the families of vertebrates, and a great number of species of invertebrates were familiarly known and named by their recognised characteristics, and these have come to be useful.

In the lowest savagery animals were killed or taken for immediate use. They were slain in the capture or they were taken to be slain. But in higher savagery the young of animals were spared and brought to the habitation for children to play with, to furnish plumage or fur for their elders to use in decorating themselves, and for the regalia of their civil and religious and military festivals. The question

[1] *Variation of Animals, &c., under Domestication,* New York, 1876, vol. ii. p. 193.

of economy obtrudes itself: how to get the best results with the least effort. There is no doubt, also, that many animals learned that they were safer under the protection of man than they were in the wilds.

The author is delighted to quote from Mr. Cushing on this point. "Both Navajo and Pueblo, but especially the latter, practise domestication. They capture young hawks and other birds—ducks and mocking birds among them—and by tethering them with strings keep and ultimately succeed in taming them. They prefer fierce animals like the hawks and, among quadrupeds, wild cats, because they are more easily captured, being less timid, and are hardier, enduring the extremely unfavourable conditions of their captivity better than the gentler creatures. They also like their voices better and their characters, as being more independent. Even porcupines are sometimes brought in, but only the badgers ever live long. The prairie dogs die of continued fright and starvation. The porcupines sooner or later prick and enrage their over familiar master, getting killed in consequence." Fright, starvation, bristling quills—too much for the youthful apostle of domestication. His wild pedagogy has been a wonderful selector, and so has the same school in all ages and all places extinguished or subdued the animal kingdom. The eagle, however, has another use not mentioned above. Its feathers are most precious appendages to the ceremonial paraphernalia.

"The moose are also easiest to tame and domesticate of any of the deer kind. I have repeatedly seen them at Churchill as tame as sheep, and even more so; for they would follow their keeper any distance from home, and at his call return with him, without the least trouble or ever offering to deviate from the path."[1]

Any American boy who has lived on the frontier knows that partridges, wild turkeys, crows, deer, foxes, 'coons, buffaloes, and many other wild species will become tame in

[1] Hearne, *Journey, &c.*, London, 1795, Strahan, p. 257.

a single generation, and some of them will breed in confinement. The author was never without some of these pets when he was a boy.

"The Campas of Peru possess a great number of tamed animals—paroquets, ourax, several species of penelopes, monkeys, ronsocos (*capybara*), and even wild boars and tapirs; and it is curious to see the mistress of the lodge when she goes, for example, to draw water from the river. If their presence around the habitation attracts at times the puma and the tigrillo, they destroy numbers of insects and small vermin infinitely more dangerous to man than the animals of the mountains. Their devotion to their masters is the more solid because it is voluntary and nothing prevents their regaining their liberty. The Campas do not eat the animals thus domesticated; on the contrary, it is not rare to see a savage woman give her breast in turn to her child and to a young monkey." [1]

The opinion that the leading domestic animals came under man's control in prehistoric times is confirmed by Ogilby's observations on the sheep, dog, goat, &c. *Ovis brachyura* is the characteristic variety of the Ugrian race; *Ovis dolichura* the appropriate breed of the Indo-Germanic nations; *Ovis platyura* is the favourite of the Semitic peoples; *Ovis steatopygia* was the original breed of Mongolic nations; and *Ovis longicauda* still continues the peculiar variety of the dark-skinned races of Asia and Africa.[2] But these are only varieties of a species that paid its fleece and flesh and milk as tribute to primitive man.

The domestication of the dog furnishes a typical example of the manner in which the process has gone on from unconscious to conscious selection in the taming of animals. It also is a refutation of those who have no good word for savage peoples.

The wolf is an uncanny creature, said to be bloodthirsty, remorseless, and deceitful. Not only so, but it practises

[1] Ollivier Ordinaire, *Rev. d'Ethnog.*, 1887, vol. vi. p. 282.

[2] *Rep. Brit. Assoc.*, 1857, pt. ii., p. 105.

co-operative murder, bringing down the stoutest mammals by mere numbers. But all the beauty, all the faithfulness, all the versatility of modern dogs were locked up in the wolf as in a mine of precious ore. These fine qualities were gradually bred out from the wolf, not bred into him. From one point of view they are indeed human productions, human inventions, just as much as the delicious music of the violin is a product of human genius. The qualities of men are mirrored in their dogs. All that the dog has and is it owes to man. In its natural state the wolf is the picture and synonym of what is rapacious in man. The latter has not learned a single virtue from the former; on the contrary, man has given no higher testimonial of his talent than the creation of such a noble animal as the dog from wolfish material.

The Eskimo in our day have preserved one of the first lessons taught the wolves. They take a sharp bit of flint or a knife-blade procured from the whalers, and fasten it to a stake or a rock and cover it with fat, which soon freezes into a ball. The incautious and hungry wolf, discovering the dainty morsel, laps the fat with its tongue until the sharp blade slices the latter and brings the blood. The taste of blood excites and infuriates the wolf, which only laps the more vigorously and commits accelerated suicide. The smell of blood and the taste thereof communicates the madness to the pack, and if at that moment all the wolves in the world could be gathered, they would be piled into a monument of mutual carnage. There is no profit in contending against such odds. The wolf is now superstitious.

The first day or two spent by Lupus in the human school must have been anything but sunshine to his soul. Hitherto he had been taken for his fur alone. Hunger on his part drove him to put his head into a noose or under a weighted log, or to move the treacherous trigger that released a stone dagger to brain him.[1]

[1] Boas, *Sixth An. Rep. Bur. Ethnol.*, Washington, 1885, p. 510, quoting Klutschak, *Als Eskimo, &c.*, Wien, 1881, p. 192.

In this primary grade, however, the wolf learned thoroughly the first lesson in domestication—the conviction of a mysterious force, the sense of having been overpowered, the ineradicable memory of a hopeless struggle. The wolf's head being larger than its neck enables its preceptor to put a collar thereon and a tether to that, against which the pupil might protest in vain. Fear also, created by the infliction of bodily pain which the creature could not escape nor revenge, completed the enslavement of the wolf.

All animals are tamed or broken in the same fashion. The author was reared on a large stock farm, and is witness to the fact that the subduing of the most incorrigible stallion or bull is first psychological. The animal is "cinched," as it used to be called—that is, it is rendered powerless. The method of breaking elephants in the East is well known. The celebrated flea-tamer puts the frisky creatures into a metal pill-box on the end of a spindle, and whirls it by means of a bow drill. When the box is opened the fleas are as docile as tortoises, and, on his testimony, they never hop again.

The next step in this lycotechny is the result of a discovery that even the native and wild instincts and habits of the wolf may be helpful to its captor. It howls at night in recognition of the approach of other wolves or wild beasts, or of man. Its refined olfactory sense discovers the sources of food unperceived by the hunter. Its fleetness enables it to run down the wounded game, which it is allowed to share with its master. It may be that in close quarters its canine teeth deliver him from an ugly foe.

One day the tired hunter discovered that on the snow the wolf-dog could pull, and that its back could bear a burden. The proverb, "to work like a dog," must have arisen in that period of history.

In Southern Canada and along the northern border of the United States a broad collar is put around the dog's neck, a girth around its body, and these are united along the shoulders by two straps. The dog is in a harness which allows perfect movement of every muscle, and yet it cannot extricate itself. Two poles are lashed to this harness, one on either side, and tied together at the upper ends, while the other ends trail behind the dog on the ground. Upon this primitive "wheelbarrow" or cart without wheels, called travois, a load of fifty pounds is lashed. In the same region a pack regularly made up is laid upon the back of the dog. Along the Arctic strip of North America and in North-east Asia, among the Eskimo and the Indians, dogs are harnessed to sledges. The amount of thought and forethought displayed in the sledge, the harness, the training and care for the animals, does great credit to the thinking powers of these savages.

Some of the Canadian dogs at the beginning of the winter, when fresh at their work for the season, are exceedingly restive under coercion of any description, and not infrequently snap at their masters, who invariably arm themselves with very strong mittens of buffalo or deer hide when harnessing a savage and powerful animal. The dogs require long-continued and most severe punishment to make them obedient to the word of command.[1]

The dog has never been able to emancipate itself from this simple but effective harness. The history of its activity must include all the tread-mills upon whose "climbing sorrow" dogs have been compelled to walk. In the most learned nation of the earth it is possible to see these creatures by hundreds harnessed with women, as they have been from the beginning dragging food for the militant class to eat. In Belgium and France also the same sights will greet the eye under the shadow of the grandest structures in the world.

The barbarous races have always welcomed the services of the dog in the care of their flocks. The shepherd's dog is a marvel of breeding and training. If the reader will watch a

[1] Hind, *Canadian Red River, &c.*, London, 1860, Longman, vol. ii. p. 93. Consult Lubbock, *Prehistoric Times*, New York, 1872, p. 557.

flock of sheep conducted through the thoroughfares of London by one of these creatures, he will scarcely be able to realise what the ancestor of the dog would have done with these same sheep many generations back.

The oldest books in Chinese, Egyptian, Indian, and the Mesopotamian languages mention dogs already bred and educated, if we may say so, out of savagery into a high canine civilisation. Révoil believes that the hunting instincts of the wild dog still survive in the hound, indeed, have not been allowed to die out, and that the great ferocity exhibited in the chase would seem a temporary recurrence to savage life and habits.[1]

The variety of acquirements among the earlier mentioned trained dogs is truly surprising, and confirms the suggestion made more than once in this book, that a multitude of the stage settings for the drama of human progress were prepared before the curtain rose upon written history.

Xenophon's *Kynegetikus* is the oldest known book upon the domesticated dog. Says Rossignol, "Among the Greeks different races of dogs were employed for different kinds of game. For hares, Castorians and fox-hounds were used; for stag hunting, Indian hounds; for the wild boar, Indian, Cretan, Locrian, and Laconian hounds. Chief attention is given to the coursing of hares. They were not to catch the hare, but to drive it into nets set at certain places. The best dogs are those with a light head and blunt muzzle, prominent black shining eyes, broad forehead; long, flexible, round neck, broad chest, straight elbows. They must be strong, well-proportioned, swift of foot, and above all they must be keen scented."[2] Then follows a number of directions for feeding, for young dogs, and for the making of specialists.

[1] Révoil, *Hist. Physiologique et Anecdotique des Chiens*, Paris, 1867, p. 394, quoted by Rossignol.

[2] For an elaborate study on the subject consult Rossignol, "The Training of Animals. A.—Dogs," *Am. J. of Psychology*, Worcester, Mass., 1892, vol. v. pp. 205–13. Reviews of Works on Dogs in Professor Brewer's Library.

"Cat and dog" are synonyms of antagonism. The cat is a tiger in blood and at heart. The dog is man's friend, chiefly; but the cat is woman's friend. The latter was invented much later in history; that is, the useful functions of the cat were developed after those of the dog.

The cat was the defender of the hoard, the house, the granary, against rats and mice and other vermin, against snakes and lizards and other reptiles. It never seized nor retrieved, nor brought home aught for man to eat. In India the cheeta or hunting leopard springs upon prey, but it still insists on devouring the game after the manner of the cat.

So, the myth relates that when in the war with Typhon the gods fled to Egypt and were metamorphosed into animals, the goddess of hunting, Diana, was changed into a cat.[1] The universal esteem in which Pasht or Bubastis was held in Egypt, pre-eminently the granary of the ancient world, is only another example of the good people of this world who have gone to live in the sky. Since in historic times the cat, in both a wild and a domestic state, has always been known in China, India, Egypt, and Greece, its first treaty of peace and amity with man goes back to those prehistoric times when men first began to lay up grain-stores for the future.

Any other one of our domestic animals would furnish an equally good example of the evolution of the art of domestication. The steps are in every case easy to follow. The first act is a kind of compromise or covenant; the second is an enslavement of the weaker minded; the third is a mutual improvement, each refining and becoming refined; the last is conscious breeding on the part of man, by which he creates, as it were, new species or varieties for a thousand uses and pleasures.

The domestication of the bee and the silkworm are marvels of human ingenuity. The bee-hunter has always a charm to readers of books of travels. In the days of slavery in the Southern States there was a class of "poor

[1] Dureau, *Ann. d. Sc. Nat.*, Paris, 1829, Crochard, vol. xvii. pp. 159–92.

whites" who eked out a living by finding bee-trees in the forests. The Africans hollow out a log, rub it with aromatic herbs, and hang it in a tree near the wild bees' nests. To unhook this log with its bees and honey and hang it in a tree nearer home is the work of a moment, and the thing is done.[1]

In Palestine the owners frequently remove their hives of bees up into the loftiest mountains as the flowers disappear from the lower regions, and put them in the woods, that the bees may gather honey from mountain thyme and other plants that bloom in autumn on those cool heights. The hives are made of plaited basket-work, formed with long, hollow cylinders, and are easily transported on the backs of mules and donkeys. The cylinders are piled up in the woods in a sort of pyramid, and covered with an old mat.[2]

The domestication of the silkworm in China was prehistoric. "In the grounds of the Imperial Palace at Peking is an altar forty feet in circuit and four feet in height, surrounded by a wall and also a temple. This is the 'Early Silkworms Altar,' in the vicinity of which a plantation of mulberry-trees and a cocoonery are maintained. It is dedicated to Yuenfei, First Wife, in her quality of the discoverer of silkworms, and annually in April the Empress worships and sacrifices to her." [3]

The latest modern efforts to change from the natural supply to an artificial basis of food through intentional selection and propagation is in the case of fishes and other aquatic products. The most primitive fishermen doubtless ate their spoils on the spot. But the savage fish-hook without barbs, and the traps and weirs and nets, could not be long in suggesting that the meal might be taken at the pleasure of the captor. So, fish-preserves, and, finally, fish-ponds are of very early date.

In Polynesian fish-ponds a circular space about ten feet

[1] See Livingstone, *Travels, &c.*, New York, 1858, p. 307.

[2] Thomson, *The Land and the Book*, New York, 1880, vol. i. p. 225.

[3] Lacouperie, *The Silk-Goddess*, London, 1891, Nutt, p. 3.

in diameter is enclosed with a stone wall built up from the bottom of a lake to the edge of the water. A small opening is left in the upper part of the wall, on the side away from the sea, and from each end of this a wall of stone extends, the two diverging as wings of a net. In this manner the creatures are intercepted on their return to the sea. Fish are usually found in these ponds, and they are excellent preserves, in which the animals are kept until they are wanted. Ellis describes a fish-pond in Tamehameha not less than two miles in circumference.[1]

Eels were a favourite fish with the Tahitians, and were often tamed and fed until they attained an enormous size. The pets were kept in large holes, two or three feet deep, partially filled with water. On the sides of these pits the eels formed or found an aperture in an horizontal direction, in which they generally remained. Ellis mentions a young chief who would give a shrill whistle near one of the holes, and bring forth an enormous eel, which moved about the surface of the water and ate out of its master's hand.[2]

The climax of this industry is the modern patent fish-jar, holding a gallon, and hatching a hundred thousand young at a time.

[1] Ellis, *Polynes. Researches*, London, vol. i. p. 138; vol. iv. p. 407. Turner, *Samoa*, London, 1884, p. 298.

[2] Ellis, *Polynes. Researches*, vol. i. p. 76.

CHAPTER X.

TRAVEL AND TRANSPORTATION.

"Look at the character of our country. Crete is not like Thessaly, a large plain, and for this reason they have horses there and we have runners afoot here. The inequality of the ground in our country is more adapted to locomotion afoot."—PLATO, *Laws*.

ATLAS and the Caryatides were men and women who went long ago to live in the skies and in the dreams of artists and architects. They once dwelt upon the earth, however, not as individuals, perhaps, but as a class of whom the artistic and the mythic types are only composite photographs.

Any day, in London or New York or Dresden or Singapore or Mexico, their descendants may be seen bowing under great burdens as of old—men carrying them upon their backs, women poising them upon their heads chiefly. In many other places these burden-bearers share with freighted ships and trains and beasts of burden the loads of commerce; but it was not so from the beginning. The commerce of the world was borne at first only on the backs of human beings. The loads were not great, the distances travelled were short, and the carrying industry was confined principally to the food-quest.

Even in those early days, conveyance was both by land and by water, and it was prosecuted for the double purpose of getting about personally and for the bearing of burdens. The former was the beginning of what is now called travel; the latter, of transportation. The vestibule train and the fast freight train, the passenger steamer and the merchant marine, are the modern representatives of those primitive methods of moving persons and things.

The interesting problem is to find out how the one became the other throughout the world, and how the more ancient methods survive into the modern time. To any thoughtful student, such immense and costly assemblages of riding and carrying devices as may be seen in a great exhibition are stimulating. So many of them are only of

FIG. 59.—Seminoles carrying the Dead. (*After McCauley, Fifth An. Rep. Bureau of Ethnology.*)

yesterday, and yet all of them are a sort of metric scale or indicator to show how far man's genius had gone in that particular year in shortening distances and lightening burdens.

In the life of primitive society the carrying of adult human beings must have been confined to the dead, the sick, or the wounded, or to the bearing of persons in authority. At any rate, in our own day, among the precipitous trails in the mountainous region of South

America, the silleteros take each a man on his back in a rude hod or chair and carry him ten miles a day.[1]

But the more primitive inventions, antedating these rude frames on which one man bears another, were devices which would enable the man himself to get about more rapidly. The exigency was never too great. Foot-gear and accessories to walking, running, or climbing were devised for the occasion. By this means the man became passenger in a coach which he himself compelled. Cold or heat, mountain or plain, open sward or volcanic slag or thorny undergrowth, only stimulated the germ of ingenuity, and gave the greater variety to what else would have been barefooted monotony.

Where the cold was intense, and the snow compact, the Eskimo walked on a snowshoe in which a few coarse meshes were sufficient, and he also, as well as the Laplander, the Finn, and the Ghiliak, put a cunning little snowshoe on the bottom of his cane.[2] The increasing size of the snowshoe and fineness of the mesh were governed by the temperature, coming southward, with the greatest fidelity, until the tier of states south of Canada were reached, and there they disappeared. Only in the elevated portions of California are they elsewhere found. In the northern regions of Europe and Asia the coarse snowshoe and the Norwegian skee prevail. But neither in Russia nor in Siberia could be found a finely meshed variety similar to those of Canada or interior Alaska. The reason is plain. Each snow-bound area where men dwelt at all, dictated the style of snowshoe to use.

The study of aids to locomotion is more interesting if we commence in the tropical regions and follow the evolution of the shoe northward.

Tropical men were practically barefooted. They wore in

[1] Columbus mentions the carrying of a cacique in his litter, *Journals, &c.*, Hakluyt, London, 1893, vol. lxxxvi. p. 119.

[2] Excellent chapter on the construction of the snowshoe, by Murdoch, in *Ninth An. Rep. Bur. Ethnol.*, Washington, 1892, 344–52, 4 figs.; also von Schrenk, *op. cit.*

some places strips of raw hide the shape of the foot, and fastened thereto by thongs. The sandal came up from the south. In other warm regions the raw hide was replaced with a coarsely woven sandal made of the toughest and most available fibre of the country—perhaps of yucca filaments as in Mexico, or of palm fibre as in other lands. Each culture area provided an excellent substance, which Necessity was not long in finding.

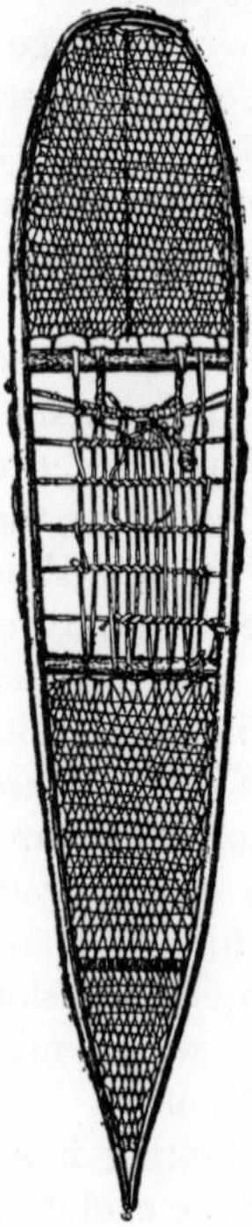

FIG. 60.—Snowshoe from Alaska. (*U. S. Nat. Museum.*)

The most primitive of foot-gear is found among the Indians of Guiana. This is a pair of sandals cut from the leaf-stalk of the aeta palm (*Mauritia flexuosa*), which is worn on very stony parts of the Savannah to protect the feet. The string which keeps the sandal on the foot passes between the great toe and the next, and when these foot coverings are much worn, the flesh between the two toes becomes callous. A very few hours' use wears out the sandals, but this does not much matter, for a new pair can be cut from the nearest aeta palm, and can be ready for use in a few minutes. They are made to the measure of the foot as carefully as though they were done by European shoemakers.[1]

Starting from this southland, where the burning or rough earth has suggested to the traveller the advisability of clothing the soles of his feet, one may hear the environment whispering new suggestions in his ear at every degree northward.

In the desert, among venomous creatures, man dons rude leggings. Among the thorns and cactus the foot was covered, the moccasin was invented, and even a long toe

[1] Im Thurn, *Indians of British Guiana*, London, 1883, p. 195.

was turned up, like the front of a snowshoe, to tread down the thorns. The Apache Indians, roving in the deserts of Arizona, were known by the style of this foot-gear. The

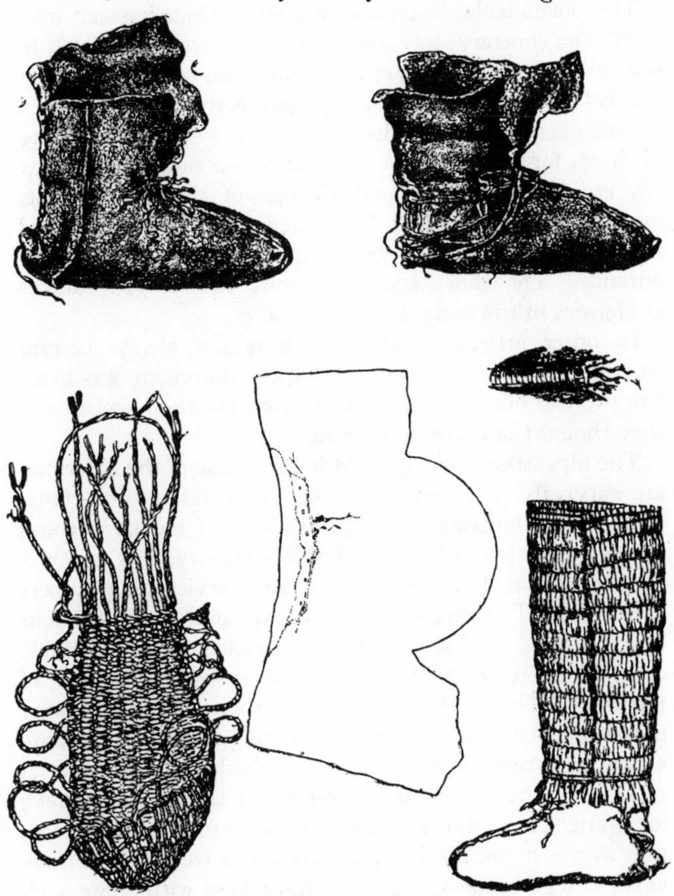

FIG. 61.—Evolution of the Top-boot, Hupa Indians, California.

covered sandal or shoe was also a helper against the cold. In the temperate zone of both continents and both hemispheres the shoe in some form came early to be prevalent.

But in the regions of the north the sandal, the moccasin, the gaiter, the buskin, and the legging were all securely sewed together into the boot.

This formidable foot-wear, whatever the chasseur may say to the contrary, was invented in the north—at least, in subarctic climes. Furthermore, the Samoyede, as well as the Eskimo, was persuaded by dame Nature to add a sock of soft grass, into which he thrusts his foot before donning his boots for a long journey. And so our modern foot-gear has required all climes and all ages of the world for its invention. Between the barefooted man and the booted Hyperborean there is a great difference of speed and endurance. The shoes and boots were the winged sandals of Hermes in the early days of travel.

In other articles of clothing it is not always certain whether the idea of utility or that of adornment was first; but there is no doubt that men protected their feet before they thought of decorating them.

The alpenstock and the gold-headed cane of the alderman are survivals of a useful accessory in travelling and transportation. Omitting the pilgrim's staff of mediæval times as being a little too far along for this inquiry, it is but just to observe that its ancestors were most serviceable members of society. The cargadores of Mexico, and all the burden-bearing tribe of Latin America, together with the noble army of coolies, everywhere carry staves for a multitude of purposes. The Pima Indian of Southern California has a notch or a fork on the top of his staff, which he may use at will as a rest for his carrying net-basket. The survival of this may be seen in the single leg of the organ-grinder's apparatus or in the coolie's staff, serving now as a cane, now as a rest for his load, and anon as a weapon, defensive against dogs or men. It must have been with some such thought in his mind that the Psalmist wrote, "Thy rod and thy staff they comfort me," serving the double purpose of supporting his flagging body and warding off his savage enemies.

Reference might also be made to stilts, which are by no means confined to the amusement of children, nor yet to the French shepherds, but may be seen in savagery as well. The Polynesians made stilts of very beautiful patterns.

Travelling upward, in the house life or along declivities and in steep places, stimulated the invention of steps cut from a log, ladders of rope or bamboo or wood. The Pueblo tribes, for safety against the roving Indians in the neighbourhood, had studied out the problem of security. Living in fastnesses of the cañons, or in adobe houses on the mesas and the plains, they rendered themselves secure by ascending to their homes on ladders and drawing these up after them. Here they defied both assault and fire.

"The road to Peten," says Morelet, "was interrupted by almost vertical descents, impassable for mules, and only ascended and descended by pedestrians, through the aid of rude ladders, formed of the notched trunks of trees placed against the rocks." [1] The inclined plane was mentioned in a previous chapter as a mechanical device for raising heavy objects. It was thoroughly studied out in the trails and mountain passes that cover the elevated portions of the earth like a network, and led up no doubt to later military engineering.

The modern elevator is in reality a vertical railway for passengers and freight. Hoisting apparatus for freight is common enough, and quite old, but, before the vertical railway was invented, passengers had to walk up into the air on steps and ladders as we have seen, but climbing was before that. Ellis tells us that a little Polynesian boy who goes up for cocoanuts, "strips off a piece of bark from a purau branch and fastens it round his feet, leaving a space four or five inches between them, and then clasping the tree, he vaults up its trunk with greater agility and ease than a European could ascend a ladder to an equal elevation. When a bunch is gathered at a time it is lowered by a rope; but when the nuts are gathered singly the boys

[1] Morelet, *Trav. in Cent. Am.*, New York, 1871, pp. 327, 420.

give them a whirling motion that they may fall on the point."[1]

The Aetas of the Philippines are prodigiously active at climbing trees, clasping the trunk with their hands and setting the soles of their feet against the trunk.[2] This is precisely the way in which men of our day mount telegraph poles. The rough bark of the tree, and the bare feet of the Philippine islander are offset on the smooth telegraph pole by spikes on the operator's boots.

Anticipatory of the modern vertical lift is the device of the bee hunters of the island of Timor. A company of natives coming upon a smooth tree, without a branch till at seventy feet from the ground, and spying three large bee combs, prepared to take them. One of them, from the stem of a creeper, made a bush-rope by which to do the climbing, and proceeded to wrap his head and body in cloths, but left his limbs free. To his girdle he had attached a long coil of small rope wherewith to let down the honey. To one end of the bush-rope a wood-torch and a chopping knife were attached. "The bee hunter," says Wallace, "now took hold of the bush rope just above the torch, and passed the other end round the trunk of the tree, holding one end in each hand. Jerking it a little above his head, he set his feet against the trunk and, leaning back, began walking up it." When he had come within a few feet of the bees he began to swing his torch. Arriving at the limb, he attached the small cord at his girdle to the comb hanging down, cut it loose with his chopper, and let it down to his companions, at the same time plying his torch. As soon as he had secured the three combs he retraced his steps to the ground, by means of his bush-rope.[3]

[1] Ellis, *Polynes. Res.*, London, Bohn, vol. i. p. 57.

[2] E. Best, *J. Polynes. Soc.*, Wellington, 1892, vol. i. p. 12; quoting Geronière, *Twenty Years in the Philipines.* Consult also Woods, *Unciv. Races*, Hartford, 1870, vol. ii. p. 33, for Australian climbing.

[3] Consult Wallace, *Malay Archipel.*, New York, 1869, p. 207.

ORIGIN OF THE MODERN "LIFT," OR ELEVATOR.
(*Scene in Alicante.*)

The Dyaks have a most ingenious method of ascending tall, smooth trees. The necessary apparatus are a number of sharpened bamboo pegs about a foot long, one or more long, slender bamboos, and cord made from the inner bark of a tree, all of these being prepared on the spot. "They now drove a peg into the tree to be ascended, about three feet from the ground, and bringing one of the long bamboos, stood it upright close to the tree, and bound it firmly to the two first pegs [one at the ground and one three feet above], by means of the bark cord. One of the Dyaks now stood on the first peg, and drove in a third about the level of his face, to which he tied the long bamboo in the same way, and then he mounted another step, standing on one foot and holding by the bamboo at the peg immediately above him, while he drove in the next one. In this manner he ascended about twenty feet, when, the upright bamboo becoming thin, another was handed by his companion, and this was joined on by tying both bamboo to three or four pegs."[1] And so on to the point desired.

The T'lingit Indians of Alaska ascended the tall trees, from which their totem post and great house logs were cut by a similar device.

Locomotion must necessarily have been largely by water at first. It was the reproach of the Choctaws living on the Mississippi river that they could not swim. But it would be very difficult to find another tribe of savages devoid of this art.

The Labrador Indians use little paddles to drag themselves quickly through the water. The tribes on the borders of Mexico, in Peru, and in several localities in the Eastern Continent, tie bundles of reeds together as floats. The ancient Assyrians are represented as buoying themselves upon inflated goatskins. Cardinal Wolsey confessed that he had ventured, like wanton boys who swim on bladders, far beyond his depth. The breaking of his high-blown

[1] Wallace, *Malay Archipel.*, New York, 1869, p. 66.

pride was true, no doubt, but the bladders used as life preservers by boys and men are difficult to burst.

On the Gulf of California there are tribes that lash two light bits of wood to a vine which they place against the breasts, exactly after the manner of the cork life-preservers. Even the eastern Eskimo at times ride on the seal-skin harpoon floats. Except in the matter of flying, the savage man solved the difficulty of going where he pleased.

FIG. 63.—Assyrian Warrior crossing river on an inflated skin.

Ellis says, " Like the inhabitants of most of the islands of the Pacific the Tahitians are fond of the water, and lose all dread of it before they are old enough to know the danger." In surf swimming they used a small board, on which they were accustomed to ride inward on the breakers. The Sandwich islanders were especially skilful with the swimming board, being able to sit, kneel, and even to stand on them when the crest of the wave was pushing shoreward." [1] In the sport called *pakaka-nalu*, the player rides the surf sitting in his canoe. The canoe poised on the inclined plain, in advance of the wave, is carried shoreward at such speed that it is possible to avoid broaching and being upset only by a delicate adjustment of forces and great skill and judgment with the paddle.[2]

But in a chapter on the beginnings of the carrying industry we are most concerned with conveniences for riding and transportation. Here, as in other arts we are to investigate the construction of each apparatus, its evolution or development, and its geographic distribution. How much of each form is due to the earth, and how much to human ingenuity will be difficult to determine. But we are at the foundation of one of the four greatest of human employments—the exploitation of the globe, the transformation of material, and the consumption of industrial products being the other three.

The first coaches were the backs of savage mothers. In the frozen north, the Eskimo woman bears her infant about in the ample hood of her warm caribou or sealskin robe. There, both men and women are compelled to wear the parka or blouse, with a hood or cowl attached, and this they may throw back at pleasure. But the mother provides a hood like a great bag, so large indeed that her babe may rest upon her shoulders and crawl about in the back of her bonnet, as in a miniature sleeping-car berth.[3]

[1] Consult Col. Lane Fox, "Early Modes of Navigation," *J. Anthrop. Inst.*, London, 1875, vol. iv. pp. 399–437.

[2] See "The Long Voyages of the Ancient Hawaiians," *Hawaiian Hist. Soc.*, May 18, 1893.

[3] Mason, "Cradles of the American Aborigines," *Rep. U. S. Nat. Mus.*, 1887, pp. 161–235; Murdoch, *Ninth An. Rep. Bur. Ethnol.*, pp. 110–120, figs. 52, 61.

Further south, as far as the borders of modern Mexico, and in the more temperate regions of South America, the vehicle in which the youthful aborigines are transported is called a papoose frame. The passenger is lashed to or in this carriage shortly after its birth, and rides about therein until his legs are strong enough to bear him up.

Fig. 64.—Eskimo Mother carrying babe in her hood. (*After Healy.*)

It is an interesting study to mark the close relationship between this primitive mode of conveyance and the environment, and also the effects which the vehicle has had on the rider. In the cold precincts of Canada birch-bark furnishes the material, making a light and warm trough in which, embedded in soft furs, the young savage takes his ride. The climate decides both the material and the method, and there is little artificial modification of the body.

On the Pacific coast of America, in that wonderful lumber belt which stretches from the Columbia river to Sitka, troughs of wood like tiny arks are excavated for the papoose, with bed of shredded cedar bark. Among not a few tribes, Chinooks proverbially, each passenger is furnished with a soft pillow which he is obliged to wear for many months upon his forehead, so attached to the sides of the cradle as to arrest the increase of the brain in front and force it to grow in the region of the vertex, pushing the skull up in that direction like a pyramid. This operation is said to have made the individual look taller, while it did not impair his intelligence. In this instance also the climate selected both the material and the style of the cradle, which, in its turn, gave shape to the body of the tiny rider throughout its life.

The Indians who treat their children thus are of several stocks. Besides this intentional modification of the skull, there is an inadvertent flattening of the occiput in all the tribes of America.[1]

In the eastern portion of the United States, the primitive vehicle was a flat board or frame, with sides of raw-hide, profusely decorated with quill-work and feather-work, and, latterly, covered over solid with bead embroidery—a dainty coach indeed. One has only to put runners beneath it and increase its size in order to have the travelling sledge in vogue in all the countries of the domesticated reindeer and the harnessed dog. The modern cradle with rockers is also the descendant of some such device, and the old-fashioned hooded cradle was not unlike it in form. The effect of this perfectly flat and straight board is visible in the limbs and carriage of the Eastern Indians.

In the Great Interior Basin of the United States and in California, wicker carriages were in universal fashion, many of them made with exquisite care. Some of them were built up from a hurdle-like wicker frame, perfectly straight, while others in form resembled the bowl of a spoon. The beds in these were differently constructed in accordance with the climate and elevation and natural resources, each region having its own unmistakable form. Where the same culture area is inhabited by more than one stock, the tribal characteristics are also marked like escutcheons upon the coach.

In one and all of these, the passenger, without being consulted, was compelled to lie on his back, to be wrapped in furs or blankets or mats, and to be kept secure in his place by being lashed therein with soft thongs or buckskin.

In the days of unchanged savagery he was borne about on the back of the mother, a soft band of buckskin attached to the top of the apparatus passing across her forehead. Occa-

[1] Porter, *Rep. U. S. Nat. Mus.*, 1887, pp. 213–235.

sionally this band was passed over the projecting knot of a limb.

> "When the wind blows the cradle will rock,
> When the bough bends, the cradle will fall,
> Down will come baby and cradle and all—"

necessarily, as they were securely lashed together.

In tropical and sub-tropical portions of both continents, where it was as much as an infant's life was worth to ride in so close a vehicle, the mother infolded it in her serape or mantle, carrying it now on her hip, now on her back, and anon on her shoulders. The women of the tropics adopted everywhere this form of carriage. It was always a marvel how the tiny creatures could keep their seats in such an insecure vehicle. A friend of the author tells of a Mohave woman who came to a barrack with her child thus mounted. Spying a rain-barrel partly full, she threw off her serape, the only garment she wore, and proceeded to bathe herself, her infant passenger sticking to her like a young lizard to a tree. Indeed, in all their travelling and working the women carry their infants upon their persons, having a half-conscious, half-automatic care of them the livelong day.

There is usually about the woman's waist or arms or shoulders a band or girdle, and the prehensile instincts of the babes soon learn to stick to this. The apes carry their children in the same manner, only they do not wear a girdle, which is an invention.

In the extreme southern regions of South America among the Araucanians, the cradle frame is similar to those in Eastern United States, and to the carrying frames of Eastern Asia; and further south, in Tierra del Fuego, there is a semblance of the Eskimo custom of infantile transportation, though the abject natives go very much more naked. So, in the American continent, from one extremity to the other, the ingenious minds of the aborigines, who were of the same race, understood the exigencies of climate and the regional resources so well as to establish in their method of carrying their children an excellent harmony.

Throughout the African continent, in Australia, among the Pacific islanders, in the Asiatic continent south of the Altai Mountains, indeed, in much of Europe, the carrying

FIG. 65.—Freight and Passenger Conveyance in one. Araucanian woman.

method of the Mohave woman before mentioned is kept up. But the little passenger attaches himself peculiarly to the mother's body in each region and tribe. In civilised Europe the child sits in front, that is, it rides on the forearm of the maternal beast of burden or of her substitute, the nurse. Mention should be made later on of the tiny carts and waggons in which the latest born of the Aryan race are dragged about. But it is not to be forgotten that in the earliest times of our species the custom of conveyance began with the transportation of infants, frequently for long distances, on the mother's back.[1]

For the better security of their babies when travelling, women are in the habit of hanging ronnd their necks a string, the ends of which have been previously fastened to the infant's wrists. The child is carried about in a *chip*, which is a sling or band made by women from the bark of the *Melochia velutina*, which is worn over the shoulder like a sash.[2]

Mention will be made later on of the methods of being carried about in apparatus of greater complexity. Leaving here for a moment the passenger primeval we may observe the most simple kinds of burden-bearing.

The millenniums of change through which human invention has passed in the transforming of a rude stick or frame to fit on a man's back, or a burden strap to fit across his forehead, or a pad to rest on his head, into the latest devices for transportation by land or by sea, constitute the history of one of the world's activities. Many volumes would be required to tell the whole story; but in this connection are to be rehearsed only a few of its opening paragraphs, and to notice how these initial efforts survive into modern culture. Here, also, we cannot stop with the thing invented, but must search for co-ordinate changes wrought in other industries.

To have some conception of the enormous amount of labour borne on human backs, calculate the weight of every mound, earthwork, embankment, fort, canal, wooden, brick,

[1] Excellent picture of Bechuana women hoeing and carrying baby at the same time in Holub, *Illustrierter Führer durch die Südafrik Ausstellung*. Prag, 1892, Otto, p. 86.

[2] Man, *Andaman Islanders*, London, 1888, Trüb., pp. 109, 180, pl. vii., fig. 25. Consult also Pokrowski, *Revue d' Ethnographie*, Paris, 1889, 50 pages with illustrations. Also *Les Nourrissons à travers les pays et les ages*, A. Collin, Paris; and Ploss, *Das Kind*.

and metal fabrication and structure on earth. These have all been carried many times and elevated by human muscle. Omitting the few heavy stones too weighty even for a combined human effort, all the freights of land trains and sea craft were first carried by men and will be lifted and borne by them again and again. The ancient field gleaner, miller and cook performed little compared with the modern farmer, roadster, freighter, warehouseman, retailer, and consumer.

If one should walk through the markets or along the docks of a great city at busy time, he would be surprised at the survivals of ancient ways of carrying existing on into our day. The human body is marvellously adapted to the greatest variety of burden bearing. Almost from the crown of the head to the foot loads may be attached. They are borne on the head, on one shoulder, on both shoulders, on the atlas, on the hips, on the thighs, on the arms, on the knees, and they are suspended from the top of the head in front ; from the forehead, resting on the back, from the neck, shoulder, breast, arms, hands, waist, and even from the knees. To suit these parts of the body there have been invented, "the milkmaid's pad," the forehead band, the porter's knot, the "Holland yoke," the Chinese yoke, the pedlar's stick, market baskets, knapsacks, burden baskets, panniers, haversacks, grip-sacks, and all the rest.

There are tribal and regional and national ways of attaching the load. What will suit the plain will not suit the mountain. What will suit meat will not be convenient for acorns, and so on to the end.

These for single carriers. Then comes co-operative carrying involving the palankeen form, the bier, the lumberman's stick held in the hands and supported on the knees. In the hill country of Hindostan as many as three hundred men have been seen carrying a menhir on a frame. The whole could not have weighed much less than fifteen tons. Every man in the street carries something, every lady has her package or her parasol. If all the loads great and small at any moment resting on human bodies could be added up, it would equal at the same moment all other loads on ships or railroad trains.

A Pullman palace train weighs from 780,000 to 1,000,000 pounds, and travels at least forty miles per hour. A pack-man among our primitive peoples cannot possibly move over one hundred pounds ten miles a day. The train, with proper relays, will move nearly a thousand miles a day. Allowing one hundred and fifty pounds for the weight of the carrier, it required four hundred thousand men to do the work of this one train. The problem is not quite so simple as that, because another great advantage is gained by the diversification of industries, and business created by the making of the track, and the train, and the commerce involved.

If all of the carriers of the world were marched in a procession in the order of the antiquity of their devices, taking into consideration the actual relationships and affiliations of certain forms, the single burden bearers, sustaining their loads without any intermediary apparatus would come first, they would be followed by single carriers who were using some sort of apparatus. All of these could be put in lines according to the part of the body involved.

The co-operative carriers might follow in accordance with the same evolutionary plan of arrangement. Before there were any means of transport over the mountains lying between Hope, on the Frazer river, British Columbia, and the Similkameen tribes, the Indians used to be employed to pack provisions on their backs. The packs were suspended by means of a band or strap passed over their foreheads, and one of them, says Mrs. Allison, packs three sacks of flour (150 lbs.) on his back, while travelling on snow shoes for a distance of sixty-five miles over a rough mountainous road, with a depth of twenty-five feet of snow on the summit of Hope mountain, over which the trail ran.[1]

Sometimes a whole family would start out on one of these

[1] Mrs. S. S. Allison, *J. Anthrop. Inst.*, London, 1892, vol. xxi. p. 306.

packing expeditions, the children, as well as their parents, each taking a load, and accomplishing the journey in six or eight days according to the state of the road. If an unusually violent snowstorm overtook an Indian while travelling in the mountains, he would dig a hole in the snow, cover himself with his blanket, and allow himself to be snowed up. Here he would calmly sleep till the snow had passed, then he would proceed on his journey.

An interesting combination of toting with packing may be seen in Jamaica on any market-day. The picture of a negro woman with an immense weight on her head, frequently over one hundred pounds, is common enough. But this same woman, still retaining the load on her head, has learned to drive a donkey in front of her to help her in her work of transportation. A halter of rope serves for bridle, and this is attached to a long line held in the driver's hand. The little creature is almost concealed beneath its panniers and bags and bundles of produce.[1]

Mr. Croffut makes an interesting statement about the Mexican porters. "In every part of Mexico have I observed the Cargadores, patiently following the trails and carrying immense loads on their backs. I recollect seeing, four years ago, near a railroad station, half a dozen of them squatting on the ground, resting. One had a sofa upon his shoulders, strapped on I could not see how; another bore a tower of chairs locked into each other, and rising not less than eight feet above his head; another carried a hen-coop with a dozen or twenty hens, and others were conveying laden barrels and various household goods. They had come, they said, from San Luìs Potósi, not less than fifty miles distant. These Cargadores will cover thirty miles a day for a week or more, going from ocean to Gulf.

"During a ride which I made over the Andes on the Mexican National Railroad, these persistent carriers were almost always in sight from the car windows, the peons and burros following each other up and down the slopes. The vice-

[1] See figure in Ward, *Jamaica at Chicago*, New York, 1893, Pell, p. 95.

president of the road, Thomas C. Purdy, said, as we watched these animated trains advancing on parallel lines, 'There is our rival. That is the only transportation company we fear. If it were not for that line, this country would treble its railroads next year, and the roads would double their profits. We are combatting the custom of centuries. Those fellows carry on their backs to Mexico the entire crops of great haciendas far over the mountains. I have sat down with a wealthy and enterprising haciendado and explained to him that we could do his carrying in a quarter of the time and for half the cost, and have seen him refuse to change, and stubbornly stick to the old method. I was never before so impressed with the tremendous force of habit.' "[1]

As transportation through the forests of Yucatan can be effected only on the backs of porters, the traveller has before him the humiliating spectacle of man reduced to a beast of burthen. The Indians, especially those of the central provinces, are accustomed to this kind of labour, which their fathers pursued before them from time immemorial, and they not only carry merchandise and baggage, but the travellers themselves, by means of a kind of chair suspended from the shoulders.[2]

" For the Indians," of Vera Paz, Guatemala, " there is no road too bad : and where no beast can keep its feet, they go and carry loads without difficulty. Herein is seen the power of habit, since these people, beginning at six years' old to carry burdens, become such active carriers as to be able to make journeys of two hundred leagues or more, without suffering, when the best mule, if unshod, becomes so lame as to be unable to move a step. I have often seen them, after having hurt themselves by stumbling, hold a burning skewer near the wound or bruise, to prevent inflammation, and start fresh on their journey after this painful treatment. When on a journey they carefully avoid drinking cold

[1] W. A. Croffut in *Am. Anthropologist*, Washington, vol. ii. p. 80.

[2] Morelet, *Travels in Cent. Am.*, New York, 1871, p. 279.

water, and quench their thirst with water as warm as it can be taken. Their ordinary food is a little roasted maize paste, called *totoposte*, which they crumble into boiling water and so eat it, or else they warm it with chile and salt. Wherever they stop they stretch themselves at full length, although it be on the stones, extending to the utmost their

FIG. 66.—Meat Vendor in the city of Mexico. (*After Holmes.*)

legs and arms, and by this means they soon regain their vigour."[1]

This will serve to emphasize what the reader has been

[1] Escobar, in *J. Roy. Geog. Soc.*, London, vol. xi. pp. 89–97, quoted in Morelet, *Travels in Cent. Am.*, New York, 1871, pp. 418–27. A good sketch of a human beast of burden in the shape of a man dragging and carrying poles is in Whymper's *Travels among the Great Andes of the Equator*.

accustomed to see all his life, but not to observe. But the frontiersman, the farmer's help, the mechanic at his ordinary employment, bows his back and harnesses himself to loads without much intermediary apparatus, the beginnings of which must now be examined.

The porter's knot is a device roughly made like a huge horse-collar, and fitting down over the head and shoulders and upper back of a man, to enable him to do his very best in carrying. With such a load as he is represented as bearing, Atlas should have worn a porter's knot. The sight of these appliances is common enough on the commercial streets of London, and in another form in Constantinople; but the author has never seen one in America. In addition to distributing the load over several parts of the body, they are padded so as to enable the carrier to take on hard boxes, furniture, and such things without bruising his flesh.

In China human beasts of burden are even now more profitable than pack animals over narrow and circuitous passes. In Southern China the long string of coolies bearing down from the hills tea-leaves in deep baskets slung on poles is a familiar sight. The transport of brick-tea from Syn-chuan into Thibet is by coolies, who bear the packages on a wooden frame strapped to their shoulders. They make a fifteen days' journey, carrying one hundred pounds of tea each.

In some places, where vehicles are used, the bridges are so narrow that the mules are unhitched and led singly, while the carts are carried over on men's shoulders.[1]

Men and women in Korea carry burdens upon frames that remind one of Europe rather than of Asia, as do other customs of the Land of the Morning Calm. The affair resembles a painter's knapsack or the framework of a wheelbarrow, without the wheel. It is hung to the bearer's shoulders by means of broad, braided ropes attached to the upper part of the apparatus, passing over and in front of the shoulders as in a knapsack, and attached by loops or

[1] Minister Denby, *J. Soc. Arts*, London, 1892, vol. xl. p. 166.

knots to the lower end of the principal pieces. From the outside of these two project two sticks outward to receive the load. This framework will hold all sorts of things, and is really a handy vehicle.

The climax of transportation after the fashion of the knapsack is reached in Monbutto land, where, Schweinfurth says, it is the fashion to pile up the hair in a high chignon. They naturally avoid exposing these artistic coiffures, accomplished with so much expenditure of time, to the danger of being crushed, and therefore carry baskets, after the manner of the panniers in Central Germany, supporting them by means of bands slung over the forehead.[1]

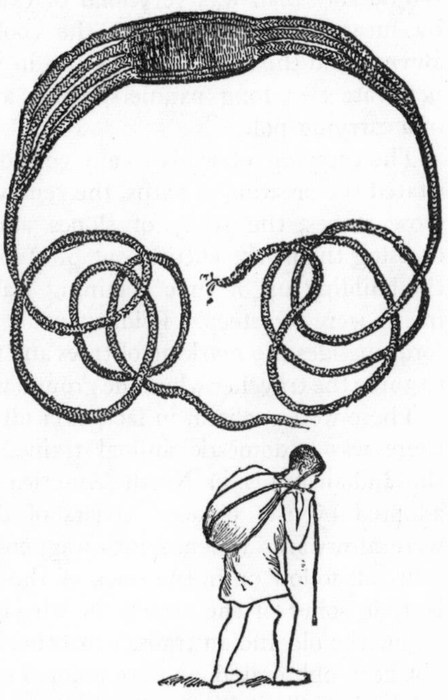

Fig. 67.—Parbuckle and Carrying Strap combined. N.E. Asia and British Columbia.

It is in connection with this forehead-band that a species of parbuckle is in use in both hemispheres. The carrier places the middle of a long line against his forehead, passes the two ends down under the pack and up to his shoulders. By pulling on the ends he rolls the load upon his back.[2]

[1] Schweinfurth, *Artes Africanae*, London, 1875, pl. xviii. fig. 15; refers to Chaillu, *Journey to Ashango*, London, p. 84.

[2] Mason, *Rep. U. S. Nat. Museum*, 1886-7, p. 287, fig. 42.

The carrying-yoke was foreshadowed in Polynesia. A pair of gourds suspended from the auamo, or carrying-stick of the Hawaiian, served for traveller's trunks, one containing food, the other clothing. They were dried and carefully cleaned, furnished with a cover, which also served as a dish and a net to hold the cover close and to form handles.[1]

The Hawaiian was very fond of carrying on his shoulder by means of the stick like the coolie's yoke. In Siam, Burma, and the Far East it is also in vogue. The Siamese perforate two long panniers, which are borne on the ends of a carrying-pole.

The carrying of loads to any considerable distance necessitated the creation of paths, the removal of brushwood and loose stones, the study of slopes and the easiest way of crossing them, the cutting out of steps here and there, and the building up of rude retaining walls where the escapements were too steep. Bridges would often be needed and fords, besides the marking of trees and the piling up of stones to guide the traveller when the ground was covered with snow.

These were realised, in fact, over all the continents before there was a domestic animal trained. Curiously enough, the Indian trails in North America were taken up and adopted by the pioneer settlers of the continent. They were afterwards widened into waggon-roads, and not a few railroads follow on in the track of the waggons. Certain it is that some of the streets in thriving cities of the west follow the old Indian trails. In other places the trails have not been obliterated, and are pointed out by antiquarians.

The term "pitching track" is applied to an Indian trail from one part of the country to another. West of Manitoba, on the crest of the ridge, there is a narrow, well-worn path, which, for many generations, probably, has been the highway of the Indians passing to the Assiniboine, through the valley of *Te-wa-te-now-sube*. This is connected with the "Ridge Pitching track."[2] The same author explains

[1] Brigham, *Cat. Bishop Museum*, Honolulu, 1892, vol. ii. pp. 7, 22.

[2] Hind, *Canadian Red River*, London, 1860, vol. i. p. 51.

how the roads are made in the Red River country for the toboggan sledges, by tramping a track first with snowshoes and following it up with the teams until it is as smooth as glass—a natural railway.

In this primitive migration there were no cross-cuts. All roads were crooked ; the trails and paths, the roads and ancient river routes are paralleled in modern times by canals winding about contours, railroads as crooked as ram's-horns, and even steamships meandering through the sea to follow their currents and the trade winds. The portages adopted by the American savages were possible where the head waters of rivers were near together, and where the buffalo-hide and the birch-bark made the light canoe possible.

The working out of highways has been along some such method as the following. By land, men went first trackless, then by trail, path, common road, highway, paved way, railroad. On the water their routes were in small streams, on rivers, land-locked seas and canals, the open sea or ocean. Parallel with these ways of conveyance the commerce and communication were first local, then regional, then continental, and finally world-embracing.

In the more favoured and cultivated regions of Mexico and Peru, the making of roads received still greater attention. For the purpose of facilitating the procuring of food for the centres of population and for moving the fighting contingent still greater attention was paid to highways.

That grading was well known to the American aborigines is attested by the approaches to the earthworks and fortifications of the mound-builders of the Mississippi valley. The Cahokia mound opposite St. Louis is one hundred feet high, and the top is reached by a graded way, which is now used for a carriage-road.

The ingenuity of savagery in this connection is best shown in the construction of bridges. A log thrown across a stream was about all that the savage mind attempted in the temperate zone. The tropical man was really the primeval bridge builder, who had great chasms to cross, and who also had around him rattan, cane, bamboo, and long pliant vines. The occasion and the material evoked the necessary intellect, as they have ever done.

A few bridges of stone were constructed by the Spaniards, some after the Conquest, and a few others have been erected by their descendants; but, as a rule, the rivers and mountain

Fig. 68.—Cliff Dwelling in Mancos Cañon, with Graded way. (*Haydon.*)

torrents are passed to-day by the aid of devices the same as were resorted to by the Incas, and at the points which they selected. In a country destitute of timber, they resorted to suspension-bridges formed of cables of braided withes, stretched from bank to bank. Where the banks are high, or where the streams are compressed between precipitous rocks, these cables are anchored to piers of stone. In other places they are approached by inclined causeways.

Three or four cables form the floor and principal support of the bridge, over which small sticks are laid transversely and fastened to the cables by vines, cords, or thongs of raw-hide.

Two smaller cables are sometimes stretched on each side as a guard or handrail. Over these frail and swaying structures pass men and animals, the latter frequently with loads on their backs.[1] It is not possible to say whether the inventors of modern suspension bridges did not copy this idea out and out without waiting for the regular processes of inheritance. In Sarawak a foot-bridge is constructed by planting two rows of long stakes in the ground alternately slanting in opposite directions, so that a small sapling laid in the fork will be horizontal and of the proper height. Each pair of stakes is lashed together at their intersection, and the bridge is further strengthened by perpendicular posts set under the footway. A pole is lashed along the top of each row of stakes to serve as a hand-rail. One of these between Paku and Serambo was a hundred feet long and nine feet high.[2]

The Dyak bridge consists of a stout bamboo for the foot-way, sustained by braces of the same material, for these engineers are well aware of the stability of the triangle. "When a stream is to be crossed an overhanging tree is chosen, from which the bridge is partly suspended, and it is partly supported by diagonal struts from the bank." In carrying a path along a precipice, the same combination of suspension and struts is used. "These bridges are tra-versed daily by men and women carrying heavy loads."

"From the landing-place to the hill a Dyak road had been formed, which consisted solely of tree-trunks laid end to end. Along these the barefooted natives walk and carry

[1] Squier, *Peru, &c.*, New York, 1877, p. 544; figs. opp. pp. 545, 558, 559. There are thousands of such bridges in Peru to-day. *Ibid.*, p. 505.

[2] Hornaday, *Two Years in the Jungle*, New York, 1885, p. 484.

heavy burdens with the greatest ease, but to a booted European it is slippery work."

"When a path goes over very steep ground, and becomes slippery, pieces of bamboo are cut about a yard long, and opposite notches being made at each end, holes are formed through which pegs are driven, and firm and convenient steps are thus built with the greatest ease and celerity." [1]

Across the Vanapa River in New Guinea, J. P. Thomson saw an ingenious bridge made of rattan. On one side of the river a large banyan-tree supported one end of the bridge fifty feet above the water. From this point it stretches seventy yards to the opposite side, where it was attached to a small tree supported by a stout post, and both of these were stayed by rattans to trees further back. The struts, foundation stringers and the rails, two on each side, were netted together by innumerable lacings of fine rattans, the whole looking like a modern cable bridge.[2]

Naturally, the carriage of passengers and freight on human backs was followed by the utilisation of animals. The study of the subject of domestication belongs to another chapter, but there are two animals in America and one in the Eastern Continent that have never done much for civilised peoples. These are the llama, the dog, and the reindeer. The last named, yielding its milk as well as its neck to the service of man, may be dismissed with a vote of thanks not only by the Board of Trade, but also by the Lapp mothers who utilised the milk of the reindeer in eking out their own,

The llama is confined to the sierras of South America, and has been used for a pack-animal from time immemorial. Its load is less than a hundred pounds, and its journeys *per diem* quite short. They are tractable animals, however, "a single drove laden with merchandise and containing from

[1] *Cf.* Wallace, *Malay Archipel.*, New York, 1869, p. 89. Figure of Dyak crossing bamboo bridge, pp. 45, 89.

[2] *Cf.* Thomson, *British New Guinea*, London, 1892, Philip, p. 90, with good page plate facing.

five hundred to one thousand head, being managed by eight or ten Indians." [1]

The freight and passengers in the Arctic region are transported overland, or, rather, over Nature's railroad track, the snow and ice, on sledges drawn by men and dogs. The parts of this operation to study in the sledge are the runners, the shoes, the cross-bars, the handles, the traction thongs, and the lashing. Concerning the dogs it is necessary to consider the domestication and training of the animal and the harness. And the economics of the apparatus involves the problem of greatest result with least effort over greatest difficulties and least natural resources. The runners are wrought from drift-wood or the bones of whales, and shod with strips of walrus ivory or whale-bone or whale's jaw, fastened on with tree-nails or lashed on with raw-hide thongs through countersunk holes. The cross-bars are longer than the sled is wide, and are not only pinned to the top of the runners, but thongs pass from their extended ends to the runners below and act as braces. The traction thongs are attached to the runners by toggles, and the thongs are joined by a loop and toggle, for this is a cold country, and knots are hard to untie. The handles of the sledge may be of drift-wood, but, wanting that, these cunning elves of the ice-land knock off the antlers of the dead reindeer, together with a piece of the skull, saw off the prongs, invert the device, lash the tips to the rear ends of the runners and the thing is done. When all is finished, and the vehicle is about to start, the Eskimo inverts his sledge, fills his mouth with water and blows it along the shoe where it freezes, fills the cavities and lubricates the surface. After polishing this surface down with his mitten it is ready for loading and progress.[2]

A Central Eskimo rarely brings up more than three or

[1] For a *résumé* of the literature on the llama and the paco see Payne, *History of America*, London, 1892, Macmillan, vol. i. pp. 317–330.

[2] An excellent account in Boas, *Sixth An. Rep. Bur. Ethnol.*, pp. 529–538, figs. 482–488.

four dogs at the same time. The young dogs are carefully nursed, and in winter they are even allowed to lie on the couch or are hung up over a lamp in a piece of skin. When about four months old they are first put to the sledge, and gradually become accustomed to pull. They undergo a deal of lashing before they are as useful as the old ones.[1]

The harnessing of the dog has reference to the character of the animals and to the nature of the work to be done. If you were to take into account the necessities in the present case they would certainly include, besides many unforeseen exigencies, the hitching of eight or ten dogs separately, provision for unhitching them instantly, the chance for the dog to jump out of his harness in a tight place, the possibility of taking a running start to set the sledge in motion, the opportunity to spread out or come together as occasion demanded, distance enough from the vehicle to keep out of the way on a descent.[2]

The old trading routes which existed among the Central Eskimo before the coming of Europeans are described by Boas. Two desiderata formed the inducements to long journeys, which sometimes lasted several years—wood and soapstone. The shores of Davis Strait and Cumberland Sound are almost destitute of drift-wood, and consequently the natives were obliged to visit distant regions to obtain that necessary material. Their boats took a southerly course, and as the wood was gathered a portion of it was immediately manufactured into boat ribs and sledge runners, which were carried back on the return journey; another portion was used for bows, though these were also made of deer's horn ingeniously lashed together.

Another necessary article of trade, soapstone, is manufactured into lamps and pots. It is found in a few places only, and very rarely in pieces large enough for the articles named. The visitors came from every part of the country,

[1] Boas, *Sixth An. Rep. Bur. Ethnol.*, p. 538.

[2] Boas, *loc. cit.*

the soapstone being dug or "traded" from the rocks by depositing some trifling objects in exchange.[1]

Tylor says, "The wheel carriage, which is among the most important machines ever contrived by man, must have been invented ages before history. . . . In looking for some hint as to how wheel carriages came to be invented, it is of little use to judge from such high skilled work as was turned out by the Egyptian chariot-builders, or by the Roman *carpentarii* or carriage-builders, from whom our carpenters inherit their name."[2]

According to the fitness of things the wheel carriage is not prehistoric. Dense population, fixed roads and great traffic or wars first called it into existence. Among the Chinese the wheelbarrow is common.[3] There the country is hilly, and the paths narrow. The waggon or cart or chariot breaks upon us in Egypt, in Mesopotamia, in the Asiatic steppes, where the whole champaign is a level road and there are no forests to impede the progress of wheels. In America the Red River cart and the lumbering Spanish vehicle are found under precisely similar environments.

No savage people have been discovered without the means of navigation. In the frozen north, among the wretched Fuegians, in Australia, boats or rafts are in daily use. The construction of these were better considered in the study of woodcraft. Here their relation to carrying passengers and to primitive commerce is to be considered. In those regions where beasts of burden were scarce boats were plentiful; they are the camels of the waters in the same sense that the latter is the ship of the desert. We shall

[1] Boas, *Sixth An. Rep. Bur. Ethnol*, p. 469.

[2] Tylor, *Anthropology*, New York, 1881, D. A. & Co., p. 193, with Egyptian figure.

[3] The old ditty—

> "The way was so long and the street so narrow,
> That he had to bring his wife home on a wheelbarrow,"

would suit exactly the prehistoric lover. Though it would more naturally be the wife wheeling the husband home.

look, therefore, for the most extended traffic by water in America and in the Indo-Pacific region.

The primitive boat was propelled by poling, by cordeling or tracking, by paddling, by rowing, and by sailing. Steam navigation is altogether a product of our century. Survivals of the other varieties may be seen in the modern sand-scow, the canal boat, the canoe, the yawl, and the catamaran. In late years important discoveries have been made of old Norse and other wrecks. This information is further increased by the ancient historians and monuments. So we come to a knowledge of the first mariners by digging up the wrecks of time, by questioning the descendants of sailors in primeval seas, by searching for some out-of-the-way arm of the ocean where generations of hardy mariners have handed down the simplest forms and names of mast and sail and rigging, by watching children playing on the beach, if haply they may reinvent some of the earliest forms of ships sailed in during the childhood of the race.[1]

This particular subject demands a volume. Every good sailor and shipbuilder knows how little the cunning savage has left for them to invent in the matters of lines and propulsion in the simpler craft.

The Australian moves his extremely rude boat by poling. In Venezuela the rivers that enter into Lake Macacaibo are navigated by shallow boats propelled by poles, the man standing on one part of the boat. All through the Southern United States the author has seen thousands of "long-boats," "scows," and even vessels with square sails pushed by negroes into and out of the shallow creeks. A clear way from stem to stern on each side enables the men to set the large end of a long pine sapling against the bottom, the smaller end against the shoulder, and to walk from stem to stern, pushing with all their might. They then walk back and repeat the process.

[1] Figures in Böhmer, *Sm. Report*, 1891. For a delightful scene in which a company of naked negroes are dragging a helpless steamer through the Nile grass, see Sir Samuel Baker, *Ismaïlia*, New York, 1875, opp. p. 53.

The Eskimo boat is a model of construction in the light of our modern craft with skins of steel drawn over frames of iron. For the construction of the umiak, which is their open freight boat, drift-wood, prepared skins of the ground seal, and abundance of stout raw-hide from the same animal or from the walrus will suffice. The framework of the umiak

FIG. 69.—The Primitive Rowlock. Eskimo lever device.

is a marvel of economy in material and of strength. After this is carefully joined together with thongs, the hide cover specially prepared is drawn over as neatly as a taxidermist finishes his specimen. The thwarts are secured by thongs, and even the rowlocks, for oars are used instead of paddles, are loops of raw-hide linked together. Into this buoyant craft the women place all the impedimenta of the family, the aged and the children, and move as occasion demands.

Dr. Rink is of the opinion that the skin-boat is the child of the birch-bark canoe, and that the more finished man's boat, kaiak, is after the same model, only, the farther the Eskimo has moved toward the open water, the more he has drawn the covering around him, until at last he sits in a well, with the gunwale fitting close to his body. In this case the utility of his craft is destroyed for freight, and becomes a passenger craft. Rather, it is turned into a man-of-war, for the deck is covered with weapons, and the manager goes forth to fight with beasts.

Again, the historical question arises whether the wonderful similarity between the graceful lines of the kaiak and those of the racing shell is a matter of gradual descent or of outright borrowing.

The paddle is older than the oar, and for rude peoples in natural streams more convenient. It has been mentioned that the Australian uses a bit of bark in bailing out his canoe, and also in forcing her along. There are other peoples who lash a flat blade to a pole for a paddle. The form cut out of the whole piece is more common. This method of propulsion is noiseless, and may be used in waters tangled with vegetation and encumbered by overhanging trees and bushes. It is universal in Australia, in the Indo-Pacific Ocean, and may be found in parts of Asia. Many pictures of Chinese, Korean, and Japanese life exhibit the fisherman using the paddle. The Eskimo, however, in their open water umiaks use the oar. The rowlocks are a model of ingenuity. The oar passes through two loops or links of stout walrus hide attached to the gunwale in such fashion that one prevents it from moving forward, the other from moving backward, but they allow perfect play in rowing. The Easter islanders employ a sculling oar, the end of which is carved to guide the water in fixed lines. There is nothing in savagery near so complicated as the ancient trireme, or the Norse boat, or the modern seine boat with oars. But with paddles the Caribs, the Haida, or the Polynesians could put those historic navigators on their muscle.

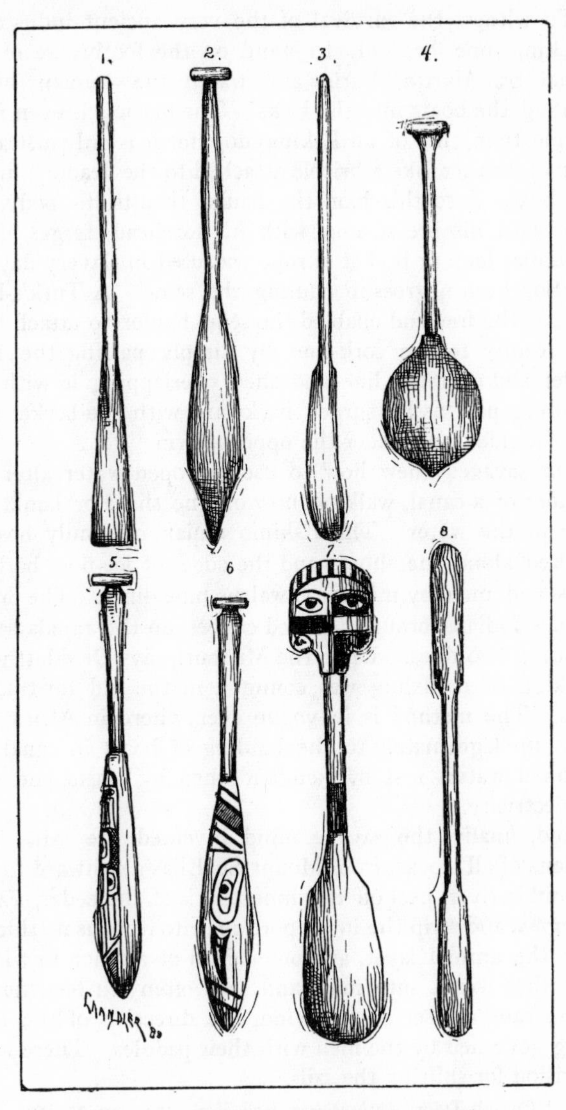

FIG. 70.—FORMS OF PADDLES IN AMERICA.
(*U. S. Nat. Museum.*)

To witness the survival of the very ancient industry of tracking, one has only to stand on the footbridge of the Canal St. Martin, Paris, and watch the human horses hauling the boats into the locks. The harness is even more simple than that of an Eskimo dog, for it is only a strap or loop of leather like a bricole attached to the leading line of the boat. Into this loop the hauler thrusts his body, and now with his breast, now with his forehead, forges along. A similar loop or becket of rope was used in slavery days by the Southern negroes in hauling the seine. A Turk's-head knot at the free end enabled the seine hauler to attach himself readily to the cork-line by simply passing the knot under and over the line and then overlapping it with the standing part as he surged backward with the becket over one shoulder, and under the opposite arm.

But savages knew how to use the open water after the manner of a canal, walking now on the shore or bank, and now in the water. The Eskimo umiak, or family boat, is tracked along the shores and the edges of ice-floes both by dogs and men by means of walrus-hide lines. The Montagnais Indians brought loaded canoes up the rapids safely by using two lines. Upon the Missouri, says Dr. Matthews, tracking or cordeling was common in the old fur-trading days. The method is in vogue everywhere in Africa, and leads up legitimately to the hauling of boats in canals or confined waters first by men, and then by beasts, and now by electricity.

And, finally, the savage mind invented the sail. The Indians of all stocks, from Mount St. Elias southward to the Columbia river, peel off the inner bark of the cedar, *Thuja gigantea*, and strip the inner portions into ribbons no thicker than the annual layer, and one-eighth of an inch in width. This they weave into mats and sails often ten feet square. These sails are set on the wind, the direction of the boat being governed by the men with their paddles. There is no provision for shifting the sails.[1]

[1] Consult Tylor, *Anthropology*, New York, 1881, pp. 252–59.

But far more successful in the use of the open water than any other savages of modern times were the Pacific islanders They made a canoe of bundles of bulrushes, in which they were not afraid to venture out of sight of land. This same craft is common in Peru. But in their dug-out and built-up canoes, made double or with outriggers, propelled by paddles and sails, they visited every archipelago in the Pacific Ocean.

FIG. 71.—Malayo-Polynesian Canoe and Outrigger, with which the Pacific Ocean was explored in the 17th century, A.D.

A little before the time when Europe was agitated on the subject of maritime discovery, canoe voyages were made by the Polynesians between Tahiti and Hawaii, a distance of twenty-three hundred miles.[1] As soon as the invention of

[1] See N. B. Emerson, *Hawaiian Hist. Soc.*, May 18, 1893; Smith, *Trans. Australian A. A. S.*, 1891; Alexander, *A Brief History of the Hawaiian People*, New York, 1891; Fornander, *Polynesian Race*, 2 vols.

the outrigger, the sail, and compressed food were perfected they sallied forth.

Associated with water travel to be supplemented by land travel is the primitive custom of establishing "carrying-places." A carrying-place is a path along which goods are transported around falls and torrents of rivers, or across country from one navigable water to another. In the former case the craft may be either hauled up through the water, or itself borne around with the goods. In the case of the birch-bark canoe, employed by the Indians of the lake region of North America, there was no difficulty of transporting it many miles.

"On these occasions only," says Hearne, "we had recourse to our canoe, which, though of the common size, was too small to carry more than two persons, one of whom always lies down at full length for fear of making the canoe top-heavy, and the other sits on his heels and paddles. This method of ferrying over rivers, though tedious, is the most expeditious way these poor people can contrive, for they are sometimes obliged to carry their canoes one hundred and fifty or two hundred miles without having occasion to make use of them ; yet at times they cannot do without them ; and were they not very small and portable, it would be impossible for one man to carry them, which they are often obliged to do, not only the distance above mentioned, but even the whole summer."[1] The bark canoe made of a single piece is found in Australia and South America, on the Columbia and the Amur river. The last named two are identical, having the bow and the stern pointed below the water-line.

How far these fur-hunting savages were from the most primitive sort of ferry ! They actually carried their ferry-boats hundreds of miles on their backs, not knowing the moment when they would encounter a lake or a river. In other words, it was cheaper to do so. There were ferries at certain points on the rivers further south, and they were

[1] Hearne, *Journey, &c.*, London, 1795, Strahan, p. 40.

common in Tibet. Catlin often speaks of being taken across the Missouri and its tributaries by Indian women in their bull-boats, made by stretching the hide of a buffalo bull over a crate made of poles. The coracle is a more improved form.

The extent and variety of ancient commerce on all the continents is attested by the occurrence of articles in graves, the materials of which are not found in the vicinity. Copper, mica, shells, curious stones, metals in the possession of historic tribes also bear witness to the diffusion of trade. Even

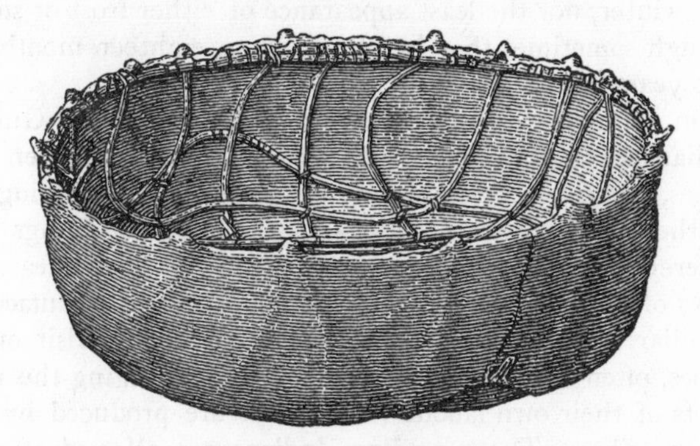

FIG. 72.—The American "Bull-boat," or Coracle.
(*U. S. Nat. Museum.*)

men belonging to peoples far distant have been met by travellers in their prospecting, and in many tribes there were laws providing that they should not be molested. This trade had pervaded Europe before Cæsar's day, and it had explored every land before its historic period.

Hearne gives a wonderful example of this extended travel. "It is indeed well known to the intelligent and well-informed part of the Company's servants that an extensive and numerous tribe of Indians, called E-ar-che-thinnews, whose county lies far west of any of the Company's or Canadian settlements, must have traffic with the Spaniards on the west side of the continent, because some of the Indians who traded formerly at York Fort, when at war with those people, frequently found saddles, bridles, muskets, and many other articles, in their possession, which were undoubtedly of Spanish manufacture. I have seen several Indians who have been so far west as to cross the top of that immense chain of mountains which run from north to south of the continent of America. Beyond these mountains all rivers run to westward. I must here observe, that all the Indians I ever heard relate their excursions in that country had invariably got so far to the south that they did not experience any winter, nor the least appearance of either frost or snow, though sometimes they had been absent eighteen months or two years."[1]

Im Thurn says that there exists among the Guiana Indians a rough system of division of labour between the tribes, and this serves not only the purpose of supplying all of them with better made articles, but also brings the different tribes together and spreads among them ideas and news of general interest. Each tribe has some manufacture peculiar to itself, and its members constantly visit other tribes, often hostile, for the purpose of exchanging the products of their own labour for such as are produced by the other tribes. These trading Indians are allowed to pass unmolested through an enemy's country.[2]

Of all the tribes on the coast the Warraus make far the best canoes, and supply them to the neighbouring tribes. In the same way, far in the interior, the Wapianas build boats for all the tribes of that district. The Macusis make ourali for poisoning arrows, the darts of blowpipes, and an abundance of cotton hammocks. The Arecunas grow, spin, and distribute most of the cotton which is used by the Macusis and others for hammocks. They also supply all blowpipes, for these are made of the stem of a palm, which, growing only in and beyond the Venezuelan boundary of their territory, are procured by the Arecunas, doubtless in exchange,

[1] Hearne, *Journey*, &c., London, 1795, Strahan, p. 40.

[2] *Cf.* Im Thurn, *Ind. of British Guiana*, London, 1883, p. 270.

from the Indians of the native districts of that palm. The Tarumas and Woyowais have a complete monopoly of the manufacture of the graters on which Indians of all the tribes grate their cassava. They are also breeders and trainers of hunting dogs. These distribute their graters and dogs through the Wapianas, who act as middlemen. The true Caribs make the best pottery, the Arawaks make fibre hammocks of a kind peculiar to them.

To interchange their commodities the Indians make long journeys. The Wapianas visit the Tarumas and Woyowais, carrying canoes, cotton hammocks, knives, beads, and European goods. Leaving the canoes they walk back, carrying a supply of cassava graters and leading hunting dogs. The Macusis visit the Wapianas to obtain graters and dogs, for which they give ourali-poison and cotton hammocks; and these in turn carry such graters and dogs as they do not require, together with their own ourali and cotton hammocks, to the Arecunas, who give in return cotton and blowpipes, or to the true Caribs, who pay in pottery. In this way travellers with news and goods pass from district to district."[1] No doubt this sort of commerce has been in vogue since the beginning of migration, scarcely any group of human beings having ever entirely lost their contact with other portions of the species.

[1] *Cf.* Im Thurn, *Ind. of British Guiana*, London, 1883, cap. xiv.

CHAPTER XI.

THE ART OF WAR.

"At no period of man's life were wars the normal state of existence. While warriors exterminated each other, and the priests celebrated their massacres, the masses continued to live their daily life, they prosecuted their daily toil."—KRAPOTKIN.

THE contemplation of the activities of primitive man in face of the greater animals is an excellent preparation for the study of war. Doubtless the motives and the actions of the wild beast were copied by men, both in their aggressive and their protective conduct. After the same fashion the weapons of warfare and those of hunting were modelled on identical notions. This parallel is painfully correct, even to the point of hunting men for food. However, the appliances in the two sets of activities are sufficiently differentiated to merit a separate treatment.

The present organisation, drill, and actions of the great armies and navies of the world are the resultants of the past efforts of leaders and students of military science. Backward in unbroken order the whole series could be traced to savagery. Not even in the lowest grade were men devoid of discipline. The first principles of defence and offence were then studied out and practised. The art of war has always engaged the greatest minds. There never was a tribe or nation that did not have its grand marshal, or generalissimo, or commander-in-chief, or war chief, whatever his title might be. For example, among the Muskhogean tribes, "the next one in dignity and power was the great war-chief. He led the army. In council his seat was nearest the mico, on his left, and at the head of the most celebrated warriors. On the right of the mico sat the second head-chief of the tribe, and below him the younger warriors of the nation."[1] This picture would be a true one for any other of the martial tribes of America, or of Africa, or of the Indo-Pacific.

The modern books on warfare divide the subject into strategy, tactics, and engineering. Strategy includes everything that is done out of sight of the enemy, and in preparation for actual fighting. By tactics is meant the actual fighting, the movement of the troops to the battle-field, their conduct and manipulation in the engagement. The engineer prepares the ground for offence or defence, and has charge also of roads and bridges. There are incidents preceding and succeeding the battle not included in these.

To obtain even a superficial view of the genius of savages in war it is necessary to examine the causes which induce them to undertake it, their modes of declaring war, the methods of recruiting and organising and drilling troops, their weapons and standards, their means of subsistence and transportation, and such other precautionary measures as they take regarding themselves, their families, and their property.

In the second place, savage tactics should be studied, their method of going to battle, of trailing the enemy, of actual fighting, their music and war-cries, captives, trophies, and neutrals.

And, thirdly, the best minds will always be found exercising themselves over the subject of temporary and permanent fortifications. These are the works of engineers in primitive times, whose remains in different parts of the world are encountered by explorers and excite the wonder of the civilised. Even the native populations can frequently give no account of their origin, only that they were erected by giants who perished long ago. The universality of such

[1] Jones, *Southern Indians*, New York, 1873, p. 11.

remains has induced many writers to defend a theory of degradation instead of progress in culture.

The most simple warfare was the duel, with or without weapons. The assassin and the duellist are survivals of the most primitive warriors. Before conflicts between communities could have existed there had to be rival communities, and even between them, if we are to believe those who know savages best, the disputes of individuals brought on blood revenge, and precipitated conflicts. It is not to be supposed that war was ever the normal occupation of any people. As now, so in all ages, war is an incident, an outbreak, a frenzy that soon exhausts itself.

Mr. Dorsey says that among the Omahas war was caused by the stealing of horses, the elopement of women, and infringement on hunting grounds. It is to be presumed that similar causes operated among other tribes. One of these causes recalls the elopement of Helen and the rape of the Sabines as cases of survival from a lower culture.[1] In aboriginal times war was occasioned by encroachments upon the standing order of things. Tribal endogamy tended to solidify each stock as against the whole world.

In Samoa war was provoked by the murder of a chief, a disputed title, or a desire for aggrandisement, and hostilities were prevented by giving up the culprit, paying a fine, bowing down in submission.[2] In more general warfare the three groups were the highway soldiers, the bush soldiers, and the sea fighters.

In Africa aggressive wars were common since historic times ; but the enslavement of negroes to bear the burdens of the world was the moving cause in most of these conflicts.

Most frequently there was no overt declaration of war, but when a great conflict was decided upon much ceremony

[1] Dorsey, *Third An. Rep. Bur. Ethnol.*, Washington, 1884, p. 312. For Mexican pretexts, see Bandelier, *Tenth An. Rep. Peabody Mus.*, Cambridge, 1877, p. 128.

[2] Turner, *Samoa*, London, 1884, p. 189.

preceded. In the aggressive tribe there were meetings, debates, laying of plans, considering ways and means. The language used on such occasions was not infrequently the national classic, incomprehensible to the laity. The author has witnessed one of these councils, and will never forget the dignity and the earnestness of some of the orators, who shrewdly guided their language by the abundance or scarcity of grunts that followed each pause.

Among the Southern Indians of the United States war was determined on by the war-chief, but his decisions were subject to the approval of the council. Subordinate to the war-chief were the leaders of parties. As soon as war was declared each warrior painted and plumed himself, provided a small bag of parched corn to eat, armed himself with a long bow and a quiver of arrows suspended from the right hip, and frequently with a formidable club made of hard wood and a spear. Thus equipped he set off from the village with a great noise and defiant shouts. The head-warrior taking the lead, was followed by the rest in single file.[1]

"The Aetas, or negrito people of the Philippines, on the death of a member of one of their tribes sally out to avenge him, and slay the first living thing they encounter as a payment. As they proceed on such an expedition they break the twigs off trees in a certain manner to warn friends off their line of march." These are among the lowest of savages, and the custom is mentioned to call attention to the primitive method of giving information.[2] Evidence is not wanting that in certain cases the enemy was notified of the coming attack, and he was warned to prepare himself for punishment.

Men were not recruited for these primitive armies either by conscription or enlistment, but each youth on arriving at an age established by usage passed through a trying ordeal, the successful endurance of which marked him as

[1] *Cf.* De Bry, *A Briefe and True Report, &c.*, Francoforti, 1590, p. 25.

[2] Gironière, *Twenty Years in the Philippines*, quoted by E. Best. *J. Polynes. Soc.*, Wellington, 1892, vol. i. p. 12.

one worthy to bear arms. The tribe was perpetually mobilised. The organisation of the male members of each for warlike purposes was adapted to its social structure. The fighting groups were kinsmen, after the manner of that people. As in all other arts, the co-ordination of structures is as complete as in the natural world.[1]

Every male of the Mexican tribe was born a warrior. When still a babe his father placed alongside of the child a small bow and some arrows in token of its future duties. There was no military caste at Tenuchtitlan or Mexico ; with the exception of children, old people, infirm or crippled persons, and sometimes priests, every one had to go to war. "No youth over fifteen years of age should remain. All had to go except children and old people." There was no standing army, the available force being composed of all the able-bodied men of the tribe of Mexico.[2]

The sources of information concerning primitive weapons and the manner of using them are the aborigines of the two Americas, the native Africans, and the mixed peoples in the Indo-Pacific region. Reference may also be made to the Arctic tribes of Asia, but neither they nor the Eskimo are given to fighting ; so their genius has been expended rather in the invention of hunting and fishing implements. Strange enough, in each of these three areas attention has been given especially to one of the three forms of wounding. The Americans are, *par excellence*, piercers ; the Africans slashers and the Indo-Pacific peoples are gifted with the club. This is not a fixed rule, however, as in each area other sorts of wounds are also in vogue. It is a fact that the true Polynesians were ignorant of the use of the bow as a weapon, that the greatest diversity of bows and arrows was in America, and that no other savages devised such a

[1] On ordeals *cf.* Catlin ; for India, Hunter's *Imp. Gazetteer of India*, vol. vi. p. 58.

[2] Bandelier, *Tenth An. Rep. Peabody Mus.*, Cambridge, 1877, p. 98, quoting Clavigero, Gomara, Torquemada, and A. Costa.

variety of apparatus for, and delighted so much in, hacking human flesh as the Africans.[1]

Ling Roth gives a list of weapons in Borneo. 1. The *Slighi*, a wooden javelin with point hardened in the fire; 2. *Apieng*, dart stems with poisoned points; 3. *Sangkoh*, long wooden spear with metal head and spud; 4. *Duku*, or *barang pedang*, a species of scimitar; 5. *Parang nabur*, or short curved sword with bone handle; 6. *Parang ilang*, the Kyan style of parang; 7. The *Katapu*, or decorated helmet of wicker work; 8. *Gagong*, or war jacket of skin; 9. *Klambi taiah*, padded or quilted cotton jacket. 10. *Trabai*, or shield.[2]

For the single warrior, offensive weapons or implements may be studied structurally or functionally. According to the former, these would be blunt weapons, edged weapons, and pointed weapons. Further on there ought to be examples which include two or even three of these in one, as the sabre bayonet, which may be used in many ways. Expressing the same ideas functionally would give us bruising, cutting, and piercing weapons.

Again, the manner of holding and using divides the same objects into hand weapons and missiles.

Hand weapons are such as do not leave the hand in doing their work. As in the case of the primitive mechanic's tools, the arms of the warrior are to be considered in their grip or handle, and in their working part. In the simple case of the stone or stick held in the hand for bruising, hacking, or stabbing, the working part and grip are practically the same; only one end is modified to suit the hand and the other to do the work. The fundamental idea in each of the three sets of hand weapons is as simple as that; but practically no people are known so rude as not to have

[1] Lane Fox, *Catalogue*, London, 1877, Eyre, &c., vols. i., ii.; Tregear, "The Polynesian Bow," *J. Polynes. Soc.*, Wellington, New Zealand, 1892, vol. i. pp. 56–59.

[2] *J. Anthrop. Inst.*, London, 1892, vol. xxii., from Papers of Hon. Brooke Low.

gone a step further. It will be seen that the handle parts of weapons may differ in length from a few inches to several feet, as in the stiletto and the Japanese long-handled sabre. They differ also in rigidity, as the slung shot and the club. They also are much varied in the adaptation to the form of the hand and to the idea of guarding and parrying, as in the totally exposed Fijian club and some types of the African swords. The working parts of weapons are subject to an infinite variety of forms dependent upon climate, natural resources, forms of defence, race proclivities, and even upon the idiosyncrasies of the enemy.[1]

The following is Adrien de Mortillet's classification of weapons modified to suit the present purpose:—

I. Bruising Weapons.

In the hand. The fist, with or without "knuckles."

With a handle. Club, flails, scourges.

Projectile. Stone bullets, blunt arrows.

II. Piercing Weapons.

In the hand. Poignard and rapier.

With a handle. The lance and the pick.

Projectile. Javelin, harpoon, thrown from the hand, with an amentum (fixed or mobile), or with a throwing stick or spear thrower (pocketed or hooked).

III. Cutting Weapons.

In the hand. The sabre with stone blades.

With a handle. The battle-axe or pole-axe.

Projectile. Boomerang, African throwing knives, bladed arrows.

"Of all weapons employed by savages the club is probably entitled to be considered the most primitive."[2]

[1] A. de Mortillet, *Rev. Mens. de l'Ecole d'Anthrop.*, Paris, 1892, vol. ii, pp. 92, 93.

[2] Lane Fox, *Catalogue*, London, 1877, Eyre, &c., p. 61.

The club is single-handed or double-handed. And in a series of them, especially from Melanesia or Polynesia, it is possible to follow minutely the thought of the maker. The plainest of the little Fijian single-handed clubs is a stick ending in a globular excrescence whose surface is regularly wrinkled. Now, in making more elaborate examples, these islanders work out similar forms, and replace the wrinkles with exquisitely carved patterns. In the same manner, the crudest two-handed club is a stock of a small tree, having a two-pronged root, on the outer margin of which is a peculiarly wrinkled appearance. But much finer clubs are carved in this same form, and, curiously enough, the carved ornamentation is always put where the wrinkled surface occurs in the crude specimens.

General Pitt Rivers's collection in the Oxford Museum is especially rich in the evolution of the club, and his catalogue is the very best treatise on the subject.[1]

In the American area the club was a complicated affair, often a compound weapon for bruising, gashing, and piercing in the most dreadful manner. The oft-repeated story of the Mexican specimens, consisting of a heavy stick grooved along the side for the insertion of blades of obsidian are more than matched by the reality in examples from the Plains tribes. The Sioux standard club is a flat piece of wood curving and widening away from the grip, and terminating in a spherical head, which in modern times carries a long spike. It is not uncommon also to see the blades of one or more butcher's knives firmly inserted along the margins. The United States National Museum possesses a great variety of these ugly weapons, designed, as the frontiersmen say, to "knock down the white man and then to brain him and cut him into mince-meat."

John Smith says that the Virginia Indians for swords often used the horn of a deer put through a piece of wood in form of a pickaxe, some a long stone sharp at both ends,

[1] Science and Art Department, South Kensington, Bethnal Green Branch Museum.

used in the same manner.[1] This is interesting because the Siouan tribes universally used the mace with head of stone sharpened at both ends, and the Pacific coast tribes in Alaska and Columbia have from time immemorial made their picks for braining slaves of horn or ivory or other hard substance "in form of a pickaxe."

The South American tribes make clubs which strongly remind one of the Polynesian varieties. Those of the Carib tribes are square in cross section, elegantly polished, and ornamented with woven cotton bands.

The African hand-weapons for striking, owing to the acquaintance of the natives with iron, are exceedingly complicated. Indeed, the plain clubs in that area are for throwing. The standard hand club is converted into a sort of tomahawk by the addition of a blade, or into a spear by the addition of a sharp spud. In the museums of ethnology may be seen now and then from Africa, plain clubs carved from the small tusks of young elephants. These have no attachments for cutting or piercing.

The ordinary hand weapon of the Dinka and other heathen negro tribes on the shores of the Upper Nile territory is the "bollong," or club, with a pointed ferrule. The Kaffir kerries are similar in shape. One of Schweinfurth's[2] examples has a disc-shaped head, and the object served as striking and thrusting weapon and for a seat. The Bechuana as well as the Zulu type is a knobbed stick or horn, useful not only at close quarters, but also as a missile in hunting, in which function it is managed with deadly precision.

The Polynesians have developed an interesting series of weapons, called paddle clubs, in which they seem to have summed up the whole story of their method of going to war. That the clubs are developed from the paddle can be seen on inspecting any series, notably those in the Oxford Museum. The whole theory of cutting, thrusting, and

[1] *History of Virginia*, Arber reprint, p. 31.

[2] *Artes Africanae*, London, Sampson Low, p. 1, figs. 3, 4, 5, 18.

striking is involved in one of them, which, on occasion, may be used also in getting a canoe out of danger or into a safe place. Indeed, many of these paddle clubs have passed into the sphere of ceremony, being elegantly carved with lace-like tracery over the entire surface.

For slashing, the American savage devised a leaf-shaped dagger-blade. A strip of soft fur wrapped about one end served for a grip ; but he also knew how to insert the tang into a haft of wood, or of antler, and secure it there with sinew lashing, gum or glue. The very same forms occur in flint over Western Europe, showing that in prehistoric times the inhabitants gashed one another's faces and bodies with short swords of stone. In countries where obsidian was abundant, blades of this material were inserted into handles, to be used as battle-axes. The Indian tomahawk is the true descendant of such a weapon.[1]

In the Polynesian area, in Easter Island specially, the axe-blades of obsidian are most savage looking. The polished stone-workers, wherever material of sufficient hardness could be procured, invented the true short sword. There is a blade of jade-like material in the United States National Museum, from Alaska, which might have been taken as a model for metal short swords. The native Africans, however, easily deserve the palm for gashing and slashing weapons. They frequently have no other function, but as often unite the office of "cut and thrust" in the same arm. Quite as much as German university students, the negro warrior prides himself upon the number of his scars. In times of peace he will hack his own flesh and retard the healing to produce ugly cicatrices. The executioner's sword in this area combines keenness of edge with the force of the battle-axe. It is the ancestor of the modern sabre. The Kingsmill islanders and other Polynesians made a dreadful slashing implement by securely sewing rows of sharks' teeth along the sides of a handle of wood. These

[1] On the Mexican *maccuahuitl*, see Bandelier, *Tenth An. Rep. Peabody Mus.*, Cambridge, 1877, p. 107.

sharks' teeth slashing weapons vary in length from a few inches to sixteen feet. They are stillettos, dirks, short swords, long swords, and poleaxes. In all the range of weapons there is nothing more blood-curdling to behold.

The dagger or dirk-knife of the African tribes is worthy of a separate chapter. Schweinfurth says that, "diffused over a great part of equatorial Africa, this weapon, which serves also for domestic purposes, forms the characteristic mark of a whole series of tribes between the Zambesi and the Upper Nile. The knives of the Balonda are not to be distinguished from those of the Niam Niam. The dirks of the South African negro tribes, and of several on the West Coast, present a contrast to the above form, being distinguished by a spear-shaped outline of the blade, suddenly becoming constricted and narrowed.[1]

The falcate edges of the Monbutto and other African swords were designed to meet an emergency. "This way of dealing hacking [picking] rather than slashing strokes was manifestly intended to wound the head, which is protected, as with a helmet, by a high coiffure, while the blow of a sabre or a sword, in our fashion, would fall ineffectually on the elastic bolster." Hence the term "pick" applied to such weapons by Pitt Rivers is eminently appropriate.[2]

This style of weapon reappears in the negroid area of the Indo-Pacific, and for the same reason, to pierce through the massive woolly coiffure. It should be compared with the Alaskan "slave-killer."

The missile weapon has developed more ingenuity in the lower grades of culture than has the hand weapon. They may be thus arranged:—

1. Hand missiles. { From the naked hand.
From a rest or an *amentum*.

[1] Schweinfurth, *Artes Africanae*, London, 1875, vol. xii. figs. 6–10. For bone daggers of the Eskimo, see Murdoch, *Ninth An. Rep. Bur. Ethnol.*, Washington, 1892, p. 192, figs. 174 and 175.

[2] Schweinfurth, *Heart of Africa*, London, 1875, pl. xviii.

2. Machine missiles. { From a sling or atlatl.
From a bow.
From a blow-tube. }

The thing projected may be a bruising weapon, as in sling-missiles and blunt arrows ; a cutting weapon, as in the trumbash or the bladed-arrow ; or, chiefly, a piercing weapon, as the whole genus of javelins, darts, arrows, and sumpitan bolts.

The grip and sudden release of the fingers enables the energy of the arm to be exploded and used in hurling the javelin. The hand-rest and the *amentum* are inventions which enable the warrior to economise the muscular effort otherwise used in grip and to employ it all in the hurling. The sling converts circular motion into rectilinear motion, adding momentum to muscular force. The projectile usually is a bruising weapon.

The Mexican atlatl or throwing-stick or spear-thrower found in Australia, Melanesia, and in America from Point Barrow to the Argentine combines muscular force with prolonged effort. It is also a convenience for a man who has only one free hand.

The bow converts the pent-up elasticity of wood or animal substance or metal into rectilinear motion. The spring-gun is a bow. The blow-tube is really the legitimate prototype of the gun. It converts the elasticity of compressed air into rectilinear motion. It is not here said that the inventors of arquebuses ever saw a sarbacan or a sumpitan, because on the theory of this book, the parentage of inventions is an intellectual one. The arquebuse and the air-gun doubtless sprung from an imaginary sarbacan, and combined the missile of the sling and of the "stone-bow" therewith. The bow and the arrow and the arbalest were its rivals. It is a hunting weapon rather than one for war.

"Nearly all savages are in the habit of throwing their weapons; even apes are known to throw stones, the North American Indian throws his tomahawk, the Indians of the Gran Chaco throw their *macana*, a kind of club, the Kaffirs and the negroes of Western Africa throw the knob-kerry and trumbashes." Grant says that the women of Faloro, East Africa, hold their knives by the tip of the blade and throw them at their adversaries. The Fiji islander throws his knob-headed club, the New Zealander his *pattoo pattoo*, and the Australian his *dowak*. Even the Franks are supposed to have thrown their *francisca*, and we learn from Blount's travels in 1634 that the Turks used the mace for throwing as well as for striking.[1]

The natives of New Hebrides were expert at throwing a stone called a *kawas*. It is about the length of an ordinary counting-house ruler, only it is twice as thick, and this they throw with deadly precision when the victim is within twenty yards of them.[2] The knob-kerry, or kiri, before mentioned, throughout South Africa, the boomerang-like rabbit-clubs of the Indians in South-western United States, and, most effective as a wounding hand-missile, the Australian boomerang deserve mention among the most clever devices in this class. The boomerang and its congeners are also quasi edge-tools, but their work is to break bones.

The pingah, or projectile of the Niam Niam, is a cutting or slashing missile of infinite diversity of shape, but in every case consists of three two-edged blades—a short, broad one at the apex, triangular or heart-shaped ; an oblong blade near the point, which is the longest of the three and is at right angles to the shaft or axis of the weapon ; a third and shorter blade on the opposite side and just above the handle, and set at an acute angle to the shaft or axis. The handle itself is only an elongation of the middle part and is wound with stout twine.

The pingah is thrown in such a way as to revolve in the plane of its blades, and no matter in what position it reaches its destination, in every instance it strikes with a sharp edge. In fact, this artistically wrought weapon is thrown only in

[1] Lane Fox, *Catalogue*, London, 1877, Eyre, &c., pls. i. and ii. p. 29.
[2] Turner, *Samoa*, London, 1884, p. 312.

extreme cases ; generally it serves as a battle-axe, the sickle-like point turned forward.

Projectiles of a similar kind, sometimes made of wood, sometimes of iron, are spread over a large part of the northern half of the continent. Among the Mohammedan negro tribes of the Soudan, from the Tchad lake to Abyssinia, flat, two-edged projectiles of wood, sickle-like and widened towards the point, are used in hunting fowls and small mammalia. Such wooden projectiles are called in the Upper Sennaar "Trumbash."

Similar to the pingah is the kulbeda of the Fundy and the Berta negroes in the Upper Sennaar. It consists of two blades, and has a wooden hilt. The longer blade inserted in the elongation of the tang is sometimes sword-like or sickle-shaped, sometimes curved to and fro in a serpentine form. The lateral blade of the kulbeda is quite short, and serves chiefly for the protection of the hilt, as a sword breaker or a guard.

In Central Sudan, projectile irons are called "Shanger-manger," and are to be seen in a great variety of shapes. The lower end of the shaft is wound with rope. As a rule, only two blades are in use.

The negroes of Borgoo, Wadai, Ennedi, and the Tibboo are fond of using them. The shangermanger of the Musgoo correspond in shape almost exactly to the kulbeda of the Berta and Fundy.

The projectiles of the Fan in equatorial West Africa exhibit the greatest similarity to those of the Niam Niam in question.[1]

The javelin thrown from the naked hand finds its highest development in the South African assagai. The Americans did not use the simple javelin. It was only in connection with the throwing-stick and throwing-strap that it found

[1] Schweinfurth, *Artes Africanae*, London, 1875, pl. xii. figs. 1–5. Consult also Lane Fox, *Catalogue*, London, 1877, pp. 28–37, pl. ii. Compare Schweinfurth, *l.c.*, pl. xviii. In this connection belongs the Hindu chakra, figured by Wood from Sir Hope Grant's collection.

place. The brown Oceanic race as well as their negroid neighbours present it in its most primitive form as a long pole of hard wood sharpened at the end. The Persian jereed is a survival of a more serious implement, gone out of date now, but quite active in European wars down to the introduction of fire-arms.

Sir Samuel Baker says: "Every man is a warrior, as the Baus are always at war. They are extremely clever in the use of the lance, which they can throw with great accuracy for a distance of thirty yards, and they can pitch it into a body of men at upward of fifty yards. From early childhood the boys are in constant practice, both with the lance and the bow and arrow."[1]

The Australians are also extremely clever at spear throwing. The most marvellous stories are told of their dexterity in this particular, as well as their coolness and agility in warding them off.

"Another means of accelerating the flight of the javelin is by means of the *Amentum*, or *Ounep*, as it is called by the Melanesian islanders. This is a thong attached to the centre of the spear in which the forefinger is inserted, and like the throwing-stick enables the thrower to continue the impulse after the spear has left the hand. . . . The amentum was used by the Greeks and Romans, and is mentioned by Virgil, Ovid, Cicero, Livy, Pliny, and other ancient writers, and is figured on Etruscan vases; it was called ἀγκύλη by the Greeks."[2] The Melanesian ounep remains in the thrower's hand, and the amentum was fastened to the javelin shaft.

In this same connection the student must not overlook the little brackets of ivory daintily carved and lashed to the shaft of Eskimo harpoons, great and small, used as "stops" for the gloved hand. Though on a hunting device the stop may just as well occur on a javelin.

According to Potter, "The most common [engine] in

[1] *Ismaïlia*, New York, 1875, p. 135.

[2] Lane Fox, *Catalogue*, &c., London, 1877, p. 40; also *Catalogue of Hamburgh*, Ethnog. Mus.

field engagements among the Greeks was a sling; which we are told by some was invented by the natives of the Balearian Islands, where it was managed with so great art and dexterity that young children were not allowed any food by their mothers till they could sling it down from the beam, where it was placed aloft. Slings resemble a plaited rope, somewhat broad in the middle, with an oval compass, and so by little and little decreasing into two thongs or reins. The manner of slinging was by whirling it twice about the head, and so casting out the bullet." [1]

In America the sling was used universally from Tierra del Fuego to the Arctic regions. And it is found diffused in Polynesia. Even in classic times it continued among the Greeks and Romans. It is not mentioned in Homer. But it was a common weapon among the Semito-Hamitic peoples, and among the Asiatic Aryans. Strutt does not know when it was introduced into England. Most of his illustrations show the slinger in pursuit of birds, but he speaks of the balistarius as a warrior. The order of invention seems to be the simple sling, then the fustibalus or staff-sling, and then the transfer of the missile to the stone bow, and the more effective arbalest.

The *fustibalus*, or staff-sling, was a common sling attached to the end of a shaft and used for heavier stones.

> "This geaunt at him stones caste
> Out of a fel *staf-slinge*." [2]

Tylor calls attention to the almost entire disuse of this weapon, but speaks of an interesting survival in the practice of the "herdsmen of Spanish America, who sling so cleverly that the saying is they can hit a beast on either horn and turn him which way they will." [3]

Among the Maoris of New Zealand, Phillips encountered a dart thrower consisting of an elastic rod and a short lash,

[1] *Antiq. of Greece*, vol. ii. bk. iii. ch. iv. p. 45.

[2] Chaucer, *Sir Thopas*, vol. i. p. 118, in *Century Dict.*, with picture.

[3] *Primitive Culture*, Boston, 1874, vol. i. p. 73.

the whole resembling a whip. A knot in the end of the lash fitted into a notch in the shaft of the dart and the latter was propelled by holding the dart in the left hand after the string was in place and giving a sudden jerk with the rod held in the right hand. The same apparatus is familiar to boys in civilised lands. But in the olden days the Maoris launched their spears at a hostile pa [village] by means of a whip sling similar to the one described.[1]

The throwing stick mentioned among the hunting instruments was with the Mexicans an apparatus for warriors also. In the codices very many pictures are given of warriors hurling javelins by means of this apparatus. Furthermore, while the forms used by the Eskimo are right-handed and carved from a single piece of wood, those in the Mexican pictures are for either hand, and the finger-holes were in bits of raw-hide or leather attached to the side of the grip. The Australian spear-thrower is of the same sort, though of different form.[2]

A separate treatise could be devoted to the bow and the arrow as exhibiting the progress of mind in the use of an elastic spring down to the fifteenth century of our era. It cannot be asserted whether the spring trap or the spring weapon is older. The bow is fundamentally an elastic rod, with a bit of wood, or fibre, or cord for a bow-string. In many places savages have not got beyond the simplest form. In Africa, universally, owing to the keenness of the poisoned arrow points of iron, the bow is weak. The negroid peoples of the Indo-Pacific make strong bows of hard wood, and in the Malayan area bamboo is the material. The Andamanese for some reason give a sigmoid shape to the apparatus. As before mentioned, the pure Poly-

[1] Coleman Phillips, *Trans. N. Z. Inst.*, Wellington, 1877, vol. x. pp. 97–99.

[2] List of authorities in *Proc. U. S. Nat. Mus.*, vol. xvi. pp. 219–222, figs. 1–6. See also *Science*, New York, September, 1893, for description of a Mexican example discovered by the author among the cliff dwellers of Arizona.

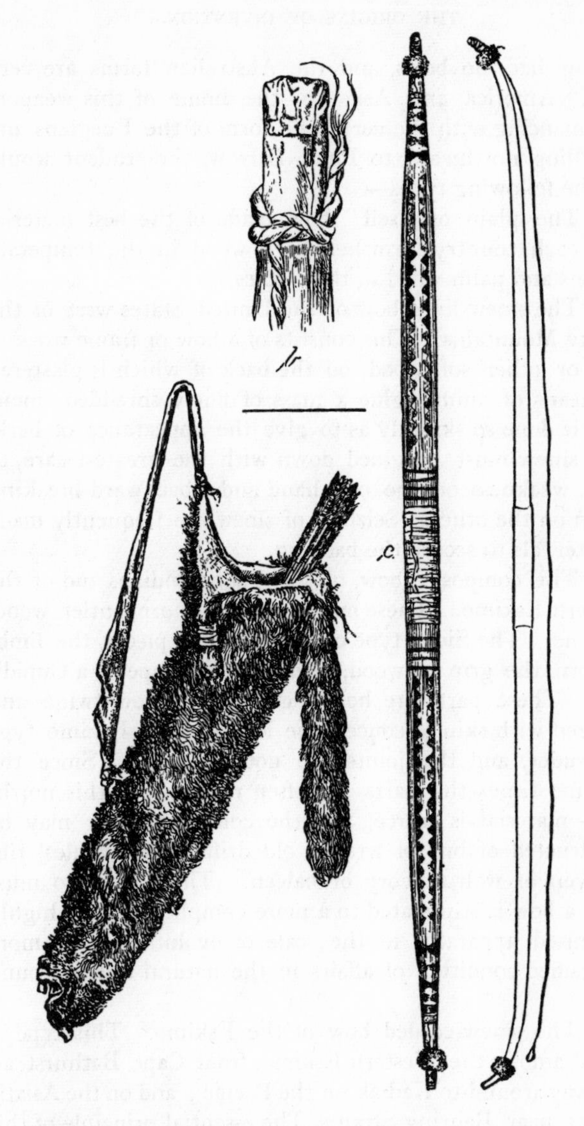

FIG. 73.—SINEW-LINED BOW AND ARROWS. ARTILLERY OF A HUPA INDIAN. (*U. S. Nat. Museum.*)

nesians had no bows, and the Australian forms are very rude. America and Asia are the home of this weapon. Commencing with the very plain form of the Fuegians and travelling northward to Point Barrow, the student would see the following types—

1. The plain or "self" bow, made of the best material that each country furnishes, hard wood in the temperate regions and palm wood in the tropics.

2. The sinew-lined bow of the United States west of the Rocky Mountains. This consists of a bow or frame work of yew, or other soft wood, on the back of which is plastered by means of animal glue a mass of finely shredded sinew. This is done so skilfully as to give the appearance of bark. The sinew must be glued down with the greatest care, to avoid weakness on the one hand and a backward breaking strain on the other. Seizings of sinew are frequently made at intervals to secure the backing.

3. The composite bow, of the Plains Indians and of the Eastern Eskimo. These may be made of horn, antler, wood, or bone. The Sioux type consists of three pieces, the limbs of horn, the grip of wood, made in the shape of a Cupid's bow. These parts are held together by sinew twine and covered with skin to conceal the joints. The Eskimo type are ruder, and the joints are not concealed. Since the whaling times the parts are often riveted. In this northland, material is scarce, and the compound bow may be constructed of bits of wrecks, old drift wood, whales' ribs or even of walrus ivory or baleen. The man who must have a bow is stimulated to a more complicated and highly organised apparatus in the scale of evolution by a more diversified condition of affairs in the natural world around him.

4. The sinew-corded bow of the Eskimo. This type is found among the Western Eskimo, from Cape Bathurst all the way around to Kadiak on the Pacific ; and on the Asiatic shores near Behring Strait. The essential principle of this invention is by means of a cable of sinew, twine, or braid, to

convert the breaking strain of a bit of drift wood into a columnar strain. In other words, it is to combine drift-wood, which is rigid and brittle, with sinew-cord, which is flexible and very elastic, so that the wood will supply the rigidity and the sinew the elasticity.[1] The cord is laid on most ingeniously, and in passing from one limb to the other is secured by half-hitches, which act like a set of fingers clasping the wood just where the greatest strain would come.

For tightening and loosening the bow, small levers of ivory with notches on alternate sides at the ends are inserted in the cable on the back of the bow at the grip. These make a half turn and then are slipped back their length. After the cable or cables are wound they are kept from unwinding by a strap of raw-hide passed through the cable two or three times and made fast to the bow. The ivory levers are then withdrawn.

The composite or built-up Asiatic bow does not in itself belong to savagery, yet it is the immediate result of compounding processes of primitive types. Indeed, the "Tatar" and the "Kung" bow of China are compounded of (1) the "self" bow as a base; (2) the separate arms of the compound bow; (3) the sinew backing or a substitute; and (4) the covering of snake-skin or bark or buckskin to conceal the joints. It is argued that this Tatar form of bow was in every-day use around the besieged city of Troy, and that the Scythians contributed this type to all the classic nations.[2]

The bows of the Monbuttoo are made of a species of Rotang, and are provided with a string of the same material, which is made and fitted on when the material is green. On the inner side of each bow is a piece of wood shaped like a weaver's shuttle and scooped out longitudinally

[1] Murdoch, *Rep. Smithson Inst.*, Washington, 1884, part ii. pp. 307–316, plates i.–xii.; and *Ninth. An. Rep. Bur. Ethnol.*, Washington, 1892, pp. 195–210, figs. 177–194. O. T. Mason, "Bows, Arrows, and Quivers of the North American Aborigines," *Smithson. Rep.*, 1893. Profusely illustrated.

[2] Henry Balfour, "On the Structure and Affinities of the Composite Bow," *J. Anthrop. Inst.*, London, 1889, vol. xix. pp. 220-245, pl. v. and vi.

toward the string. This not only serves as a protection for the hand against the recoil of the sharp-edged string, but as a receptacle for poison, to be used on the spot.[1]

The bow has not been known as a weapon among the brown Polynesians in historic times. Its occurrence as a toy in one place and as a ceremonial object in another may point to prehistoric use, but the fact remains that, while the negroid peoples around them carried the arrow especially to a high degree of perfection, the brown race discarded the apparatus of the archer altogether.[2] A type of bow from this area has a groove along the back, which would seem to be an element of weakness rather than of strength. By one it is alleged to be a survival of the crease in the bamboo bow, by Captain Cook as a place for the arrow, and by Moseley as a resting-place for a cord like the Eskimo sinew-backed bow.[3]

As an arm or weapon the arrows of the Americans were of the simplest kind, consisting of shaft, feather, and head of stone. The last named was often barbed to prevent its being withdrawn. As to poisoning their arrows, there is conflicting testimony, with the preponderance against the systematic use of deadly substances: this remark is not true of the darts from the "blow-tubes." The collectors of American arrow-heads are wont to divide them into sagittate and lanceolate, alleging that the former were for men, the latter for beasts. This lacks confirmation, but the two styles of head did prevail widely.

But, for refinement of cruelty that cannot possibly be surpassed, the war arrows of the negroid peoples take the lead, whether on the continent of Africa or in the Indo-

[1] Schweinfurth, *Artes Africanae*, London, 1875, pl. xix., fig. 23; pl. xx., fig. 7. Compare wrist guard of the Tinneh Indians, which is a bit of wood the shape of a bridge on a violin, attached to the bow and not to the shooter's wrist.

[2] See the subject well worked out by E. Tregear, in *J. Polynes. Soc.*, 1892, vol. i. p. 56.

[3] Balfour, *J. Anthrop. Inst.*, London, 1889, p. 241.

Pacific isles. The Papuan and Fijian arrows are barbed with human bones, the thorns of the stinging ray, and every other diabolical thing that would serve the purpose. In Africa, with the recollection of some such prehistoric cruelty or in imitation of some murderous shrub, the blacksmiths make ragged the heads of the arrows with thorn-like projections, which lacerate on entering the body, and which cannot be removed.

"The arrows of the Monbutto have broad triangular and spatulate heads of iron, furnished along the shank with thorn-like barbs in endless variety. These are said to inflict at short distances much worse wounds than the sharp-pointed arrows. It is important to remember that all originally lanceolate heads acquire an obtuse or spatulate shape by repeated grinding." [1]

Not all African arrows are feathered. The Monbutto feathering is produced by inserting a narrow strip of skin into a longitudinal slit at the base of the arrow, so that the long hair may protrude. The seizing is of bark.

The Bongo arrow-shaft is of cane or cut out of a piece of wood, and made about as thick as the cane. The tang of the ragged head is inserted into the shaft, cemented and bound with bast. The nock is wrapped thickly with the same substance (*Grewia mollis*). The cane shafts are cut so as to have a knot near the head, to secure a break at that point and render the wound more dangerous. The bows are made of bamboo or tough wood, and the string is twisted from a vegetable fibre (*Crotalaria cannabina*).[2] In poisoned arrows the tang of the head is wrapped between the ragged barbs with bast soaked in the juice of *Euphorbia venefica*.

Makrigga is the name of a very thorny shrub (*Randia dumetorum*) which seems to have been in the eyes of the Bongo armourer as a model in making his jagged, and toothed, and thorned lance and arrow-heads. These thorns are produced on the corners of the quadrangular tang by

[1] Schweinfurth, *Artes Africanae*, London, 1875, p. xix. fig. 21.

[2] Ibid., pl. vii.

chiselling the metal in a red-hot state, and, considering the rudeness of the tools, cannot fail to excite the highest admiration. The shaft is of bamboo or light wood, and as a counterpoise to the head, it is set at the lower end with bands or spuds or wrappings of metal.[1]

Of the Mittoo and Madi arrows, Schweinfurth says, they are distinguished by the number and variety of the barbs, and manifest a truly diabolical ingenuity in the invention of means to render a wound as dangerous as possible. These heads, too, as in the Bongo arrows, are fastened by means of bast, just above a node in the cane, so as to make them break off more easily and adhere to the wound. The tang is always four-sided, and the barbs, now shorter now longer, are cut out from the edges of this tang.[2] The endless variety of these barbs would of course baffle any surgeon. The only redeeming consideration in the premises is that among peoples who take no prisoners of war, the more speedy death is the more beneficent.

The blow-tube, called Zarabatana and Pucuna in the Western Hemisphere, and Sumpitan in Southern Asia, completes the series of projectiles. Its office is that of hunting rather than in killing men, and, therefore, a more detailed account of its manufacture and use will be found elsewhere (chapter viii.).

The modern representatives of these implements are the fowling-piece on the one hand, which keeps up the old traditional function of bird slaying especially, and, on the other hand, the man-slaying series beginning with the pistol and ending with the rifle, cannon, and dynamite gun.

Among hunting weapons, those for capture and retrieving occupy a prominent place. But in war, among savages especially, little ingenuity is expended on this point. It is true that the Australian lover steals upon his sleeping flame and twists his spear into her matted hair, just as the darkeys in the Southern States wind the opossum out of a hollow tree or steal cotton from the warehouse through a knot

[1] Schweinfurth, *Artes Africanae,* London, 1875, pl. viii. [2] Ibid., pl. x.

hole. This is, however, a peaceable matter, though it frequently issues in bloody conflicts. The barbs on piercing weapons are not for retrieving the foe, but to ensure his death by preventing the withdrawal of the point.

The use of poison was resorted to quite generally in savage warfare. Leaving the secret administration through treacherous and false friends to more cultured peoples, the primitive warrior did not hesitate to lay the fatal dose on his cutting, and especially upon his piercing weapons. De

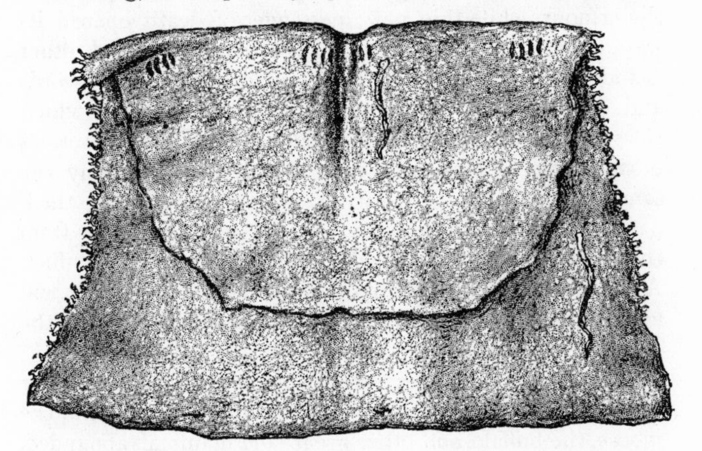

FIG. 74.—Elk-Skin Armour, West Coast of America.

Mortillet holds that prehistoric man was a poisoner. It is well known that the Botocudos of South America, the Bushmen in Africa, and the Negritos of Asia make up for their weakness in other regards by their insidious chemistry.[1]

The defence of the body or of the home is quite as important to the warrior as his weapons. The history of invention as applied to war has been the record of alternate advances in this line, and in overcoming defence. The

[1] Mortillet, "Empoisonnement des Armes," *Rev. Mens. de l'Ecole d'Anthrop.*, Paris, 1891, vol. i. pp. 97-106.

Spaniards conquered the American natives simply because the bullet when it was discharged and did hit its mark would penetrate anything the Indians opposed to it, while the Spanish armour would resist the showers of arrows rained upon it. The English archers long discussed the question of laying their bows aside, because they could discharge three dozen deadly arrows while the arquebusier was getting his missile started. But for a long, long time the shaft had to go searching for the joints and weak places in the armour, while this new messenger of death opened its way through walls of steel. The Kingsmill and other islanders arm themselves with horrid rapiers and swords and battle-saws of cocoa wood, along the sides of which rows of sharks' teeth have been sewed. It makes one's blood curdle to look at them. But it is in precisely the same islands that the warrior arrayed himself in thick panoply of cocoa-nut fibre that would tear the teeth from these dreadful dogs of war whenever they came into conflict. A more detailed description of savage defensive armour will follow, the design here is to emphasize its importance in the current of invention.

The American savages were acquainted with body armour when they were first encountered. Wherever the elk, the moose, the buffalo, and other great land mammals abounded, there it was possible to cover the body with an impervious suit of raw-hide. Such armour is to be seen in many museums. The Eskimo and his Asiatic neighbours shielded their bodies with plates of ivory, and with armour made of encapsulated rings of raw-hide precisely after the style of the Japanese. Rods of wood laid parallel were woven together and fitted to the body by North American and Asiatic tribes. In Mexico cotton armour was worn. "Sometimes they went to war without any other protection, but in most cases the warrior wore a frock of quilted cotton, about three-quarters of an inch thick up to one and one-half inches. This was the cotton armour subsequently adopted by the Spaniards under the name of

Escuapil. Sometimes even the limbs were encased in such quilted protection.

"Warriors of merit inserted their heads into wooden forms, intermediate between masks and helmets, imitating heads of ferocious beasts." [1]

Quite as important as the armour of the warrior was his shield, rendering him ambidextrous. Only those who have seen Australian savages using the parrying shield, or one of the American aborigines warding off arrows with his disc of raw-hide, have any conception of the efficiency of this

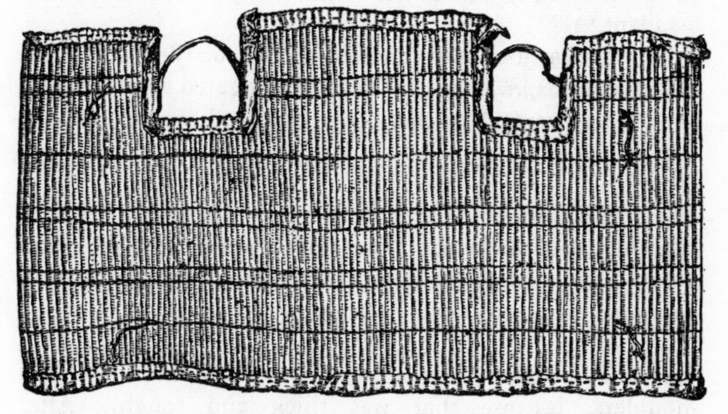

FIG. 75.—Slat Armour from California. (*U. S. Nat. Museum.*)

apparatus. The general notion is of a target simply opposed to the arrows of the enemy, but this idea would be dissipated after seeing a Shoshoni and an Apache trying their best for twenty minutes at a range of only six feet to shoot an arrow into each other's unarmed bodies. The knack and dexterity, the enormous energy put into the operation, excite the spectator to such a pitch that he forgets the element of death involved.

[1] Bandelier, *Tenth An. Rep. Peabody Mus.*, Cambridge, 1877, p. 109; Squier, *Nicaragua*, vol. ii. p. 347; Dall, *Third An. Rep. Bur. Ethnol.* p. 93. D. Hough finds Japanese greaves among T'lingit armour.

The inhabitants of Drummond's Island, in the Kingsmill group, use a cocoa-nut club pointed at both ends for warding off spears. The Hottentots warded off stones with their kirri-sticks. The Dinka negroes use a club in the left hand to push aside the spears of the enemy.

The shield, according to Lane Fox, was developed from the parrying-stick. The gradual widening of the stick or club in the centre, with the addition of a contrivance to cover the hand until at last the long narrow shield is produced, developed into the broad shield constructed to cover the body from the thrusts and missiles of an assailant.[1]

The Dinkas used parrying-sticks made of the wood of the *Diospyrus mespiliformis*. They are elongated spindle shape, the bulbous central portion being hollowed out, and having a grip for the hand. The Dinkas also use a parrying-bow, apparatus in vogue elsewhere, but often mistaken for an offensive weapon.[2] There is a slight difference between parrying and shielding. The one diverts the point of a flying missile and lets it go on; the other receives it, arrests its momentum, and takes its blows.

The ancient armourer among the Sioux made shields of the buffalo hide, choosing the part over the neck and shoulders, because that was thick and tough. After removing the hair by sweating the hide, he stretched the latter over a pit containing a mass of reeking coals. The heat shrunk the material, and made it hard and impervious to any arrow. The disc of raw-hide is common everywhere in the area of the great mammals, and worn on the left hand, it received the club, the arrow, or the slashing blade.

In the modern army nothing is more prominent than the flag, the standard, the corps badge or symbol. It was so in mediæval and classic times. The soldier fought about his

[1] *Cf.*, Lane Fox, *Catalogue*, p. 6.

[2] Schweinfurth, *Artes Africanae*, London, 1875, Sampson Low, pl. i., figs. 13–16. *Cf.* Wood's *Africa*, fig. 11.

standard, and sought no greater glory than to die with it folded in his arms. The savage had the same pride. His flag was his totem ; it was painted on his shield and on his body. He had a sign language which would reveal his clanship, even if he could not speak. When he went out in his canoe, there was no mistaking his affiliation. He had no banner that he held aloft, but every one who "ran could read" his brotherhood on both stern and bow of his war boat.

"The Polynesians in their war canoes had some distinguishing badge of their district hoisted on a pole—a bird, a dog, a bunch of leaves. Land forces had certain marks on the body by which they knew their own party. It seems that this means of recognition could be changed from time to time, being blackened cheeks one day, marks on the breast another, and a shell suspended from the neck a third."[1] This countersign is in contrast with the American Indian scheme, for the identification of the tribe and the clan was always by the same token. The Mexicans had their standards mounted on a staff, and bent in such a way to the shoulders of the bearer as not to hinder him from fighting, and yet it could not be captured without hacking him to pieces.[2]

Drill is also of great importance in keeping up the skill of modern armies. Of the Plains Indians Colonel Dodge says that they spend a considerable portion of their time in drilling. This applies especially to their history since they became possessed of horses, but the very same evolutions could be practised on foot.

"There seems to be no fixed system of tactics, each chief instructing according to his own peculiar ideas. There are no ranks, no units of command, but there are words or signals by which the same evolutions are repeatedly performed. The whole band will charge *en masse*, and

[1] Turner, *Samoa*, London, 1884, p. 191.

[2] Icazbalceta, doc. i., Mexico, 1858, *Torquemada*, vol. i. p. 525 ; Zelia Nuttall, *Arch. and Eth. Papers*, Peabody Museum, vol. i. pp. 10, 11.

without order on a supposed position of the enemy. At a word it breaks and scatters like leaves before a storm; another signal, a portion wheels, masses, and dashes on a flank, to scatter at another signal. The plain is alive with circling, flying horsemen, now single, lying flat on the horse or hanging to his side, as if to escape the shots of a pursuing enemy, and now joined together in a living mass of charging, yelling terror.

"The remarkable control of the chief is exercised by signals. Wonderful as it may seem, the orders are given on a bright day with a piece of looking-glass held in the hollow of the hand. In communicating at long distances their mode of telegraphing is equally remarkable. Both the signalling and telegraphing are modifications and extensions of their sign language. All are offspring of a necessity growing out of the constant wariness instant to a life of peculiar danger."[1]

All the operations of the Baris are conducted by signals given by the drum. In early morning, shortly before sunrise, the hollow sound of the big drum is always heard giving the signal, by a certain number of beats, to milk the cows. The women and young men then commence, and when the operation is completed the drum beats again, and the large herds are driven to pasturage. The signal is repeated in the evening. Should an enemy attack the country the sheik's big drum gives the alarm by a peculiar series of beats. In a few seconds this alarm will be re-echoed by every drum throughout the villages.[2]

In Melville's *Typee*, recounting a four months' residence in the Marquesas, he tells of a primitive mode of signalling. "The word 'botee' was vociferated in all directions, and shouts were heard in the distance and growing louder and nearer at each successive repetition, until they were caught up by a fellow in a cocoanut tree a few yards off. This was the

[1] Dodge, *Plains of the Great West*, New York, 1877, p. 369.

[2] Sir Samuel Baker, *Ismaïlia*, New York, 1875, p. 134. Drum language is common both in Africa and Polynesia.

vocal telegraph of the islanders, by which condensed items of information could be conveyed in a few minutes from the sea to their remotest habitation—eight or nine miles." [1]

The superstition which condemns every scalped warrior to annihilation is the primary cause of a drill peculiar to the Indians, namely, stooping from a horse going at full speed and picking up objects from the ground. At first, light objects are selected. These are exchanged for heavier and more bulky ones, until some individuals attain such wonderful proficiency as to pick up, while going at full speed, the body of a man and swing it across their horses. This is generally done by two men working in conjunction.[2]

Of the Mexicans Bandelier writes: "There were no regular times set for military practice, but every twenty days there occurred a religious festival, at which the warriors skirmished, showing and practising their skill in handling arms. . . . When in 1743 the tribe of Tlalilulco agreed upon attacking Mexico, they practised beforehand with as much secrecy as possible. Setting up posts of hard wood, they beat against them with their swords and clubs; they sped arrows and threw darts at thick wooden planks; and, lastly, they went out into the lake and shot at birds flying." [3]

Each group was its own quartermaster and commissary, partly carrying its subsistence, but for the most part relying on pillage. It has been said, indeed, that for the first time in the history of war the army of Frederick the Great was wholly independent of the country invaded. Likewise there was little need of quartermaster. Every warrior furnished his own tent, rolled himself in a robe of skin, and slept upon the naked ground, as hundreds of thousands of brave men have done in historic times.

[1] *Typee*, Harper Bros., 1852, New York, p. 119. The man in the tree for looking out and signalling was common in America.

[2] Dodge, *Plains of the Great West*, New York, 1877, p. 369.

[3] Bandelier, *Tenth An. Rep. Peabody Mus.*, Cambridge, 1877, p. 101.

Transportation was on the backs of warriors, what little was needed, when the travelling was by land ; but on the water they fared better, for canoes were always at hand to transport both troops and equipments. It is easy to understand, therefore, that the more complicated methods of fighting were invented by peoples who possessed the means of water transport, and of preserving compact food. When there were no means of transportation the fight had to be brief, and consisted for the most part in a band of warriors falling upon unsuspecting enemies asleep, or engaged in some peaceful festivity.

A "trail" is a succession of marks left on the ground by anything moving to a definite end, as a trail of troops, an Indian trail, a deer trail. Trailing, or following trails, is second nature to the Indian. He is taught from childhood to read every mark on the ground, to tell what made it, its age, and all about it of interest or importance to himself. To these are added a thorough knowledge of the habits of animals of any kind, and a pair of eyes exquisitely sharpened by constant practice.

When anticipating pursuit, the Indian will resort to all ruses, keep as much as possible on rocky ground, mount a high hill, only to go down again on the same side. Getting into the bed of a brook, he will keep along its channel for miles, going out and coming in again, doubling on his track.

Indians travel by "landmark." A good trailer will tell from the general appearance of the country what landmarks an Indian is travelling by. When the pursued resorts to ruses the pursuer loses no time in painfully tracking him through all his windings, but goes at once to where he thinks the Indian will pass. There he looks for the trail, and finding it, pushes on. The pursued may spend several hours in making a devious trail, which the astute pursuer will jump over in as many minutes.[1]

As a rule the Indian relies upon surprises, upon the effect

[1] Dodge, *Plains of the Great West*, New York, 1877, chap. xxxix. See the chapter in full.

of a sudden and furious dash, accompanied with unearthly yells to demoralise his enemy and render him a sure prey. In this he has no superior, nor can he be excelled in the spirit with which he follows up a first successful effort, nor in the remorseless vigour of his pursuit of a flying foe.

If two hostile bands nearly equal in number should meet on the plains a prolonged contest at long range is sure to ensue. This goes on until one party shows signs of weakness and gets away the best it can.[1]

The Indians never receive a charge, and very rarely meet one. When charged the portion immediately in front of the charging force breaks and melts into individual Indians, while the bands on either side close in to harass the flank and rear. The broken Indians, wheeling in circles, form on the flanks to attack and break again when charged. Should the attacking force become scattered its defeat and destruction are almost sure.[2]

An excellent and detailed account of the apparatus and the processes of savage warfare, written by one who is thoroughly conversant with the language of the tribe about which he writes, is that of J. Owen Dorsey, on the war customs of the Omaha. No item of detail is omitted concerning the preparation This description includes the going out of small war parties, the start, the secret journey, and the method of procedure. It also embraces the setting out of large war parties, the feasting, the government of these, the order of march, the songs and dances and encampments, and the behaviour of those who stay at home.[3]

War was not carried on by the Siouan tribe as it was by the nations of the Old World. They had no standing armies, no general who holds office for life or for a given term. They had no militia ready to be called into the field

[1] Dodge, *Plains of the Great West*, New York, 1877, p. 371.

[2] Ibid., chap. xxxv. The plan of campaign among the Indians is well worked out in this otherwise unfavourable book.

[3] Dorsey, *Third An. Rep. Bur. Ethnol.*, pp. 312–333 ill.

by the government. Military service was voluntary in all cases, from the private to the commanders, and the war party was disbanded as soon as home was reached. They had no wars of long duration, in fact, wars between one Indian tribe and another seldom occurred, but there were occasional battles. This was for want of transport and compact food.[1]

When near a hostile town or in the vicinity of the spot where a meeting with the enemy was anticipated, the most profound silence and circumspection was observed. A sudden attack, a fearful succession of wild yells, an indiscriminate massacre, and the demolition by fire of the habitations of their enemies, and then a hasty return with captives and bloody trophies of the pillage and butchery, these constituted, as a general rule, the sum total of a successful military excursion. "Their manner of warres," says Thomas Hariot, "among themselves, is either by sudden surprising one an other most commonly about the dawning of the day or moone light, or else by ambushes or some suttle deuises: Set battels are very rare, except it fall out where there are many trees, where eyther part may haue some hope of defence, after the deliuerie of euery arrow, in leaping behind some or other."[2]

Says the Gentleman of Elvas: "The Indians are so warlike and nimble that they have no fear of footmen, for if these charge them they flee, and when they turn their backs they are presently upon them. They avoid nothing so easily as the flight of an arrow. They never remain quiet, but are continually running, traversing from place to place, so that neither crossbow nor arquebuse can be aimed at them. Before a Christian can make a shot with either the Indian will discharge three or four arrows, and he seldom misses his object. Where the arrow meets with no armour it pierces as deeply as the shaft from a crossbow. Their bows are very perfect; the arrows are made of certain canes,

[1] Dorsey, *Third An. Rep. Bur. Ethnol.*, p. 312.

[2] *A Briefe and True Report*, *&c.*, Francoforti, 1590, De Bry, p. 25.

like reeds, very heavy and so stiff that one of them, when sharpened, will pass through a target. Some are pointed with the bone of a fish, sharp, like a chisel; others with some stone, like the point of a diamond; of such the greater number, where they strike upon armour, break at the place where the parts are put together; those of cane split, and will enter a shirt of mail, doing more injury than when armed."[1]

A public declaration of war was made by planting arrows along the pathway leading to the principal village of the enemy. They were also able, by means of ignited tufts of dried moss and grass attached to the heads of their arrows, to set fire to the thatched cabins located in the fortified towns of their adversaries.[2]

Wherever thatched roofs or stockades were prevalent they were attacked by fire. The North American tribes were extremely careful to keep a good wide space burned away from their stockades or ditch banks.

A custom existed among the Plains Indians when fighting called "giving the *coup*." When a foe has been struck down in a fight the scalp belongs to him who shall first strike the body with knife or tomahawk. This is the *coup*. If in a *mêlée* a warrior kills an enemy he, in order to secure his proper recognition and reward, must rush at once on the prostrate body and strike his *coup*. Otherwise, says Dodge, the enemy being in full flight, a brave and skilful warrior who presses on adding victim to victim, returns to find his scalps at the girdles of laggards.[3]

The Andaman islanders are ignorant of the most elementary rules of warfare. Should a dispute arise, a general personal conflict ensues, after the manner of a

[1] *Narratives of the Career of Hernando de Soto, &c.*, trans. Buckingham Smith, New York, 1846, p. 26.

[2] Jones, *Southern Indians*, New York, 1873, p. 18, quoting *Brevis Narratio*. Also Adair, *Hist. of the American Indians, &c.*, London, 1775, p. 377, *et seq.*

[3] Dodge, *Plains of the Great West*, New York, 1877, p. 389.

street brawl. The wounded are not cared for, and unless speedily removed receive the *coup-de-grâce*. They do not mutilate the slain. In case of more systematic attacks the assailants steal upon their enemies, availing themselves of natural cover, but they take no further precautions, or devise stratagems, to conceal their trail. The favourite time of attack is the break of day, when the enemy are asleep, or at a late hour, when they are at the evening meal. No captives are taken, except children, who are frequently carried off and adopted into the conquering tribe.[1]

In Dyak warfare the men cluster around their chief and are indifferent to the fate of others so long as the chief escapes. Similarly relatives cluster together. They carry away the dead and wounded when possible, at least they sever the head and bury it in the forest. War is declared at a great feast, and the plan of campaign agreed upon. Notice to get ready is given by sending a spear around from village to village. The women prepare the bags of provisions, the men put the canoes in order. They take nets for fishing, and dogs for hunting by the way. The men furbish their arms, sharpen their weapons, and decorate their helmets and war jackets. As long as the men are away their fires are lighted on the small fireplaces just as if they were at home. The mats are spread and the fires are kept up till late in the evening, and lighted again before dawn, so that the men may not be cold. The roofing of the house is opened before dawn, so that the men may not lie too long, and so fall into the hands of the enemy. The omen birds are consulted. There is no attempt at order in going until the proposed landing-place is reached. A camp is then formed and guarded, the canoes are hauled up, and the neighbourhood explored.

On a given day the march commences, each one shouldering his pack and stepping out in Indian file—the guides ahead, and closely followed by a few of the hardiest, boldest, and most experienced men at their heels. This line reaches

[1] *Andaman Islanders*, London, 1883, Trübner, p. 135.

many a mile if the war party be a numerous one. Surprise, a sudden rush, fire created by javelin torches hurled into the thatch, and bloody duels for heads, constitute the action.

The defence of the Dyak consists in palisades, wattle, and *chevaux de frise* of spiked bamboo. The waterside, the landing-places, and the paths leading to the villages, as well as the foot of each ladder, are all spiked. Pits are also dug in the pathway. Women and treasures are concealed on the hills and in the forests. Decoys and ambushes, blockades of streams and paths by falling trees are commonly resorted to.[1] A careful study of this description reveals the entire art of war in embryo. It will be noticed that only time and distance, as well as the complexity of apparatus and methods have been modified as the world progressed. The art of fortification and annoyances to marching were well developed by the Dyaks. And this brings us to consider that matter.

In Samoa there was in each district a certain village called the "advance troops." It was their province to take the lead in fighting. The boundary between villages was the battle-field. Women and children were moved off. Wives of chiefs often went to the field, carrying their clubs or some part of the armour. The chiefs and heads of families united in some central spot, and whatever they decided on the young men endeavoured to carry out. Stockades were thrown around the villages where war parties were assembled. Their favourite tactics were of the surprise and bush skirmishing orders. "Their heroes were the swift of foot, like Achilles or Asahel; men who could dash forward towards a crowd, hurl a spear with deadly precision, and stand for a while tilting off with his club other spears as they approached him, within an inch of running him through."[2]

[1] Ling Roth, Low's "Natives of Borneo." *J. Anthrop. Inst.*, London, 1892, vol. xxii. pp. 52–59.

[2] Turner, *Samoa*, London, 1884, chap. xvii.

Schultze says that the Australians of South Finke river murder their enemies by stratagem, waiting and spying them by night.[1]

This is the testimony of the best observers concerning warfare among the lowest people everywhere. Living from hand to mouth, and lacking social organisation and drill for any purpose, the men must contend single-handed, and the fight must be bloody and brief.

In addition to the weapons of defence and offence among savages, for the single warrior, there began to be devised, even in primitive times, appliances of the same sort for the corps, the family, the village, the tribe, in short, community offence and defence, especially the latter.

As we have seen, men went out in squads against squads, by concerted action set fire to villages, and on the water they manœuvred their fleets of war canoes. The action in such cases is co-operative, organised life against the same.

However, savages have few engines of attack on land. But this lack is atoned for by their land defences. There are tribes so low down as not to wear any personal defensive apparatus, but none who do not know how to protect their villages by means of some kind of stockade or platform or earth wall. In some places, as we have seen, they set up ugly splinters in the paths.

There was not a scheme for entrapping animals that was not improved upon to catch the unwary foe. Especially in sedentary tribes the permanent villages were made difficult of approach.

The mediæval caltrop and the modern abattis are quite anticipated and surpassed by the Dyak *tukah* and *ranjan*. This device is simply a strip of bamboo, large or small, as the occasion demands, sharp at both ends, and stuck in the ground wherever an enemy may be passing. Around villages the whole *chevaux de frise* is thus constructed. Every pathway, landing-place, and the foot of each ladder is thus guarded. In the shallow beds of streams these

[1] Schultze, *Trans. Roy. Soc. S. Austral.*, 1891, vol. xiv. p. 221.

dreadful splinters are set up to impale the feet of the men who have to tumble out of the canoes to haul them over the rapids.[1]

The Mango negroes, on the Maringa river, Africa, set up slender bits of bamboo about fifteen centimètres long in their path to cover a retreat. They always carry an abundance of these with them, and when they have to flee they dip the sharp ends of the splints in poison, making a shallow cut around to enable them to be easily broken. The pursuers thrust their naked limbs against the poisoned points and are severely wounded.[2] They choose strategic points for their villages, where bluffs or marshes or watercourses will shield them from sudden attack. In every condition of land surface and natural resources, new inventions meet changing conditions. The American savages were especially ingenious in this regard. Here their villages were located near some spring of water, and surrounded by a fence of logs set close together on end, and pierced at intervals for archers. At the base of the stockade the earth was heaped up to increase the security. In another place a bluff or tongue of land, with precipitous sides, was rendered more secure by a continuous fence or wall of stone extending for miles. At the opening of these fortifications on the land side were ramparts and gateways covered by mounds of earth, so that an enemy would be compelled to enter single file. Not far away, on prairie lands, huge mounds were erected with steep sides, to whose tops the people could fly in hours of danger, while in the south-west of the United States great communal houses were built with outer walls solid. Entrance was by ladders to the first stage, whence the people descended into the ground floor or ascended to

[1] Ling Roth, Low's "Natives of Borneo," *J. Anthrop. Inst.*, London, 1892, vol. xxii. p. 59.

[2] Allaire, *Ann. de la Prop. l. Loi*, Paris, 1892, No. 389, p. 263. Excellent portrait of native, p. 262. The Rev. O. F. Cook brought to the United States National Museum baskets containing hundreds of these "path-splinters" from the Gola and Mandingo area in West Africa.

the stories above. Besides these, pueblos were frequently erected on the extremity of a mesa, where nature had built up an indefinite number of stories below. If the same good friend also provided a canopy of mountain above, the inhabitants had only to lay off the pueblo on some shelf or cave floor and they were secure from marauders. The log, or living tree stockade, the fortified bluff, the refuge mound, the cliff dwelling, the pueblo once invented, it is only necessary to replace the elements of their composition with more durable material, or with more elaborate details to write the history of fortification. The more civilised Mexicans and Central Americans and Peruvians wrought in stone the same ideas, for until 1492 nothing more deadly than an arrow, or a javelin, or a club, or a stone-bladed sword or battle-axe, had ever assailed human life in the western hemisphere.

The Veiburi people, in British New Guinea, owing to incessant raids made upon them, were compelled to establish themselves on the bank of a stream in the midst of high and large trees; here the village was constituted of two houses on the surface and eleven in trees. These aerial dwellings are constructed in the highest trees, one hundred feet above the ground, and approached by means of almost perpendicular ladders, constructed of long spliced saplings lashed eighteen inches apart by cross-bars at every fifteen inches. These houses, supplemented by detached platforms, are stocked with food and weapons of defence, and constantly occupied by their owners, who are so intimidated by the raids of their slayers that they leave their dwellings no longer than they can possibly help for procuring food.[1]

This tree fortress is a tropical device. In those portions of Guiana and Venezuela where the ground is submerged a portion of the year, the natives naturally escape to the trees, have developed an arboreal life, and find their only needed defence therein. The latter country received its name

[1] Thomson, *In British New Guinea*, London, 1892 Philip, &c., p. 51, excellent fig. on p. 52.

FIG. 76.—ANCIENT WATCH TOWER AND FORTRESS IN NEW MEXICO, AND A MODERN SCOUT KEEPING A SHARP EYE ON THE DESCENDANTS OF THE BUILDERS. (*After Jackson.*)

from the circumstance of the natives living on what the navigators called piles.

The best-known example of water defence was the pile structures of Southern Europe and Switzerland in prehistoric times.

The Polynesian places of defence were rocky fortresses improved by art—narrow defiles or valleys sheltered by projecting eminences—passes among the mountains difficult of access, yet allowing their inmates a secure and extensive range and an unobstructed passage to some spring or stream.

Sometimes the natives cut down trees and built a kind of stage or platform, projecting over an avenue; upon this they collected piles of stones and fragments of rock, which they hurled down on those by whom they were attacked.[1]

The Kyans, of Borneo, when they make a camp, strew dead leaves outside the fence, so that no one, not even a dog can approach without being heard. Punans make their camp in a circle, each hut facing a different direction, so as to prevent surprise.[2]

The Hawaiians had a curious contrivance to protect the house from invasion. No locks were known. A heavy stone was suspended over the door in such a way that a person entering after the trap was set would probably be crushed to death.[3]

The fate of the captive in the wars of savages is intimately connected with four words of awfully ominous import in history—torture, cannibalism, slavery, and sacrifice.

The American Indian was addicted sparingly to cannibalism. Slavery in the Pacific region and torture in the eastern slope was the usual fate of the captive. Sacrifice, as will be seen, is a higher idea, and had its evolution after slavery. "The Indian," says Dodge, "does not claim to do murder in the name of his religion. He does it because he

[1] Ellis, *Polynes. Res.*, vol. i. p. 313.
[2] Ling Roth, *J. Anthrop. Inst.*, London, 1892, vol. xxii. p. 56.
[3] *Cat. Bishop Mus.*, Honolulu, 1892, vol. ii. p. 32.

likes it, because his savage instincts and vindictive temper impel him to it." [1]

But this would hardly satisfy the modern ethnologist. In the Indo-Pacific, where large mammals were unknown, cannibalism was pre-eminent. Africa developed slavery, with the other ideas of secondary importance. Americans were most gifted in torture, with outcroppings of the other ideas. The sacrifice of human victims to the gods, to serve them as food, or as objects of vengeful torture, or as slaves to wait on them, must in any event succeed the acts which it apotheosises. The fact remains that torture and cannibalism and slavery and sacrifice had their roots in savagery, and their most refined differentiations were then developed.

War was carried on in Mexico largely for the procurement of human victims, their religion demanding human sacrifices at least eighteen times a year. Every important event, like an improvement of the "teocalli, and especially the installation of a new war chief of the highest degree, had to be celebrated by a special butchery of men—and these victims had to be obtained through war. Therefore the well-known Mexican custom on the battle-field, to look more to the capture than to the slaying of their foes." [2]

"Some of the men of Beit 'Abdel Hady had attacked the villages south-east of Carmel, had burnt the houses and driven off the cattle and flocks. But what most excited the wrath, especially of the women in that region, was the report that the raiders had abused and even killed women and children. During the civil wars that desolated Lebanon in 1841 and again in 1845, women were not molested even in the heat of battle. I have repeatedly seen those of both parties hastening with jars of water for the relief of their friends who were either wounded or suffering from thirst, and they were neither insulted nor molested." [3]

"In the French and Indian wars of North America," says

[1] *Our Wild Indians*, Hartford, 1883, p. 524.

[2] Bandelier, *Tenth An. Rep. Peabody Mus.*, Cambridge, 1877, p. 128.

[3] Thomson, *The Land and the Book*, vol. ii. p. 167

Ellis, "the custom which soon came in, to soften the atrocities of Indian warfare by the holding of white prisoners for ransom, was grafted upon an earlier usage among the natives of adopting prisoners, or captives." "In their earlier conflicts with the whites, the Indians generally practised indiscriminate slaughter. . . . In the raids of the French with their Indian allies, upon the English settlements prisoners taken on either side came gradually to have the same status as in civilised warfare, and to be held for exchange."[1]

The war paraphernalia of nations still absorb a large part of their industries and their genius. As one examines with great care the armoured war-ships, the built-up guns on pneumatic carriages, the elaborate fortifications, it strikes him that the world has gone a long way from the Polynesian war canoe, the Carib *pucuna*, and the mound-builders' work in the Ohio valley. But we can but marvel at the voyages of the Polynesians, the cleverness of the built-up blow-tube, and the astounding patience displayed in the erection of fortifications by hand, containing millions of cubic feet of earth.

The way in which war is engendered and determined upon, the preparations and precautions therefor, the apparatus used by land and by water, the actual conflict, the atrocities and cruelties, the conduct after the engagement, all these are foreshadowed so far in savagery as often to be characterised as among its relics. The best schools even now charge high for tuition, and war has been a costly teacher of men. It seems almost to have been necessary in the pioneer days of human struggle, when beasts and men were arrayed to decide thus who should be master.

In looking through the museums of Europe and America for the material proofs of inventive genius, the student finds no other class of objects more highly organised for the co-operation of intelligent action. In the refinement of

[1] G. E. Ellis, in Windsor, *Narr. and Crit. Hist.*, Boston, vol. i. pp. 287, 289.

the thought involved, in the growing complexity of the mechanical elements and their movements, in the co-ordination of great numbers of men, in the material and political and ideal rewards or ends to be attained, war, at least in primitive times, stands forth pre-eminently as an incitement to the genius of invention and discovery.

CHAPTER XII.

CONCLUSIONS.

THE principles I have sought to illustrate in this book may be briefly summed up.

Invention is indigenous in the nature of man. The first being on this earth worthy of that name was an inventor. The only moment in the life of an individual or a people in which the distinction of true humanity may be worthily bestowed on them is that in which something new is added to the stock of knowledge or experience. When men or nations originate they live and grow ; when they cease to do that they decay and die. This has been true from the beginning.

Invention is stimulated by human wants for :—

1. Food, nourishment of all kinds.
2. Shelter for the person, or clothing.
3. Shelter for the family, or habitation.
4. Rest, recuperation.
5. Locomotion on land and on the water.
6. Delight of the senses.
7. Knowledge, the explanation of phenomena.
8. Social enjoyment, leading to co-operative life in many directions.
9. Spiritual satisfaction.

From this point of view inventions are not only things, but languages, institutions, æsthetic arts, philosophies, creeds, and cults.

All invention is based on change :—

1. In the materials and thing invented.

2. In the apparatus and processes employed.

3. In the mind of the inventor.

4. In society.

This change is in both structure and function, and proceeds from simple to complex and compound in all the particulars named above. It is also always a change from the natural to the artificial, as Payne has well emphasized in his *History of America*. The true destiny of man is to subdue the earth, and to dress and to keep it.

The changes in things or in the powers and results of inventions have followed some such law of evolution as the following :—

1. In most primitive life inventions were natural objects unchanged in form or material used for their natural function, as thorns for piercing, or teeth of rodents for chisels.

2. Natural objects slightly modified in structure to better their performance of the same function, as the same things hafted and sharpened.

3. Natural objects little changed in form to perform a new or different function, as stones for hammers, sticks for weapons.

4. Natural forms or structures copied in a variety of materials for a multitude of functions, as gourds imitated in wood, basketry, and clay.

5. Natural objects or materials changed in form to perform a diversity of functions. This is the most prolific series, and has been ever growing in complexity.

6. Change of motive power, as man, elastic spring, weight, beast, wind, running water, steam, chemical and electric power.

7. Imitation of human activity by machinery.

8. Multiplication of man's power through mechanical powers, as the inclined plane, wedge, roller, wheel, wheel and axle, pulley and screw.

9. Co-operative apparatus, demanding a corps of men and performing more than one function.

Psychical changes include the following series :—

1. Perception, noticing the relation of cause and effect in the natural world, and making up the mind to produce the same results with the same means.

2. Happy thought, imagining that the same result may be differently achieved.

3. The combination of mental activities, discovering relation of one invention to another, resulting in machines.

4. Purposeful invention for its own sake ; predetermined invention.

5. Co-operative invention.

The change of reward has been from individual to international.

1. Beginning with a man making an invention, manufacturing it with his own hands, and putting it on the market, granting himself letters patent and exclusive use for life, gaining to himself the highest pleasures and applause in the world, and being apotheosised at death.

2. Ending with a world-involving and world-benefiting invention like the telephone, in which the inventor is enriched and all mankind brought into relation through one central office of thought.

This evolution is from immediateness to remoteness; from materiality to ideality ; from individuality or personalism to plurality or sociality ; from egoism to altruism.

The change in society has been along the same lines.

1. The ability to fish, hunt, glean, build, weave for more than one person made social groups possible.

2. The differentiation of special activities created centres of activity. The first activities carried the actor to the source of material supply, mineral, vegetal, animal. The last and highest activities takes the natural supply to great, artificial centres of invention and co-operative machinery, from the natural source to the artificial civic centre.

At once it will be seen that this group of social effects

is the result of inventions in all sorts of apparatus and machinery, and *pari passu* men have invented languages, organisations, cities, international exchanges, and laws. Society has invented itself.

In each culture-area of the earth such styles of invention have been elaborated as to confer upon the people thereof their local or tribal traits. The doings and sayings and even the bodily appearance of the peoples of the earth are composite photographs of all that they have been thinking out along the paths or contours laid down by nature.

Finally, in contemplating the exalted position to which acquired knowledge and experience have brought the favoured race, we are apt to forget how many have helped to place them there. The many patents and inventions now on the earth are only a "handful to the tribes that slumber in its bosom."

It is a well-established fact in biology that the humblest creature is just as important a link in the chain of creation as the highest mammal. The higher forms are so well known, and so little has been found out concerning some of the more lowly creatures, that the naturalist is very glad to leave the ninety and nine and go into the wilderness to seek the one that is lost.

The devices of pristine man are the forms out of which all subsequent expedients arise. The fire-sticks of savages are the earliest form of illumination by friction. The tribulum is the modern thresher with stone teeth. The kaiak furnishes the lines of the swiftest racing boats. The sewing-machine makes no new loops. Warfare is still cutting, bruising, or piercing. All art lines and geometry were born in savagery. Society even can never change in organisations and motives. Our most precious maxims antedate literature. The whole earth is full of monuments to nameless inventors.

INDICES.

—o—

INDEX OF AUTHORS.

INDEX OF SUBJECTS.

THE M.I.T. PAPERBACK SERIES

1 **Computers and the World of the Future** edited by Martin Greenberger
2 **Experiencing Architecture** by Steen Eiler Rasmussen
3 **The Universe** by Otto Struve
4 **Word and Object** by Willard Van Orman Quine
5 **Language, Thought, and Reality** by Benjamin Lee Whorf
6 **The Learner's Russian-English Dictionary** by B. A. Lapidus and S. V. Shevtsova
7 **The Learner's English-Russian Dictionary** by S. Folomkina and H. Weiser
8 **Megalopolis** by Jean Gottmann
9 **Time Series** by Norbert Wiener
10 **Lectures on Ordinary Differential Equations** by Witold Hurewicz
11 **The Image of the City** by Kevin Lynch
12 **The Sino-Soviet Rift** by William E. Griffith
13 **Beyond the Melting Pot** by Nathan Glazer and Daniel Patrick Moynihan
14 **A History of Western Technology** by Friedrich Klemm
15 **The Dawn of Astronomy** by Norman Lockyer
16 **Information Theory** by Gordon Raisbeck
17 **The Tao of Science** by R. G. H. Siu
18 **A History of Civil Engineering** by Hans Straub
19 **Ex-Prodigy** by Norbert Wiener
20 **I Am a Mathematician** by Norbert Wiener
21 **The New Architecture and the Bauhaus** by Walter Gropius
22 **A History of Mechanical Engineering** by Aubrey F. Burstall
23 **Garden Cities of To-Morrow** by Ebenezer Howard
24 **Brett's History of Psychology** edited by R. S. Peters
25 **Cybernetics** by Norbert Wiener
26 **Biological Order** by André Lwoff
27 **Nine Soviet Portraits** by Raymond A. Bauer
28 **Reflexes of the Brain** by I. Sechenov
29 **Thought and Language** by L. S. Vygotsky
30 **Chinese Communist Society: The Family and the Village** by C. K. Yang
31 **The City: Its Growth, Its Decay, Its Future** by Eliel Saarinen
32 **Scientists as Writers** edited by James Harrison
33 **Candidates, Issues, and Strategies: A Computer Simulation of the 1960 and 1964 Presidential Elections** by I. de S. Pool, R. P. Abelson, and S. L. Popkin
34 **Nationalism and Social Communication** by Karl W. Deutsch
35 **What Science Knows About Life: An Exploration of Life Sources** by Heinz Woltereck
36 **Enzymes** by J. B. S. Haldane
37 **Universals of Language** edited by Joseph H. Greenberg
38 **The Psycho-Biology of Language: An Introduction to Dynamic Philology** by George Kingsley Zipf
39 **The Nature of Metals** by Bruce A. Rogers
40 **Mechanics, Molecular Physics, Heat, and Sound** by R. A. Millikan, D. Roller, and E. C. Watson
41 **North American Trees** by Richard J. Preston, Jr.
42 **God and Golem, Inc.** by Norbert Wiener
43 **The Architecture of H. H. Richardson and His Times** by Henry-Russell Hitchcock
44 **Toward New Towns for America** by Clarence Stein
45 **Man's Struggle for Shelter in an Urbanizing World** by Charles Abrams
46 **Science and Economic Development** by Richard L. Meier
47 **Human Learning** by Edward Thorndike
48 **Pirotechnia** by Vannoccio Biringuccio
49 **A Theory of Natural Philosophy** by Roger Joseph Boscovich
50 **Bacterial Metabolism** by Marjory Stephenson
51 **Generalized Harmonic Analysis and Tauberian Theorems** by Norbert Wiener
52 **Nonlinear Problems in Random Theory** by Norbert Wiener
53 **The Historian and the City** edited by Oscar Handlin and John Burchard
54 **Planning for a Nation of Cities** edited by Sam Bass Warner, Jr.
55 **Silence** by John Cage
56 **New Directions in the Study of Language** edited by Eric H. Lenneberg
57 **Prelude to Chemistry** by John Read
58 **The Origins of Invention** by Otis T. Mason
59 **Style in Language** edited by Thomas A. Sebeok
60 **World Revolutionary Elites** edited by Harold D. Lasswell and Daniel Lerner